Recent Developments and Current Practices in Odor Regulations, Controls and Technology

TRANSACTIONS

Recent Developments and Current Practices in Odor Regulations, Controls and Technology

Papers from an International Specialty Conference

Edited by
David R. Derenzo
Derenzo and Associates, Inc.
Livonia, Michigan

Alex Gnyp
University of Windsor
Windsor, Ontario Canada

Air & Waste Management Association
Pittsburgh, Pennsylvania

TR-18
Recent Developments and Current Practices in
Odor Regulations, Controls and Technology

A Peer-reviewed Publication
(A&WMA Transactions Series, ISSN 1040-8177; No. 18)

ISBN 0-923204-05-9   1991

These papers were originally presented at an International Specialty Conference at Detroit, Michigan, October 23-26, 1989. The Conference was sponsored by the Air & Waste Management Association's Odors Committee, EE-6, and Solvents, Odors and Gases Committee, AE-2, and hosted by the A&WMA Michigan Chapter.

Additional copies of this book may be purchased for $75 (Association members $50) from the Order Fulfillment Clerk, Air & Waste Management Association, P.O. Box 2861, Pittsburgh, PA 15230.

CONTENTS

I. Odor Regulation

II. Odor Controls and Environmental Systems

III. Odor Measurements, Modeling and Technology

IV. Odor Measurements, Modeling and Technology

V. Odor Emission and Control Technologies at Waste Water Treatment Plants

PREFACE

The last specialty conference on odors was held over ten years ago. The Transactions of the papers presented at this conference were published by the Air Pollution Control Association and still serve as a valuable reference. The ten year period represents a relatively long time frame when compared to the frequency of other air pollution related conferences. Therefore, the need for the International Specialty Conference held in Detroit, Michigan from October 23 to 26, 1989 on "Recent Developments and Current Practices in Odor Regulations, Controls and Technology" was very evident to the individuals and committees involved in its development. The response given the request for papers by prospective authors, and number of people that attended the conference confirmed the need for a forum to transfer new information on odor related issues.

The United States Environmental Protection Agency's definition of air pollution does not include odors. Therefore, no federally promulgated regulations for odor control exist. This regulatory structure leaves odor control to the individual state and local air pollution control agencies. With these agencies reporting that over 50% of the air pollution complaints received are related to adverse odor impacts, it is apparent that the issues surrounding odor evaluation and control are very important. A review of the odor control programs developed by these regulatory agencies finds that a variety of technical procedures are used to measure and evaluate odorous emissions.

The European Community has conducted much work on odor measurement techniques, controls and the standardization of evaluation methods. A large portion of this work has its origins in the study of odorous impacts related to agricultural activities. This experience provides an important basis for establishing an acceptable approach to the investigation and control of odorous emissions for all types of sources.

Issues concerning repeatability and reproducibility of odor measurements are not complete and uncertainties still exist in the use of published odor threshold data. These uncertainties have been recognized by the American Society for Testing of Materials (ASTM). It has recently indicated that studies on acceptable odor measurement techniques are very necessary.

The Editors of these Transactions are confident that the papers provide a comprehensive summary of the state of odor evaluation procedures, measurement methods and control technologies, and will be a valuable reference resource. It is hoped that the materials in this publication will encourage continued work on our understanding of odors.

David R. Derenzo
Alex Gnyp
Technical Program Chairmen

ACKNOWLEDGEMENTS

The specialty conference programs organized through the National Chapter of the Air & Waste Management Association (A&WMA) are important tools in the transfer of data and information. To complete the organizing and planning functions of the Odor Specialty Conference, many efforts were required. The Technical Program Chairmen acknowledge the contributions made to the success of this program.

Thanks are due to the overall planning committees that formalized the structure of the conference. These key individuals contributed to making the conference a very successful event. Dr. Peter Warner's experience in the field of odor analyses, and in conference planning were very important in solidifying the conference structure. Messrs. Sol Baltimore, Kevin Dunn, Tom Wackerman and Howard Murray spent many hours in securing appropriate accommodations, finalizing the conference budget, arranging exhibitor facilities and promoting the conference.

Appreciation is due to all of the authors, session chairmen, and peer reviewers who provided significant contributions to make the Odor Specialty Conference technically strong and an important tool for future odor studies.

The Fall Conference of the Michigan Chapter of the A&WMA was held immediately after the Odor Specialty Conference. Much coordination between the Odor Conference planning committees and those of the Michigan Chapter was required. Messrs. Van Mauzy and Tom Weeks, with support from their employers, provided the on-site registration services.

The success of the Odor Specialty Conference is in part due to the efforts provided by the Air & Waste Management Association. The experience and participation of Mr. Steve Stasko and Ms. Sharon DeAndrea in coordinating the conference activities helped greatly to ensure a high quality meeting.

David R. Derenzo
Alex Gnyp
Technical Program Chairmen

CONFERENCE ORGANIZING COMMITTEE

General Conference Chairman
Peter Warner, Wayne County Air Pollution Control Division

General Conference Vice Chairman
Richard Kruse, Duall Division, Met-Pro Corporation

Technical Program Chairmen
David R. Derenzo, Derenzo and Associates, Inc.
Alex Gnyp, University of Windsor

Exhibition
Tom Wackerman, Applied Science and Technology, Inc.

On-site Registration Chairman
Van Mauzy, Wayne County Air Pollution Control Division
Tom Weeks, Clayton Environmental Consultants, Inc.

Facilities Chairman
Kevin Dunn, Petro-Chem Processing

Special Arrangements Chairman
Sol Baltimore, Environmental Communications

Publicity Chairman
Howard Murray, Wayne County Air Pollution Control Division

Michigan Chapter
Chuck Hersey, Chairman

EE-6 Odor Committee
Richard Duffee, Chairman

AE-2 Solvents, Odors and Gases
James Berry, Chairman

SESSION CHAIRS/PEER REVIEW PANEL

Session I
David Benforado, 3M Company, St. Paul, MN
James Frazier, National Academy of Science, Washington, DC

Session II
John Swanson, Swanson Environmental, Farmington, MI
Perry Lonnes, Interpol, Minneapolis, MN

Session III
Alex Gnyp, University of Windsor, Windsor, Ontario, Canada
Carl St. Pierre, University of Windsor, Windsor, Ontario, Canada

Session IV
David R. Derenzo, Derenzo and Associates, Inc., Livonia, MI
Peter Warner, Wayne County Air Pollution Control Division, Detroit, MI

Session V
Richard Kruse, Duall Division, Met-Pro Corporation

THE EDITORS

Mr. David Derenzo is president of Derenzo and Associates, Inc., an environmental consulting company specializing in analyses related to air quality issues.

Mr. Derenzo received a Bachelor of Science degree in Mathematics from the University of Massachusetts in 1975 and a Master of Science degree in Atmospheric Science from the State University of New York at Albany in 1978. He has worked for a regional transportation planning agency and an environmental consulting company. In 1984, Mr. Derenzo joined the Wayne County Air Pollution Control Division as Assistant Director of Technical Services where he provided and supervised analyses to support the agency's engineering, permitting and enforcement activities.

Mr. Derenzo is a member of the Air & Waste Management Association (A&WMA), and a member of various local technical committees. He has made presentations at environmental conferences and published associated materials.

Dr. Alex Gnyp is the coordinator of the Environmental Engineering program in the Department of Civil and Environmental Engineering at the University of Windsor. He received his B.A.Sc. (1953), M.A.Sc. (1958) and Ph.D (1959) degrees from the University of Toronto. To bridge the gap between the academic and industrial worlds, he serves as a senior consultant to Clayton Environmental Consultants, Ltd.

Dr. Gnyp's research and professional activities involve:
* Sampling, analysis and control of odorous emissions
* Stationary source sampling and assessment
* Analysis, design and operation of air pollution control equipment
* Assessment and improvement of indoor air quality

Dr. Gnyp is a member of the Association of Professional Engineers of Ontario (APEO), the A&WMA and the A&WMA Odor Committee, EE-6.

AN OVERVIEW OF THE
CONFERENCE TECHNICAL PROGRAM

Introduction

The Conference on Recent Developments and Current Practices in Odor Regulations, Controls and Technology program consisted of Keynote and Luncheon speeches, and five technical sessions. The Keynote address was presented by Mr. Richard Duffee who outlined "Odor Technology: Present and Future Directions". The Luncheon speech was presented by Dr. Thomas Getchell. He discussed "Odor Detection and Analysis: A Biological Approach".

Of the 37 technical papers presented at the Conference, 33 are included in these Transactions. The contents of these papers are outlined in the following text.

Session I - Odor Regulation

The driving force for the analysis and control of air contaminants resides in the implementation of appropriate regulations. Not since David Benforado conducted his initial survey of the odor regulations used by air pollution control agencies in the United States has any detailed follow-up work been conducted. Mr. Benforado revised his initial work with new survey information. With odors being responsible for over 50 percent of the complaints registered with many control agencies, these organizations are implementing regulatory approaches to address public concerns. Two papers in this session provide specific approaches to local and state control of odorous emissions. Much work has been conducted in Europe on the evaluation and regulation of odors. Key issues that have been investigated deal with odor measurement techniques, repeatability and reproducibility of results, and standardization of methods. Three papers in this session provide information on the activities being conducted in Europe.

Session II - Odor Controls and Environmental Systems

Many different technical approaches are being used to control odorous emissions. There are six papers in this session. An engineering perspective of odor control is presented in one paper. The best control approach must define the constituents in the odorous emission that need to be reduced to a desired level, and consider cost and associated design parameters. Two papers present information on the principles of biofiltration, its odor removal effectiveness and use as a supplement to standard odor control practices and mechanisms. Information on the use of isolation distances to ensure that an odor is not perceived at a specified location is presented in a paper. Little work has been conducted on the characteristics of wastes from odor removal operations. A paper presents information on the properties of fully exhausted carbons used to remove sulfide odors in waste water treatment plants. The use of a new monolithic packing versus conventional random dumped packing for odor control is compared in a paper.

Sessions III & IV - Odor Measurements, Modeling and Technology

The need for odor regulation and control depends on the strength of the emission source and the extent of impact on the surrounding environment. Therefore, it is necessary to conduct odor measurements and analyses to assess odorous impacts. Two sessions on odor measurements, modeling and technology were held at the Conference. The two sessions produced a total of 16 papers.

Odor impact data are generally obtained from studies on the affected communities. The importance of developing background data on the community for use in verifying a perceived odor problem is presented in one paper. Odor threshold values are used to determine when concentrations of an identified material can cause problems. Four papers in the two sessions discuss errors in measurement methods that incorporate chance, difficulties in establishing health related standards, uses of threshold data by regulatory agencies to determine potential community odor impact problems and a "Decismel Scale" for comparative measures of odor levels and sensitivities. Much work has been conducted on odor measurement techniques; however, one paper presents information that indicates that odor panel selection and training may have the greatest effect on measurement results. Community surveys provide a mechanism whereby the extent and levels of odorous impacts can be quantified. Two papers discuss investigations of odor annoyance experienced by local populations and methods of defining the area surrounding an emissions source. Extensive data bases on the potentials of materials to generate odorous emissions do not exist. The results of testing conducted on automotive coatings to establish potential odor emissions is presented in one paper. The results of extensive sampling at three different sewage treatment plants and 28 different locations in the sewage collection system are presented in another paper. Some regulatory agencies identify the use of specific odor test methods to be used in making measurements. Data that compare odor measurements made with syringe and dynamic olfactometry methods are presented in a paper. An update on the recent investigations being conducted by the ASTM Task Group E-18.04.25 on Sensory Thresholds is also presented.

Much odor impact work utilizes atmospheric dispersion models to determine the effect of emissions on the community. A paper compares results obtained from the use of the ISCST Gaussian dispersion model, the U.S. EPA's INPUFF integrated puff model and a fluctuating plume puff odor model. Communities with numerous sulfur sources can have odor problems. A paper discusses the procedures used to sample aggregate reduced sulfur gas with a mobile unit.

SessionV - Odor Emission and Control Technologies at Waste Water Treatment Plants

A significant amount of work has been conducted on the control of odors from waste water treatment plants. This session had nine papers that describe a wide range of odor control perspectives. Experiences in the use of mist scrubbing, carbon adsorption, packed towers, vapor phase reactions and biological degradation systems are presented in this session by equipment manufacturers, consulting engineers and regulatory agency personnel.

Due to the nature of the material presented by equipment suppliers, a complete peer review of these papers was not conducted.

David R. Derenzo
Derenzo and Associates, Inc.
Livonia, Michigan

I. Odor Regulation

Survey of States for Odor Information
David M. Benforado, 3M
James A. Frazier, National Academy Sciences

ABSTRACT:

The National Academy of Sciences (N.A.S.) Committee on Odor from Stationary and Mobile Sources, prepared a state of the art report on odor in 1979. In order to compile background data for that report, the committee sent a questionnaire to all state and selected local agencies, to obtain available information on odor complaints, sensory data on odors and related court cases.

The authors sought to assess the status of available information on odors ten years later in 1989, by conducting a follow up survey using the same questionnaire as used in 1979. The questionnaire was sent to 73 state and local agencies. The results of this new survey are reported in this paper.

BACKGROUND:

In 1979, the NAS published a state of the art report prepared by the Committee on Odors from Stationary and Mobile Sources. (1) That report covered a study of the effects of odors on public health and welfare, odor measurement and control technologies, and the costs of odor abatement strategies. The report is still an excellent resource for anyone getting involved in odor measurement and control.

The committee concluded that establishment of federal ambient-air-quality or emission standards would confront various conceptual and technical difficulties as follows:

> The adverse effects of odors on people are variable, and our knowledge about the effects is incomplete. Thus it will be difficult to define standards that will be widely accepted.

Although odor perception can be assessed by psychophysical methods, and some odorous substances can be measured by modern instrumental methods, the two sets of methods are difficult to relate to each other; and are costly and time-consuming.

Features were identified which should be incorporated into any approach made to establish standards, should the U.S. EPA attempt to develop odor standards in spite of the problems. Refer to the report. (1)

In recognition of the difficulty that would confront any attempt to establish odor standards, the kinds of studies needed to be done were also identified. See the report for listing of these studies.

The NAS report was used by the U.S. Environmental Protection Agency (U.S.EPA) to prepare their 1980 report on the Regulatory Options for the Control of Odors, (2) in response to a Congressional mandate that appeared in the 1977 Amendments to the Clean Air Act. The U.S.EPA report evaluated the need for national ambient air standards for odors, and what control strategies might be appropriate for odor abatement.

The conclusions from the U.S.EPA's evaluation was that federal regulatory involvement does not appear to be warranted (in terms of trying to establish federal odor standards) for the following reasons:

1. The available odor data are not sufficient to support the establishment of primary and secondary air quality standards.

2. Techniques used to measure odors are considered generally inadequate for regulatory purposes.

3. Reliable procedures for relating ambient odor levels to the extent of community annoyance do not exist.

4. Community tolerance or odor annoyance levels vary widely.

5. Use of best control technology for new or existing sources of odors under section 111 of the Clean Air Act also has its problems:

a.   It would require controls nationwide, even though the problem may only be local..

b.   It does not guarantee community odor nuisance levels would not be exceeded, especially when odor is due to fugitive sources, or multiple sources in close proximity.

c.   Regulating all odor sources would require an inordinate expenditure of Federal, State and Local agency resources.

6.   State and local odor control procedure relying on nuisance rules appear to be generally adequate, and more cost effective than a national regulatory program.

In regard to odor regulations, it was learned from Mr. William H. Prokop, consultant, that the National Renderer's Association (NRA) in 1982 published the results of a Survey of State and Local Odor Control Regulations. Those interested in additions and revisions which have occurred since the NRA survey should contact Mr. Prokop.

INTRODUCTION:

As background for the 1979 National Academy Science (NAS) report, the committee sent a questionnaire to all state and selected local agencies, to obtain available information on odor complaints, sensory data on odors, and related court cases.

The authors sought to make a comparison of earlier results with those ten years later.

A letter similar to the l979 letter was sent to 72 state and local agencies, asking for their response to the following same three questions asked in l979:

1.   Does the agency have summaries of complaints about odors that identify the odor sources and or the number of people affected?

2.   Does the agency have sensory data and/or measurements that were used to determine the source, intensity, and quality of the odors?

3.   Does the agency have summaries of court cases regarding public exposure to odorous substances?

Time constraints and other priorities precluded the same level of follow-up effort used by the authors in preparing the NAS report. A limited amount of follow up was done with those agencies who did not respond in a timely manner.Thus, the present response rate to the questionnaire is markedly reduced from that of 1979.

The authors would be glad to make the original responses available to anyone interested in pursuing the survey in further detail.

RESPONSES TO THE QUESTIONNAIRE:

A list of all agencies contacted and a brief overall summary of the results of the survey is presented in Table 1.

Detailed sheets prepared from those few agencies that sent data, is presented in Tables 2 - 10. The one page sheets provide examples showing the type of information received and available from these agencies:

Table 2: Maine - The Maine Department of Environmental Quality does not routinely summarize odor complaints including the number of people affected. However, the complaint file in the computer data base was reviewed and a summary of odor complaints to the Air and Water Bureaus was provided. Table 2A shows typical data available on complaints to the Water Bureau from 1986 - 1988. Table 2B shows typical data on some of the complaints received by the Air Bureau in 1988.

Table 3: New Jersey - The New Jersey Department of Environmental Protection does compile citizen odor complaint information on a monthly basis. Table 3 shows one of the monthly summary reports for August 1988, which is typical of the data available from the agency.

Table 4: Puerto Rico - The Commonwealth Agency of Puerto Rico does keep summaries of odor complaints including the number of people affected, and sensory data and court cases.

The summaries have not been received as of press time. However, the authors thought it would be useful to include a copy of the Commonwealth's Agency letter response as Table 5.

Table 5: Tennessee - The Tennessee Division of Air Pollution Control does keep summaries of odor complaints and sensory data from measurements performed by the Field Services Program. There have been no court cases regarding public exposure to odors.

The agency has a standard operating procedure for handling complaints. In Table 5 A is presented a Complaint Investigation Form, and in Table 5B if the Odor Complaint Observation/ Assessment Form used by the agency personnel in evaluating a complaint.

Table 6: Wisconsin - The Bureau of Air Management, Wisconsin Department of Natural Resources, can provide summaries of odor complaints, but does not have sensory data, or summaries of court cases.

Some of the typical data available from the 1988 Complaint Tracking Report is presented in Table 6A. The key to the headings is shown in Table 6B.

Table 7: Michigan - Wayne County - The agency keeps complaint cards which identify the character of the odor and the source. A sample card is shown in Table 7A. The number of households affected is recorded when the agency takes action on a complaint. An example of the data available is shown in Table 7B.

The ASTM Syringe Method is still used by the agency, and the data is available in the files. Good Engineering Practice is used as basis for evaluating complaints and is defined as an odor threshold measurement of 150 odor units.

The agency is currently involved in litigation with several companies. A summary of the status of each case is available from the agency.

Table 8: Tennessee - Chattanooga-Hamilton County - The agency does maintain summaries of major complaints which are available for review. Chemical analysis is used for evaluating hydrogen sulfide emissions which result in complaints.

There were a total of 193 odor complaints and 204 open burning complaints in 1988.

There have been no court cases in 8 years. However the agency has had a few administrative cases, and summaries are available from the agency.

All new sources must use reasonable and proper technology. BACT must be used for any new volatile organic compound (VOC) source.

The agency also monitors indoor odor complaints. An example of the summary of indoor complaints is shown in Table 8.

Table 9: Pennsylvania - Philadelphia - The agency does not try to measure odors. The inspector makes a determination subjectively if an odor complaint represents a nuisance in the community.

Court cases have been relatively few. Administrative orders and penalties are emphasized instead.

A summary of odor complaints and violations by category is presented in Table 9.

Table 10: Virginia - The agency maintains a file on individual odor complaints. A generic complaint form is used as shown in Table 10.

The agency does not measure odors, but makes a subjective determination if an odor source responsible for a complaint is a nuisance.

There have not been any court cases, but public hearings are conducted as needed, which frequently lead to corrective action.

GENERAL COMMENTS:

The response to the survey was very disappointing. Due to the low percentage of response to the I989 questionnaire, direct comparisons with 1979 cannot be constructed. Only 9 agencies provided data indicating activity in responding to odor problems, 25 agencies did not respond at all, and the rest of the agencies provided a brief comments that has been summarized in Table 1.

In comparison,the numbers of responses to the original questionnaire in 1979 were 18 agencies that provided data on odor problems and only 7 agencies did not respond.

It is difficult to draw conclusions from such limited data. However, it might be speculated that not very much has improved in regard to the resources available to deal with odor problems, nor to systematically tabulate the odor problems by type or categories of odorants and their sources.

Further, It appears that odor complaints still account for the majority of complaints agencies receive.

Most agencies apparently still handle odor complaints on a case-by-case basis, using nuisance statutes and negotiation to effect an abatement of the odors. This may be the most cost effective manner of dealing with the single source local odor type problems.

References:

(I) "Odor from Stationary and Mobile Sources," Committee on Odors from Stationary Sources, National Research Council, National Academy of Sciences Washington D. I979."

(2) "Regulatory Options for the Control of Odors', George H. Wahl, Jr., U.S.EPA report 450/5-80-003, Research Triangle Park, N Feb 1980.

TABLE 1. STATE SURVEY RESULTS

STATE	ODOR COMPLAINTS SUMMARY AVAILABLE YES	NO	DATA IN FILE	SENSORY YES: YES	WHO MEASURES	DATA NO *NO SUMM	COURT CASES SUMMARY YES	NO	COMMENTS
Alabama		No				No		None	Not involved in court cases-nuisance only
Alaska		No				No		None	
Arizona-State		No	Yes			No		No	No record court cases
Arizona-Pima		No	Yes			No		No	Some data in files-will compile for a fee
Arkansas	NR								
CA-State		No				No		No	Data in local districts
CA-Bay Area			Yes	Yes	Agency	No*		No	Keep files-Have specific odor rule
CA-San Diego	NR								
CA-Santa Bar	NR								
CA-So.Coast	NR								
Colorado		No	Yes	Yes	Agency	No*		No	No cases in recent years
Connecticut	NR								
Delaware	NR								
District Col	NR								
Florida-State		No				No		No	District handles
Florida-Dade	YES					No		No	
Georgia		No				No		No	No odor regulations
Hawaii		No	Yes			No		No	No court cases
Idaho	NR								
IL-State	NR								
IL-Bedford P	NR								
IL-Chicago	NR								
IL-Will Cty	NR								
Indiana		No	Yes			No		No	1 case in 1967-livestock odors-dev. regulation
Iowa		No				No		No	Rescinded odor rule
Kansas		No	Yes			No		No	Maintain info and actions in files
Kentucky		No	Yes	Yes	Kentucky	No*		No	
Louisiana	NR								
Maine			Yes			No		No	No court cases-See Table 2
MD-State	Yes		Yes		Chem. Analy.	No		No	No court cases
MD-Baltimore		No	Yes			No		No	No court cases-nuisance approach
MA-State		No				No		No	
MA-Boston		No				No		No	Over 3-4 complaints/yr-handled case-by-case
MI-State		No	Yes			No		No	
MI-Wayne			Yes	Yes	Agency	yes	yes		
Minnesota		No	Yes	Yes	MWCC				Odor data confidential
Mississippi	NR								
MO-State				Yes	Agency		Yes		Court case in 1980
MO-Springfield	NR								
MO-St. Louis		No	Yes			No		No	No court cases
Montana	NR								
Nebraska		No				No		No	But whole series of livestock cases
Nevada		No	Yes			No		No	Limited # of complaints
New Hampshire	NR								
New Jersey	Yes		Yes	Yes	Trained Invest			No	See Table 3
New Mexico-St		No				No		No	No regulation
NM-Albuquerque	NR								
New York	NR								
North Carolina		No	Yes			No		No	Uses odor evaluation form
North Dakota		No	Yes	Yes	Agency			No	30 complaints/year. Scentometer
Ohio-State		No				No		No	Handled case-by-case
Ohio-Cincinnati	NR								
Ohio-Cleveland		No	Yes	Yes	Varies			No	
Oklahoma	NR								
Oregon-State		No	Yes			No		No	Scentometer not effective-now use nuisance
Oregon Lane Cty	NR								
PA-State		No				No		No	
PA-Philadelphia	Yes		Yes			No		No	See Table 9 and text
Puerto Rico	Yes		Yes	Yes	Agency			Yes	See table 4 Rule 420
Rhode Island	NR								
South Carolina		No				No		No	No reg. Case-by-case
South Dakota		No				No		No	Handled locally by Nuisance
TN-State	Yes		Yes	Yes	Agency			No	No court cases See Table 5
TN-Chattanooga	Yes		Yes			No		No	See Table 8 and text
Texas	NR								
Utah		No				No		No	Local Gov't. -nuisance rule
Vermont		No	Yes	Yes	Agency	No*		No	Air quality modeling
Virginia		No	Yes			No		No	See Table 10 and text
Washington	NR	No				No		No	Nuisance statute
W. Virginia	Yes								
Wisconsin						No		No	See table 6
Wyoming		No	Yes	Yes	Agency			No	Scentometer but no success.

TABLE 2A. MAINE – COMPLAINTS TO WATER BUREAU

Odor Complaints Received by the O & M Section
Water Bureau – Augusta – May 1986 to September 1988

Date	Type	Location	Impact
5/86	Treatment plant	Mexico	Local
5/86	Treatment plant	Wilton	Local
6/86	Agriculture (manure)	University of Maine	One
6/86	Street pipe odor	Woolwich	One
7/86	Treatment plant	Kennebunkport	Local
7/86	Sludge spreading	Farmingdale	Local
8/86	Sludge spreading	Falmouth	Local
8/86	Surface water, marsh	Wells	One
8/86	Treatment plant (pump station)	Farmington	One
10/86	Treatment plant (sewer line-cellar)	Jackman	One
10/86	Treatment plant	Brewer	Local
11/86	Surface water, stream	Waterville	One
11/86	Industrial	Freeport (Eastland Shoe)	One
2/87	Sewer odor in cellar	Gardiner	One
2/87	Surface water odor	Boothbay	One
2/87	Treatment plant	Gardiner	Local
6/87	Treatment plant	Scarboro	Local
6/87	Industry or treatment plant discharge odor?	Champion (Bucksport)	One
7/87	Dead fish	York	One
7/87	Surface water, stream	Bristol	One
7/87	Surface water (lake)	Smithfield	One
8/87	Treatment plant	Lincoln POTW	Local
8/87	Treatment plant	Hartland	Local
8/87	Dead animals?	Turner	One
8/87	Leach field	Augusta	One
8/87	Treatment plant	Gardiner	Local
8/87	Landfill	Augusta	One
8/87	Surface water (street pipes)	Yarmouth	One
8/87	Storm drain (sewage odor)	Portland	One
9/87	Surface water (brook)	Portland	One
11/87	Sewage smell	Wells	One
6/88	Treatment plant	Rumford POTW	Local
7/88	Treatment plant	East Wind (St. George)	One
7/88	Treatment plant	Gardiner	Local
7/88	Treatment plant by leach fields	Sebago Lake	One
8/88	Agricultural	Gray	Local
9/88	Treatment plant	Robinson	Local

TABLE 2B. MAINE - COMPLAINTS TO AIR BUREAU

Number	Type of Complaint	Source	Town	Complainant	Date Received
1159	Open Burning	Newport Town Dump	Newport	Anonymous	06-15-85
1131	Tires	Old Town Dump	Old Town	Bangor Staff	04-15-85
1135	Tires	Old Town Dump	Old Town	Bangor Staff	04-28-89
1136	Tires	Old Town Dump	Old Town	Bangor Staff	05-04-88
1145	Incinerator	Orland Consolidated School	Orland	Anonymous	05-23-88
1196	Soot	Penobscot Energy Recovery Co.	Hampden	Marcel Whitney	09-02-88
1117	Noise/Smoke	Penobscot Energy Recovery Co.	Orrington	Ackley, Carl	01-06-88
1129	Smoke	Penobscot Energy Recovery Co.	Orrington	Ackley, Carl	04-14-88
1130	Smoke	Penobscot Energy Recovery Co.	Orrington	Judd, Pat	04-15-88
1132	Odor/Smoke	Penobscot Energy Recovery Co.	Orrington	Webster, Linda	04-21-88
1133	Soot/Odor	Penobscot Energy Recovery Co.	Orrington	Kenny, Randy	04-26-88
1143	Odor	Penobscot Energy Recovery Co.	Orrington	Webster, Linda	05-18-88
1146	Odor	Penobscot Energy Recovery Co.	Orrington	Webster, Linda	06-02-88
1147	Odor/Noise	Penobscot Energy Recovery Co.	Orrington	DeSanctis, Elsie	06-02-88
1148	Odor	Penobscot Energy Recovery Co.	Orrington	Kenny, Randy	06-03-88
1151	Odor/Smoke	Penobscot Energy Recovery Co.	Orrington	Bouzan, Laura	06-08-88
1152	Odor/Noise	Penobscot Energy Recovery Co.	Orrington	Adams, Seth	06-08-88
1153	Odor/Noise	Penobscot Energy Recovery Co.	Orrington	DeSanctis, Marge	06-08-88
1155	Odor	Penobscot Energy Recovery Co.	Orrington	Adams, Seth	06-13-88
1156	Odor	Penobscot Energy Recovery Co.	Orrington	Daeuth, Helen	06-14-88
1158	Odor/Noise	Penobscot Energy Recovery Co.	Orrington	Residents of Orrington	06-16-88
1154	Odor/Noise	Penobscot Energy Recovery Co.	Orrington	Linda Webster	06-17-88
1165	Noise/Smoke	Penobscot Energy Recovery Co.	Orrington	Helen Daeuth	06-21-88
1171	Odor/Litter	Penobscot Energy Recovery Co.	Orrington	Carl Ackley	06-24-89
1172	Odor/Smoke	Penobscot Energy Recovery Co.	Orrington	Helen Daeuth	07-01-89
1176	Odor/Noise	Penobscot Energy Recovery Co.	Orrington	Helen Daeuth	07-14-88
1181	Odor	Penobscot Energy Recovery Co.	Orrington	Helen Daeuth	07-21-88
1163	Odor/TSP	Penobscot Energy Recovery Co.	Orrington	Carl Ackley	07-22-88
1184	Noise	Penobscot Energy Recovery Co.	Orrington	Pat Judd	07-28-88
1187	Plant Emissions	Penobscot Energy Recovery Co.	Orrington	Carl Ackley	08-15-88
1190	Odor	Penobscot Energy Recovery Co.	Orrington	Linda Webster	08-16-88
1195	Odor	Penobscot Energy Recovery Co.	Orrington	Helen Daeuth	08-29-88
1197	Odor/Noise	Penobscot Energy Recovery Co.	Orrington	Jeff Toutillotte	09-10-88
1214	Odor	Penobscot Energy Recovery Co.	Orrington	Borns, Margaret	11-29-88
1180	Smoke	R. Andersons Dump/burning tire	Steuben	Ruth Faulkinghan	07-20-88
1167	Dust	Rodney Cummings gravel pit	Alton	Corrina Doucette/L. Feero	06-21-88
1172a	Dust	Sewer Construction	Milbridge	Snowdeal, Mary	07-05-88
1174	Smoke	St. Josephs Hospital	Bangor	Tom Bocze	07-08-88
1168	Dust	Steego Auto Pts/Sandblasting	Old Town	Thomas Nadeau	06-21-88
1194	Smoke	Town Dump	Otis	Fire Chief E. Austin	08-23-88
1192	Smoke	Town Dump	Sorento	Anonymous	08-23-88
1163	Dust	Traffic from road	Enfield		06-17-88
1161a	Dust	Traffic on Road	Enfield	Stevens, Heidi	06-16-88
1162	Dust	Traffic on Road	Enfield		06-17-88
1123	Soot	Ultrapower	Jonesboro	Cox, Connie	02-10-88
1202	Odor	Unknown	Bangor	Barry Goodell	09-20-88
1142	Open Burning	Worcesters Landfill	S. W. Harbor	Anonymous	05-17-88

11

TABLE 3. NEW JERSEY ODOR COMPLAINT INFORMATION ENFORCEMENT OPERATIONS

SUMMARY OF CITIZEN COMPLAINTS

New Complaints Received	Complaints Investigated	Complaints Referred for Investigation	Total Complaints Resolved
369	203	104	309

COMPLAINTS RECEIVED DURING PERIOD BY COUNTY AND BY TYPE OF COMPLAINT

County	Number	County	Number
Atlantic	1	Middlesex	15
Bergen	27	Monmouth	72
Burlington	24	Morris	3
Camden	10	Ocean	4
Cape May	11	Passaic	20
Cumberland	2	Salem	10
Essex	19	Somerset	12
Gloucester	20	Sussex	0
Hudson	25	Union	6
Hunterdon	6	Warren	18
Mercer	64		

Type of Complaint	Number
Odor	320
Smoke	14
Particulate	30
Open Burning	6
Emergency Response	6
Noise	3
Code Violation	3
Other Miscellaneous	5

COMPLAINTS RESOLVED DURING PERIOD

Verified Complaints	17
Complaints Not Verified	252

TABLE 4. RESPONSE FROM PUERTO RICO

COMMONWEALTH OF PUERTO RICO / OFFICE OF THE GOVERNOR

Environmental Quality Board

Mr. James Frazier
Institute of Medicine
National Research Council
2101 Constitution Ave. NW
Washington, D. C. 20418

> RE: ODORS FROM STATIONARY AND
> MOBILE SOURCES
> NATIONAL ACADEMY SCIENCES,
> 1979 REPORT

Dear Mr. Frazier:

Reference is made to your letter dated August 17, 1988 related to the subject matter.

The following are the answers to the questions contained in the odors survey:

1- This Commonwealth Agency keeps summaries of odors complaints including information of the number of people affected.

2- We keep sensory data. This data includes measurements for intensity and quality of odors. The sources and origin of odors are also determined.

For those measurements we have developed a <u>Procedure and Method for Perception and Evaluation of Precense of Odors in the Atmosphere.</u> Basically this Procedure and Method consist of a Committee or Group of persons who evaluate the quality and intensity of the odors using a uniform strategy already developed. The persons who mostly of the time are our engineering and technicians staff have already received a special odor detection and evaluation training which qualify them to be eligible to constitute what we call Objectable Odors Detectors Committee.

You are receiving an Spanish version of this Procedure and Method. Feel free to translate it and in case your prefer us to do it, please advise.

OFFICE OF THE BOARD: 204 DEL PARQUE ST. CORNER OF PUMARADA / MAILING ADDRESS: P.O. BOX 11488.
SANTURCE, PUERTO RICO 00910 / TELEPHONE: 725-5140

13

TABLE 4. RESPONSE FROM PUERTO RICO (page 2)

Mr. James Frazier

October 19, 1988
Page 2

3– Yes, we have summaries of court cases related to persons who have been exposed to odors, specially the objectionable ones.

The Regulation for the Control of the Air Pollution of Puerto Rico contains Rule #420 which is the one which contemplates the presence and evaluation of objectionable odors. This rule definitions is included in the Exhibit attached to the Procedure you are receiving.

So far we have been using this Method and Procedure, the results and experiences have been very positive and practical.

We would appreciate you evaluate and judge this procedure based on other techniques you have knowledge these surveys received from other sources.

For more information your might need or be interested, please contact Mr. Luis S. Matos, Senior Scientist, Air Quality Area, EQB, Tel. (809) 722-0077.

We really appreciate be included in your mailing list.

Cordially yours,

Santos Rohena Betancourt
Chairman

TABLE 5A.   TENNESSEE DIVISION OF AIR POLLUTION CONTROL
COMPLAINT INVESTIGATION FORM

Received Via   Received By   Reference Number _____

Phone _____   Name _____   ACCR [ ][ ][ ]
Letter ____(Copy   Office_____
        Attached)   Field
                                                  Office [ ][ ]
Date:_____

Complaint Log Number _____   County [ ][ ]

File Reference Number _____   Date
                                                  Received [ ][ ][ ][ ]
Complainant _____Phone _____

Address_____County_____ Investigator [ ][ ]

Directions_____   Pollutant

_____   ___01 Smoke
                                                   ___02 Dust
Referred to _____of_____Field Office   ___03 Odor
            Empl Name   ___04 Soot
        on_____   ___05 Fly Ash
                Date   ___06 Fumes
                                                   ___07 Acid Mist
     *   *   *   *   *   ___08 Other

Complaint Against_____   Effects

Address_____County_____   ___01 Soiling
                                                   ___02 Eye Irritation
Directions_____   ___03 Respiratory
                                                          Irritation
_____   ___04 Nausea
                                                   ___05 Plant Damage
Nature of Complaint_____   ___06 Material Damage
                                                   ___07 Reduced
_____   Visibility
                                                   ___08 Other
Action Taken (include time, date and results)
                                                   Weather
_____
                                                   ___01 Clear
_____   ___02 Cloudy
                                                   ___03 Fog
_____   ___04 Rain
                                                   ___05 Snow
_____

                                                   NO. OF
                              ELAPSED TIME   PEOPLE
SIGNATURE_____DATE_____ FROM RECEIPT_____ AFFECTED_____
        FIELD INVESTIGATOR   DAYS

NOTE: THIS FORM MUST BE RETURNED TO THE NASHVILLE OFFICE BY THE FIELD
       OFFICE SUPERVISOR AFTER THE COMPLAINT HAS BEEN INVESTIGATED.

Please Route To: Chief of Field Services

(In East Tennessee Route thru: GOW)   PH-2920
                                                        APC 4/87

15

ODOR COMPLAINT OBSERVATION/ASSESSMENT FORM
TENNESSEE AIR POLLUTION CONTROL DIVISION

Date_____   Time_____

PART I

1. Name and Address of Complainant _____
    _____
    _____
    _____

2. Source of Odor (if known) _____
    _____
    _____

3. Description of Odor   _____
    (see Odor Quality   _____
    Evaluation form)   _____
    _____

4. Intensity (strength) Chart
       0 No odor 1 Slight odor 2 Moderate Odor 3 Strong Odor
       4 Very strong odor

       Frequency, Duration and   DATE   TIME   INTENSITY
       Intensity of odor (attach   _____
       separate sheet if necessary) _____
       _____
       _____
       _____

5. Describe Weather Conditions:
       Was sky overcast?   YES_____   NO_____
       Was it raining or snowing?   Rain_____   Snow_____
       Wind Speed: Still_____ Low_____ Moderate_____ High_____
       Wind Direction _____

6. Does complainant claim that odorous conditions have caused or
    contributed to illness of any member of household?
                     YES_____   NO_____
    Is a physician's statement available for documentation?
                     YES_____   NO_____
    Is a physician's statement attached?
                     YES_____   NO_____

7. Describe Nature of Illness _____
    _____
    _____
    _____

16

ODOR COMPLAINT OBSERVATION/ASSESSMENT FORM
TENNESSEE AIR POLLUTION CONTROL DIVISION

PART II

AIR POLLUTION FIELD OBSERVATIONS

1. Observation   Location   Times   Intensity   Description
    Point   Code
    1   _____   ____to____   ____   _____
    2   _____   ____to____   ____   _____
    3   _____   ____to____   ____   _____
    4   _____   ____to____   ____   _____
    5   _____   ____to____   ____   _____
    6   _____   ____to____   ____   _____

    Intensity (strength) Chart:
    0 No odor   1 Slight odor   2 Moderate odor   3 Strong odor
    4 Very strong odor

    On Reverse, Draw a map indicating observation points,
    distances to source, plant boundaries, wind direction,
    location of complainant, and other area odor sources.

2. Wind speed and direction: _____   Cloud cover _____
    Data obtained from: _____ (use VEE manual method)
                                                    rain _____
                                                    snow _____
    Temperature _____   Humidity _____

3. Suspected source of odors   _____
    (Include company name and   _____
    number, emission point   _____
    description, odorous   _____
    chemicals and any other   _____
    information.)   _____

4. Remarks _____
    _____
    _____
    _____
    _____
    _____
    _____
    _____
    _____

Name _____   Signature_____
Title_____   Date_____

TABLE 6A. WISCONSIN – 1988 COMPLAINT TRACKING DATA

DIS FORM TYPE	COMPLAINT ON	DATE REC'D	COMPLAINT BY	INVEST. DATE	SRCE. TYPE	# DAYS TO INVEST.	COMMENTS
SED 4229 O/S	ADVANCED ENGINES DEVELOPMENT CORP.	05/02/88	ROSEMARY BLASCZYK	05/02/88	I	0	DIESEL TYPE ODORS - WHITE SMOKE
SED 4142 0	AFW FOUNDRY	08/10/88	DEBORAH SCHULTZ	08/17/88	I	7	FOUNDRY ODORS
SED 4023 0	AIRPORT SPUR	06/20/88	WILLIAM PELZEK	06/23/88	R	3	NO ACTION AT THIS TIME
SED 4279 0	ALDRICH CHEMICAL	04/08/88	BILL BARTMANN	04/13/88	I	5	STRONG ODOR
SED 4260 0	ALDRICH CHEMICAL	03/22/88	ANON	04/08/88	I	17	GAS ODOR
SED 1887 O/S/OT	ALICE PLOPPER	07/18/88	BUILDING	07/20/88	R	2	BURNING
SED 3753 O/S/OT	ALLIED POWER	07/13/88	LLOYD CHRISTENSEN	07/18/88	I	5	COMPANY WILL BE MOVING IN SEPTEMBER
SED 4369 O/S	ALLIS CHALMERS	06/27/88	DALE CROUSE	06/27/88	I	0	BURNING WOOD ODORS, DUST
SED 4195 O/S	APARTMENT	07/28/88	EDWARD HOFFMAN	08/02/88	R	5	DARK BLACK SMOKE FROM INCINERATOR
SED 1882 O/S	APARTMENT BUILDING	08/10/88	ANONYMOUS	08/12/88	I	2	ODOR AND SMOKE
SED 4218 O/S	APARTMENT HOME INCINERATOR	05/27/88	JAMES KINGBEIL	06/08/88	R	12	SMOKE FROM INCINERATOR
SED 4314 0	ARTISTIC LANDSCAPING	04/15/88	JUDY KINZLE	04/18/88	CH	3	MANURE SMELLS
SED 4315 0	ARTISTIC LANDSCAPING	04/14/88	DIANE PRONDZINSKI	04/18/88	C	4	PIG MANURE ODOR
SED 4118 O/S	ASTOR APARTMENTS	08/16/88	EDWARD HOFFMAN	08/16/88	R	0	SOOT-UNBURNT PARTICLES IN AIR
SED 4104 0	BESSON AUTO BODY	07/01/88	ARLENE BOMAN	/ /	CH	0	SPRAY PAINT ODORS
SED 4379 O/S	BON-AIRE APARTMENTS	06/02/88	DAVID BABCOCK	06/22/88	R	20	BAD ODORS
SED 4177 0	BREYER BROS.	07/08/88	SHIRLEY ARNOLD	07/14/88	A	6	RESOLVED BY INDUSTRIAL WASTEWATER
SED 3847 0	BREYER BROS.	07/08/88	EARL HEDER	07/14/88	A	6	RESOLVED BY INDUSTRIAL WASTEWATER
SED 3848 0	BREYER BROS.	/ /	LINDA DAY	07/14/88	A	0	RESOLVED BY IND. WASTEWATER
SED 3849 0	BREYER FARM	/ /	JANET HEDER	07/14/88	A	0	RESOLVED BY INDUSTRIAL WASTEWATER
SED 4114 O/S	BRIGGS & STRATTON	07/26/88	ANGIE JERMAN	07/27/88	I	1	ODORS
SED 4106 0	BRIGGS & STRATTON	07/15/88	JIM BENSON	08/04/88	I	20	BURNING ODORS
SED 4148 0	BRIGGS & STRATTON	06/23/88	RUTH STEMPSKI	06/23/88	I	0	CHEMICAL ODORS
SED 4390 O/S	BRIGGS & STRATTON	06/20/88	JOSEPH PAWALKA	06/23/88	I	3	SULFUR ODORS - SMOKE
SED 4082 0	BRIGGS & STRATTON	06/15/88		06/21/88	I	6	SOLVENT ODORS VERY BAD TODAY
SED 4304 O/S	BRIGGS & STRATTON	06/01/88	JOYCE POZEN	06/07/88	I	6	GAS FUMES, BLACK SMOKE
SED 4346 O/S	BRIGGS & STRATTON	05/24/88	MR. & MRS. SAUER	05/25/88	I	1	ODOR AND SMOKE
SED 4337 O/S	BRIGGS & STRATTON	05/20/88	ANONYMOUS	05/24/88	I	4	HAZE AND ODORS
SED 3722 O/S	BRIGGS & STRATTON	04/05/88	ANON	04/05/88	I	0	SMOKE AND ODOR

KEY TO 1988 COMPLAINT TRACKING REPORT

Heading	Description
DIS	District office where complaint was received.
FORM	Complaint Form number
TYPE	Type of complaint (e.g., O = Odor, S = Smoke, D = Dust, OT = Other)
COMPLAINT ON	Name of company or source of complaint
DATE RECD	Date complaint was received at the District Office
COMPLAINT BY	Name of Person registering the complaint
INVEST. DATE	Date the district office investigated the complaint
SRCE. TYPE	Type of source (e.g., G = Government, R = Residential, CM = Commercial, I = Industrial, CN = Construction, A = Agricultural, OT = Other)
# DAYS TO INVEST.	Number of days to investigate the complaint (from the day the complaint was received)
COMMENTS	Self explanatory

Bad odors right now.

| 1 NOTICE ACTIVE | 3 FURTHER SURVEIL. | 5 CORRECTIVE MEAS. | 7 REFER OTHER BUREAU | 9 NO JURISDIC |
| 2 NOTICE ISSUED | 4 NO CAUSE | 6 COND. CORRECTED | 8 INCORRECT ADDRESS | 0 INSPECTION |

▲ PREVIOUS COMPLAINTS   CAR NUMBER AND TIME   INSPECTOR'S SIGNATURE   DATE
▼ COMPLAINANT   PHONE

▲ ADDRESS   STREET   COMMUNITY   ZIP
WCHD-APC 9-07.0 COMPLAINT RECORD   WAYNE COUNTY HEALTH DEPARTMENT, AIR POLLUTION CONTROL 9

▼ LOCATION ADDRESS   STREET   COMPLAINT NUMBER▶ 02541   AREA AND GRID
27140 Princeton.   16-2351
Environmental Waste Control Inc.   Inkster   48141
▲ COMPANY AND/OR OCCUPANCY   COMMUNITY   ZIP
● COMPLAINT, WHEN   ▼ DATE   TIME   TAKEN BY   INSPECTOR
OCCURRED, PROBABLE
CAUSE AND SOURCE   3-15-88   8:00AM   Valet/LS   R.Elliott

The stink is rancid again down there, if there is any
way to sign a formal complaint I would like to sign
formal complaint agains them this morning too.

Companies and Relative Numbers of
Complainants and Complaint Situations

Company	Odor Situation	Complainant Households Involved in Agency Action
Wayne By-Products	Court action to obtain temporary restraining order resulted in consent decree and installation of control equipment.	100
TAS Graphics	Litigation related, newly installed control system is still being evaluated.	2-5
Usher Oil	Presently installing a new scrubber.	6
General Oil	Reactive scrubber with recently improved efficiency has greatly improved odor complaint situation. Modification of processes are now in progress.	15
Valassis Printing	Installation of a catalytic lead control system has eliminated odor problems.	12
Quaker Chemical	Process changes are being made to improve emissions from tall oil processing.	3-4
Mazda	Automotive paint spray lines result in solvent emission. An odor survey is in progress to determine relative magnitude of sources prior to installation of control equipment.	20-30
Chem-Met Services	A comprehensive control system of enclosures and scrubbers has been installed to capture all volatile emissions from treatment of waste chemicals using alkali causes of occasional emissions are being pursued.	20-30

The Bureau did not begin detailed recording of indoor odor complaints until October of 1988.

10-17-88   <u>elderly woman complained of smell in her apartment that left a bitter taste in her mouth and burned her skin-</u> The Bureau employee was unable to detect an odor. The woman had recently moved into the apartment and the carpets had been cleaned just prior to occupancy. Powdered boric acid had been sprinkled on the carpet to prevent pests.   It was suspected that the woman was sensitive to either some ingredient in the carpet cleaning preparation or the boric acid. The occupant covered the carpet with another and the problem seemed to lessen.

10-27-88   <u>exhaust fumes penetrated wooden flooring and irritated employees in business above-</u>   A Superfund cleanup requiring the use of bobcats and forklifts was taking place utilizing the basement. The Bureau employee tested and found high levels of CO in the business.

11-8-88   <u>odor in trailer near refrigerator that smelled like "something had died"-</u>   The occupant had suffered many serious illnesses since she had lived in the trailer and felt something might be "wrong" in the trailer that was causing her illnesses. Bureau employees could not find the source of the odor and the occupant chose to move.

11-10-88   <u>woman thought an odor in her house was making her ill-</u> Her health had steadily declined in the past few years. She seemed acutely sensitive to odors. Air fresheners were placed throughout her house because she felt the odor was so potent. She moved out of her house because her eyes burned and she became nauseated whenever she entered the house.   The Bureau employees did detect a chemical "pesticide" type odor.   The husband said that he had used chlordane on the house many years ago and the summer before.   Air monitoring for chlordane and heptachlor was done by the Bureau. Neither pesticide was found in significant amounts. The family decided to sell the house.

12-8-88   <u>several office units in a commercial complex complained of a strong glue smell-</u> A Bureau employee discovered an open container of industrial type glue in the warehouse behind one of the businesses.   However, after the container was sealed and removed, the smell seemed to reoccur occasionally.   The Bureau went to the location a second time but did not find a source of the odor.

22

TABLE 9. PENNSYLVANIA - PHILADELPHIA
UPDATE OF ODOR PROBLEMS
7/1/87 - 6/30/88

Category	Complaints	Violations
Petroleum Refining	0	3
Chemicals	61	19
Rendering	2	3
Metal Smelting	53	14
Incineration & Refuse Dry	0	18
Sewage Treatment	30	43
Metal Fabricating	62	29
Restaurants & Food	55	33
Transportation	9	4

TABLE 10.   VIRGINIA – COMPLAINT REPORT

LOGGED IN _____

D.A.P.C.   - REGION V
COMPLAINT REPORT

FILE UNDER _____

DATE _____ TIME _____ RECEIVED BY: PHONE ___; MAIL ___; PERSONAL CONTACT ___ .

COMPLAINANT'S NAME _____

ADDRESS _____

_____ PHONE _____

SOURCE (NAME & LOCATION) _____

_____

_____

DISTANCE COMPLAINANT IS FROM SOURCE _____

NATURE OF PROBLEM (SMOKE, ODOR, DUST, ETC.) _____

_____

TIME OF OCCURRENCE _____

IS IT GOING ON NOW? _____

HOW OFTEN DOES IT HAPPEN? _____

ADDITIONAL COMMENTS _____

_____

_____

RECEIVED BY: _____

ACTION TAKEN: _____

_____

_____

_____

_____

I have entered a brief statement of the action taken in the complaint log. _____

INVESTIGATED BY _____ DATE _____

NEW JERSEY'S APPROACH TO ODOR PROBLEMS

Leo Beck
Bureau of Enforcement Operations
Division of Environmental Quality
New Jersey Department of Environmental
West Orange, New Jersey

Vanessa Day
Bureau of Enforcement Operations
Division of Environmental Quality
New Jersey Department of Environmental Protection
West Orange, New Jersey

ABSTRACT

This paper discusses New Jersey Administrative Code
7:27, Subchapter 5 and Odor Complaint Guidelines
utilized by the New Jersey Department of Environmental
Protection. The Odor Complaint Guidelines were
implemented to ensure consistent enforcement by state
investigators. An odor intensity scale was developed
by the Department to provide a standardized description
of odor intensity. A case study involving a New Jersey
composting site is utilized to demonstrate the
procedures used to investigate odor problems. In this
case as well as many others, the Department was upheld
in its determination by the New Jersey Administrative
Courts.

INTRODUCTION

"Odor pollution is becoming a major factor in the total air pollution problem. Because odor is so easily noticed, complaints about obnoxious odors are frequently received by local air pollution control authorities." Nuisance odors are a growing concern, and odor regulations are among the more difficult enforcement problems since quantitative standards are difficult to define. During 1988, the New Jersey Department of Environmental Protection handled a total of 3752 odor complaints either directly or through county and regional agencies.

The State of New Jersey successfully addresses odor problems by enforcing a broadly-worded odor code, Subchapter 5. To ensure fair and consistent handling of all odor cases, the New Jersey Department of Environmental Protection has implemented guidelines for odor investigations. This approach has been upheld in decisions of the New Jersey Administrative Courts.

THE NATURE OF ODORS AND PUBLIC PERCEPTIONS

Although progress has been made in reducing air pollution problems, the Department continues to face an ever-increasing number of odor complaints. The public is more aware of contaminants in the air they breathe. They are concerned about short-term and long-range health effects. It should be recognized that public outcry against "toxic odors" can often be described by the terms "unpleasant" or "irritating," rather than life threatening.

Odors are very subjective, with responses varying greatly from individual to individual. Odor descriptions from the general public tend to be unreliable. How pleasant or unpleasant an odor is perceived, is often a matter of association. On a recent odor investigation, neighbors were interviewed and asked to describe an odor which the complainant had characterized as "an irritating, sickening chemical odor." During interviews with other complainants, one described the odor as "hamburgers" and "rather pleasant," while another neighbor defined it as the odor of "pickled herring." The actual odor source turned out to be a food processor who was making 1000 gallon batches of barbecue sauce that day, whose main ingredient was cooked onions.

SUBCHAPTER 5

In New Jersey, the regulation used to prosecute
polluters who cause odor problems is New Jersey
Administrative Code 7:27, Subchapter 5. This
regulation prohibits "air pollution." Air Pollution is
defined as "the presence in the outdoor atmosphere of
one or more air contaminants in such quantities and
duration as are, or tend to be, injurious to human
health or welfare, animal or plant life or property, or
would unreasonably interfere with the enjoyment of life
or property." Subchapter 5 further states "no person
shall cause, suffer, allow or permit to be emitted into
the outdoor atmosphere substances which shall result in
air pollution." The key phrase here is "unreasonably
interfere with the enjoyment of life or property." It
is obviously subject to a certain amount of judgment
and interpretation. According to the Department
guideline, to cite a Subchapter 5 violation the
Department must be able to verify and prove that a
person (as defined in the regulation) has caused or
allowed the release of air contaminants to the outdoor
atmosphere which had an effect on people or the
environment. Odor is an air contaminant as defined by
the New Jersey Superior Court in the Department of
Health v. Owens-Corning Fiberglas Corporation legal
decision.

Guidelines for the Enforcement of Subchapter 5 Complaint Investigations

To ensure fair and consistent enforcement, the
Department has developed guidelines since odor problems
vary from case to case. The guidelines provide
flexibility to the investigator if the situation
warrants, therefore a deviation will not necessarily
invalidate the investigation.

Complaint investigations are the Department's highest
priority. The Department considers timeliness an
important factor in the resolution of odor problems.
Odor investigations are conducted in accordance with
these guidelines.

1.   When complaints are received by the
     Department, they are immediately forwarded to
     the Regional Enforcement Officer. He will
     then assign the complaint to an investigator.

2.   The investigator should immediately contact
     the complainant to ascertain the current
     situation. The investigator should obtain
     the following pertinent information to assess
     the problem:

a. Is the problem currently occurring?
b. A description of the odor which includes the nature, intensity and duration.
c. The suspected source.
d. Any physical effects incurred by the complainant.

3. If a complainant identifies a suspect facility or source, the investigator should quickly review the facility's files. Problematic facilities in an investigator's assigned geographical area should be familiar to the investigator, therefore reducing or eliminating the necessity for this review.

4. The investigator should proceed to the complainant's location and record the arrival time. In order for a Subchapter 5 violation to be processed the investigator must verify the presence of the odor at the complainant's residence or work place. When odors are verified the complainant will be requested to complete a Statement of Complaint form (Attachment 1). The complainant must be advised that he or she will be needed to testify in court if a violation is cited. If the complainant does not testify, the Subchapter 5 violations may be subject to substantial challenge during the administrative hearing process. The complainant's identity will be kept anonymous until the administrative hearing is held.

5. After the odor is verified in the presence of the complainant, the investigator proceeds to conduct a 360 degree odor survey of the suspected source. When an odor is detected the investigator records the following information: the characteristics of the odor and weather conditions including wind direction and speed. Also, any physical effects on the investigator or the complainant caused by the odor should be noted. The intensity of the odor should be evaluated in accordance with Table 1 and recorded.

The odor intensity Scale developed by the Department is not meant to be used as a scientific measure to determine what odor constitutes an enforceable complaint situation. Rather it is a factor that if taken into consideration with the nature of the odor and the duration will help the inspector make a judgment that the complainant's enjoyment of life and property

was unreasonably interfered with A low - scale
strawberry odor may be considered acceptable over short
intervals scattered throughout the day, but may be
judged 1 when it persists for an 8-hour shift.

TABLE 1.

SCALE/DESCRIPTION	ODOR INTENSITY DESCRIPTION
0	Odor not detectable.
1 - Very Light	Odorant present in the air which activates the sense of smell but the characteristics may not be distinguishable.
2 - Light	Odorant present in the air which activates the sense of smell and is distinguishable and definite but not necessarily objectionable in short durations. (Recognition Threshold)
3 - Moderate	Odorant present in the air which easily activates the sense of smell, is very distinct and clearly distinguishable and may tend to be objectionable and/or irritating.
4 Strong	Odorant present in the air which would be objectionable and cause a person to attempt to avoid it completely, could indicate a tendency to possibly produce physiological effects during prolonged exposure.
5 Very Strong	Odorant present which is so strong it is overpowering and intolerable for any length of time and could tend to easily produce some physiological effects.

CASE STUDY (TOWNSHIP OF PISCATAWAY DPW)

The Township of Piscataway hosts the Kilmer Composting
Project which is a joint effort of six Middlesex County
municipalities. The Project was started in 1984 to

reduce the amount of waste being sent to County
landfills. A low-technology process is utilized at the
Kilmer Composting site in which leaves are transported
to the site during the fall months, placed in windrows,
wetted and over time, turned with heavy machinery. The
next fall season, the leaves are sufficiently
decomposed to be shredded and distributed as compost.

During the fall of 1986, several municipalities decided
to deliver numerous plastic bags containing leaves.
When leaves were delivered to the site in plastic bags
an additional debagging step was needed before
composting. In the fall of 1986, the Kilmer Project
was the largest in the State. The Kilmer Project
Management Team had not anticipated how many bags would
arrive at the site. Thousands of plastic bags
containing leaves were delivered but manpower was not
available to debag them. In June 1987 a debagging
apparatus was purchased to cope with the problem. Some
of the bags from the fall of 1986 were opened on June
23, 1987. By that time the composting process had
become anaerobic and acidic, resulting in foul and
noxious odors being emitted. Several complaints were
filed. Employees at the Middlesex Water Company, the
Kilmer Project's nearest neighbor, complained of itchy,
burning eyes and of an odor of rotted vegetation
throughout the day.

An inspector was dispatched to the area of the
complaint. The inspector conducted a 360 degree
neighborhood odor survey. Odors were detected on the
complainant's property nearest to the source. Figure 1
is a diagram of the location where odors were detected.
The odors were classified by the inspector as # 4 on
the Odor Intensity Description Scale (Table 1). The
wind conditions were noted by the inspector as 2 to 5
miles/hour north-northwest.

Several complainants filed a Statement of Complaint.
As a result of the investigation, a Field Record of
Violation was filed. The New Jersey Department of
Environmental Protection issued an Administrative Order
and assessed a penalty of $2000.00.

The Township filed for a hearing to contest the order.
The basis for the appeal was whether public policy
should require a public agency to pay a penalty. The
Township was engaged in large scale leaf composting for
the public good. Since the process was relatively new,
the Township faced many unknowns such as the effects of
opening a large number of plastic bags containing
decomposing leaves all at one time. The Department's
position was the air pollution created was no less so
because the violator was a public agency. The lowest

possible penalty was assessed by the Department for this type of violation, because it was already taken into consideration that the violation was unintended and did not recur. The judge in the case agreed with the Department and ordered that the $2000.00 fine be paid.

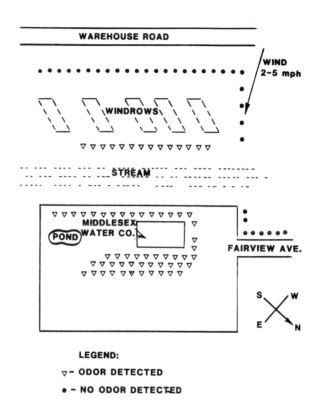

FIGURE 1:   INSPECTOR'S ODOR INVESTIGATION MAP

CONCLUSIONS

Since standard methods such as odor panel testing and the establishment of standardized sensory scales are not suitable for field investigations, the New Jersey State Department of Environmental Protection has attempted to standardize its odor enforcement by issuing procedural guidelines and an "Odor Intensity" scale, to be used by its field personnel in odor complaint investigations. This approach has proven successful in withstanding legal challenges of odor

violations in the State's Administrative Court System
and subsequent legal challenge before the Appellate
Division of the Jersey Superior Court.

REFERENCES

(1)   Cheremisinoff, Paul A. and Richard A. Young.
      Pollution Engineering Practice Handbook. Ann
      Arbor, Michigan: Ann Arbor Science Publishers,
      Inc., 1981.

(2)   Department of Environmental Protection, Division
      of Environmental Quality vs Township of
      Piscataway, Department of Public Works. Initial
      Decision, OAK DKT. NO. EEQ 2661-88, Agency DKT.
      NO. 88-63.

(3)   Department of Environmental Protection, Division
      of Environmental Quality vs Township of
      Piscataway, Department of Public Works. Final
      Decision, OAL DKT. NO. EEQ 2661-88, Agency Dkt.
      NO. 88-63.

(4)   Hesketh, Howard E. and Frank L. Cross. Odor
      Control Including Hazardous/Toxic Odors.
      Lancaster, Pennsylvania: Technomic Publishing
      Co., Inc., 1989.

Q-098 (1/89)

New Jersey Department of Environmental Protection
Division of Environmental Quality — Environmental Enforcement

STATEMENT OF COMPLAINT

Name _____

Address _____

Phone _____ Age _____

Address Where Employed *(if applicable)* _____

Lived (Worked) at Above Address About _____[month(s)/year(s)]

Nature of Complaint _____

_____

_____

Source of Complaint _____

Recently Noticed On _____ At About _____

Distance from Facility to Home (Work)   Approximately _____ Feet

This Condition has been continuing for about _____ (Days, Months, Years)

Describe Activity and where problem was noticed _____

_____

_____

_____

Physical Effect(s) and/or Condition(s) _____

_____

I have written the above statement and it is true. Further, I am aware that I may have to present testimony at an adjudicatory hearing pertaining to this statement of complaint.

_____
Signature

_____
Date

Witness of Signature:

_____
Signature

_____
Date

33

Introduction and Related Practical Aspects of Odour Regulations in the Netherlands

A. Ph. van Harreveld
Project Research Amsterdam BV

Abstract

Odour annoyance remains a high priority environmental problem because of its immediate effect on population. Regulatory tools are urgently required. In the Netherlands zoning around stables was the first form of regulation. As a general approach characterisation of sources using odour threshold measurements, combined with dispersion models and odour concentration standards seems the most straightforward and practicable regulatory approach. In the Netherlands this approach has been used and met with acceptance, in spite of the existing inaccuracy of olfactometry. In 1984 a provisional odour concentration standard has been published. Improvement of repeatabilty and reproduceability of olfactometry is an imperative pre-requisite for regulation based on these measurements. This implies that inter-laboratory differences must be reduced from presently common values of factor 20–30 to within a factor 2–3. Standardisation and quality assurance structures are required to achieve the desired performance of olfactometry. Recent experiences with advanced, computer controlled olfactometers, such as the Olfaktomat, indicate that the performance required as a precondition for odour regulation on the basis of source characterisation using olfactometry can be achieved on a routine basis. The developments in the Netherlands of the past fifteen years can be considered a model for the developments that could take place in the U.S. and some European countries, as a result of continuing and increasing complaints about odour nuisance as an indicator of general environmental annoyance.

1 Odour nuisance and introduction of standards in the Netherlands

1.1 Introduction

To create the need for odour regulations in a society certain conditions must be present:

1. A high degree of industrial and agricultural activity at such a scale that odours are released in significant amounts.

2. A high population density, which forces industry, agriculture, recreation and housing to be close neighbours.

3. A high level of wealth and education, with matching expectations where quality of life is concerned.

4. A political setup that weighs interests of citizens on a similar scale as those of larger economic interests.

All these factors are prominently relevant to Dutch society. The population density is high, at 436 per square kilometer. The country is wealthy, with a per capita income (1985) of 18900 Dutch Guilders[1], and this wealth is distributed very evenly. The average level of education is high and people are well aware of their political and legal rights. The industrial sector is large in volume, and highly diversified. The agricultural sector is economically very important and highly intensified (Holland is the third largest exporter of agricultural products of the world, measured in value, after the U.S. and France[2]).

In this situation about every citizen with a nose and a telephone turns into an odour monitoring station. When odour nuisance occurs it can be established instantly, as the complaints are phoned in to the relevant authorities. In Holland the number of complaints, phoned in to environmental complaint registration lines is considerable, with a maximum of 68.2 complaints per 10,000 people in the industrial Rijnmond area [1].

The need for odour regulation became apparent and acknowledged by the authorities around 1975. Greater importance was given to environmental issues by the public, and most registered complaints were about odour annoyance. Research showed that 15% of the Dutch population experiences odour annoyance from time to time, while 5% of the population perceives considerable annoyance due to odours[3]; the equivalent of 700,000 considerably annoyed potential voters. The problem at the time, however, was that no tools were available to tackle the problem. Odour measurement was not available in a straightforward format. Chemical analysis was generally of little use in the most common cases where two or more chemicals caused the odour. Practicable means for odour emission

---

[1] 1 US$ = 2.18 Dutch Guilders, August 1989
[2] 1987 data from Dutch Economical Information Service (EVD)
[3] data Central Bureau of Statistics, 1987

abatement, such as the now widely used compost filters, were at that time largely unavailable or not common knowledge.

Now, after considerable effort and research, tools are available to characterise, quantify and regulate sources of odour annoyance and to reduce the annoyance using a range of methods.

1.2 A first regulatory step: zoning for stables and sewage treatment plants

The first step in regulatory effort was therefore a simple, straightforward and highly practical approach. The Health Inspectorate decided to tackle a widespread class of problems, that gave rise to a large number of conflict situations: stables for intensive rearing and fattening of pigs, poultry or veal calves. These stables need a permit according to the Nuisance Law, and are usually located very near to residential areas. Also, with increasing migration from the city to suburban and agricultural areas, the non-agricultural neighbours became more numerous and more demanding where nuisance was involved. A number of health inspectors decided to go out in the field to collect observations at a number of sites. This led to a zoning guideline [2] that prescribed a minimum distance between a stable and residential housing. The width of the buffer zone was determined by the number of animals that were housed in the stable only (see figure 1 from [2]). The pig became the reference animal, and other animals were expressed in 'pig units' using conversion factors as indicated in table 1 from [2]. This practical approach and the conversion factors were later supported by research, using sensory measurements (olfactometry) and chemical analysis as parallel tools. This research led to the conclusion, among others, that one 'pig unit' is the equivalent of an emission of 2 odour units per second.

This zoning regulation was and is applied with satisfactory results. Later, in 1977, sewage treatment plants, another common source of odour annoyance, were also included in a zoning guideline.

1.3 Questions remained to be answered before setting a general regulatory framework

For industrial sources the situation was more complex. Here, each situation was different, and needed careful study. A quantitative approach to regulation was urgently required. A number of approaches and possible tools for assessment of odour nuisance situations were considered and studied, in a number of systematic studies that were mainly initiated and funded by the Ministry of Housing, Physical Planning and the Environment (Ministry VROM[4]), from about 1982 onwards.

---

[4]Ministerie van Volksgezondheid, Ruimtelijke Ordening en Milieu

The following approaches were considered:

1. Assesment of odour nuisance using field panels

2. Assesment of odour nuisance using large scale population panels

3. Assesment of odour immission levels, using emission measurements and dispersion models

   (a) Chemical analytical approach
   (b) Approach using olfactometry

These approaches were considered with a view to providing regulatory tools.

1.3.1 Using field panels

The use of expert field panels that 'sniff out' the situation around a source received the least attention. Although this method was used to formulate the zoning guideline for stables, the general attitude among experts was that a good impression could be obtained of the maximaum distance at which the source can be perceived, but that it did not provide the sort of quantitative description that could be used objectively in specific cases.

1.3.2 Using population panels

A great deal of effort was put into research that was aimed at the development of a measure of odour annoyance using population panels [3]. These panels, recruited from residents usually living around large scale and or complex sources, would be asked to go out at a specific time every week to assess the odour nuisance in the ambient air at their residence. Although the results are very interesting, this method does not seem to lead to an effective regulatory tool. It does take into account the annoyance of perceived odours, but it fails to distinguish sharply and in a quantitative manner in most cases. Only in situations involving very large, complex sources the method can indicate general trends. The method seems to be to cumbersome to provide a useful regulatory tool for specific sources. In practical terms it is not being used as a regulatory instrument.

1.3.3 Using source characterisation with chemical analytical methods, combined with dispersion modelling

The chemical analytical approach was tempting. Analytical procedures were familiar, and the interpretion of dispersion was equally well established. In regulation of air quality, emission standards based on acceptable immission concentration levels, using source characterisation combined with calculations with dispersion models, is a very common approach.

There was (and is!), however, a major problem: it turned out to be virtually impossible to predict odour using the results of gas chromatography, even when combined with mass spectrometry. Odours are often determined by trace concentrations of a large variety of chemicals, very smelly even at levels around or well below detection thresholds. Even if the most powerful detection capabilities would be able to grab the vast number of components in emissions that can be perceived at trace concentrations (sub-ppb), a more fundamental problem would remain. The combined sensory perception of mixtures of even as little as two or more odorants turns out to be unpredictable. The odour is rarely an additive result of the concentrations in the mixture. A good scientific parallel is found in psycho–physical research into taste of sugars, where the effect of three or more sugars in a mixture still seems to defy proper modelling. Furthermore, these problems turned out to be insurmountable, and this approach, after extensive attempts, was finally abandoned as a practical approach in applied odour research in Holland.

1.3.4 Using source characterisation with olfactometry, combined with dispersion modelling

Source characterisation using olfactometry could hardly be considered a tempting option. The measurement of odour thresholds was complex, often leading to widely differing results. Published odour thresholds for chemical substances are silent but clear testimony to these analytical pitfalls. Olfactometers, although they were widely used in the mid-seventies, were experimental, the procedures for sampling and measurement equally so. Repeatablity was often unknown and reproduceabilty was shown in various inter-laboratory comparitative studies to be well below acceptable levels. Results differed between laboratories with factors 10–1000. Differences within a factor 20–30 are not uncommon. The *source characterisation→dispersion model→immission regulatory* approach was however the most straightforward, and therefore the approach using olfactometry as a main provider of basic data remained by far the most popular, in spite of the analytical problems. It provides quantitative data about specific situations, that can be used in a way familiar to most air quality regulatory agencies.

As a consequence, great effort was put into the improvement of olfactometric methods and equipment. The need for standardisation was felt. In a first step to such a standardisation, a guideline published by the Ministry VROM in [4] the aim was stated: differences between olfactometric laboratories should be no more than a factor three. Later the objective was set even more stringently to differences no more than a factor two [6].

1.4   1984: A provisional odour concentration standard

In 1983 an 'interim odour concentration standard', was tentatively formulated by the Ministry VROM [5]. Investigations around existing sources had indicated that odour annoyance seemed to be at acceptable levels when the LTFD dispersion model indicated that an hourly average concentration of 1 ou·m^{-3} was not exceeded more than during 0.5% of yearly hours (the 99.5% isopleth).

After consultation the Ministry VROM in 1984 published the *provisional odour concentration standard* or OSCN[5]. This standard was not yet legally binding, but appeared in a long term policy document, the Indicative Long–term Programme for Air Quality 1985–89 [7].

The provisional standard was formulated as follows:

For existing installations, an hourly average concentration of 1 odour unit per m^3 may not be exceeded for more than 2% of time at locations with domestic dwellings.

This corresponds to the area within the 98-percentile isopleth. For new installations a more stringent criterium was chosen:

no more than 0.5% of yearly hours an hourly average concentration of 1 odour unit per m^3 may be exceeded if domestic dwellings are present.

The isopleths must be calculated using a national dispersion model, based on a Gaussian plume model (narrow plume approach), the LTFD model[6]. Low cost software for this model on PC is available, in a commercial package developed by Project Research Amsterdam.

The impact of the provisional OSCN standard was quite noticeable. The demand for a straightforward, quantitative method for regulatory purposes turned out to be considerable. As soon as this regulatory tool was available, even provisionally, a large number of municipal and provincial authorities started to base their (very legally binding!) Nuisance Law licences on the OSCN. Appeal cases at the State Council, the highest court for administrative appeals, were from that time onwards judged with this sorely needed guideline in hand. Thus, although not included in a law, a de–facto odour concentration standard has been applied since 1984.

1.5   Odour concentration standard part of legal framework

The provisional OSCN standard has been applied in hundreds of cases. The government is aiming to make an odour concentration standard, along these lines, into Law as a General Administrative Order[7][6].

To provide the lawmakers with a feel for the practical impact of this standard, and to underpin the final form of the standard, an inventory is now being made of all cases in which the OSCN has been applied, in one form or another. These cases run into the hundreds. My firm, carrying out the inventory for the Ministry VROM, expects to report the results of this study in December 1989. The way in which to include the odour concentration standard in the legal framework that can be used to regulate odour nuisance is expected to be determined by the Minister[8] in the course of 1990.

---

[5] Ontwerp Stank Concentratie Norm
[6] Long Term Frequency Distribution model
[7] Algemene Maatregel van Bestuur (AMVB)
[8] Secretatry for the Environment

2 Practical aspects surrounding introduction of the odour concentration standard

The regulatory and legal framework looks quite established, quite straightforward. However, the legal tools far exceed the tools that are available to the consultant to provide data on which these regulatory decisions are based. In an inter–laboratory study, in which twelve Dutch laboratories involved in olfactometry participated, the diffences between laboratories proved to be very dissapointingly large: up to a factor 1,000 [8]. This was quite a shock, as most of these laboratories produce data that are used in regulatory processes. Sometimes very large investments are required from industries to comply with the OSCN standard. The data on which these decisions were based turned out to be of insufficient reliability, in some cases. The demand for analysis had evidently exceeded supply. Olfactometry, an analytical technique at that time still in its research stage, found itself suddenly in the regulatory arena.

It was not surprising, in these circumstances, that industry expressed their doubts concerning the impending legal implementation of an odour concentration standard, when the olfactometric measurements on which decisions were to be based were apparently not reliable. It was equally predictable that the authorities, in this case the Ministry VROM, took the initiative to improve the standard of olfactometry, and achieve the aim stated previously, to bring reproduceabilty between laboratories within a factor 2–3.

2.1 Standardisation of olfactometry in the Netherlands

At the initiative of the Ministry VROM, the Netherlands Normalisation Institute (NNI) created a sub–committee 'Standardisation Olfactometry' of their established Committee on Air Quality 390 146[9]. This sub–committee was entrusted with the task of formulating a standard for olfactometry, that would reduce the differences between olfactometry laboratories to within a factor 2–3. In this committee all those who were actively involved in olfactometric analysis were invited, as well as representatives from the NNI, Ministry VROM and the Netherlands Metrology Service, recently renamed to Nederlands Meet Instituut (NMI)

The work of the subcommittee on standardisation of odours has led to the publication of a Dutch Provisional Standard[10] for olfactometry in mid 1990, known as NVN2820. This standard, available from the author in English translation, is supposed to become a firm standard after a three year revision process.

The standard contains a number of procedural requirements and recommendations, summarized in table 2. Furthermore a number of requirements have been defined for the instrumentation to make and present diluted odours to panelists. These requirements are very stringent, and as far as I know the most stringent in existence anywhere in world of olfactometry. Also procedures and

---

[9]Structured along guidelines ISO/tc 146
[10]Nederlandse Voor Norm

equipment for sampling of odours, an underestimated but serious and frequent source of error, will be included in the standard. Finally quality assurance procedures will be formulated. These include procedures for instrumental calibration of olfactometers, as a measure of internal quality control. Most importantly, however, quality assurance procedures for the entire analytical method have been formulated, and the structures that are required to effectively set up internal *and* external quality assesment are now being prepared.

2.2   Quality assurance of olfactometry

Internal quality control is meant to enable olfacometer operators to improve their own performance. External quality assurance gives the end users of olfactometry results with information about their reliabilty. This information also creates a strong impetus for competing laboratories to strive for improved performance.

The structure for quality assurance consists of the following elements:

1. Internal quality assessment

   (a) Facilities and methods for tracer calibration of dilution apparatus used in olfactometry.
      i. suitable analytical method for tracer method, sensor with known response characteristics over five decades.
      ii. suitable reference materials for tracers used.
      iii. evaluation procedures.

   (b) Facilties and methods for integral quality assessment of repeatability of total method (including olfactometer, panel, operator, procedure, odour room, clean dilution air).
      i. Suitable and stable odourous reference materials for regular use in control measurements, i.e. butanol and/or H_2S calibration gases.
      ii. Procedures to carry out internal evaluation of results.

2. External quality assessment.

   (a) A structure and organisation that can provide odourous reference materials, traceable to primary standards, and provide samples with concentrations unknown to the olfactometer operators at regular intervals.

   (b) A structure and organisation that can evaluate results of blind testing and give feedback to the participating olfactometrists.

   (c) A method to provide results concerning reproduceabilty and repeatabilty of olfactometry laboratories to end users of these services.

(d) A system to encourage a tendency to improve quality and to exclude those facilities that do not match up to the standard that can be considered achievable, such as regular publication of findings.

A structure along these lines is now being set up, with the Netherlands Measurements Institute (NMI) in the pivotal role of providing the reference materials, traceability, data processing and reporting.

An important stimulus to participate in the costly programme is that compliance with the standard, including external quality assessment, will be required by the authorities if a laboratory provides data for official use.

The external quality assurance programme, running since January 1990, involves monthly measurements of four samples, two of n-butanol and two of H_2S, both in a high and a low concentration range, comparable to concentrations of odourants in commonly encountered emissions (10^1 to 10^5 of dilutions to threshold). After ten of these data sets have been produced, the repeatabilty and reproduceabilty of a participating olfactometer can be established. These data will be published at regular intervals, in order to inform those who commission and use olfactometric data.

2.3   Recent developments

The standardisation and the apparent upgrading of the demands on olfactometry have led to a number of developments. Generally it was felt that the dilution equipment was the cause of a lot of trouble and variability, caused by lack of repeatabilty of dilution levels, memory effects, high risk of operator mistakes, ad–hoc pre–dilution methods necessary because of limited dilution range etcetera.

This has led to development of new, radically different types of equipment for olfactometry. Our company developed an olfactometer, the **olfaktomat**[11] that was designed not only to solve instrumental problems, but to control the entire operational protocol and control of the measurement, inluding the panel procedure, using integrated microprocessor control and software. This virtually excludes operator mistakes and -bias. All knowledge we had acquired in seven years of doing odour measurements has been built into this software.

---

[11]olfaktomat is a registered trade mark of Project Research Amsterdam bv

The main aim however was to get the dilution apparatus right:

dilutions range	4–100,000
repeatability	better than 10%
setting dilutions	within 30 seconds
control principle	Fully software controlled, no buttons, no dials
memory effects	absolutely minimized by avoiding any measurement component in sample flow

Since early 1988 five olfactometers of this type have been operational, three at consulting firms in the Netherlands and two at a governmental research institute in the United Kingdom[12]. Well over a thousand threshold measurements have been performed. Our own research has demonstrated that the repeatability of **olfaktomat** measurements are within a reasonable range. Preliminary results of comparitative measurements with these five instruments used at different laboratories, using 50 ppm n-butanol calibration gas, show very encouraging results, that indicate that the aim of olfactometry with a reproduceability of within a factor 2–3 can be achieved, on a routine basis.

In table 3 threshold values for n-butanol are presented that were obtained in the first half year of monthly measurements in a round robin that is to last 2–3 years. Twelve olfactometers in the Netherlands now participate in this inter-laboratory test that was also mentioned in the previous paragraph. The concentrations of the reference materials varied over a wide range and were unknown to the operator performing the analysis. The threshold for n-butanol based on these measurements is 184 μg·m^{-3} with a 95% confidence level of 71–498 μg·m^{-3}.

3 International perspective

3.1 Towards European traceabilty (1992!)

As a scientist from one of the smallest countries in the European Community I cannot resist adding some international perspectives. Several countries in the EC are formulating odour standards as regulatory tools [6]. In the EC the advent of the unified market in 1992 stresses the need for harmonisation of methods and standards. More important is the creation of structures around the existing standards offices that would ensure traceabilty of olfactometric data. A quality assurance program as that now being implemented in the Netherlands would be very useful to these ends, if it were to be extended to a European scale. The availability of traceable calibration gases plays a crucial role in this aspect. The Standards Offices of the member states are the logical providers of traceability.

---

[12] AFRC Institute of Grassland and Environmental Research, IGER, Hurley, Maidenhead

3.2 The U.S. perspective

It would seem that in the U.S., while being at the forefront ten years ago, environmental scientists have almost given up on olfactometry and odour standards based on it. The demand for regulatory tools, however, is considerable. The public awareness of the value of a clean environment continues to increase sharply, as is illustrated in figure 2 from [9]. A trend towards more localised environmental action among citizens living around industries has been observed [9]. These groups tend to look at the Nuisance Law as a useful tool to get a legal grip on the environmental performance industries. As it happens the one air quality factor *anyone* can measure plays a role in this nuisance law. Therefore, in my opinion, a development parallel to the regulatory process that occurred in the Netherlands can be expected to take place in the U.S., with increasing demand for quantitative assessment, regulation and abatement of odours.

4   Conclusions

1. The environmental annoyance caused by odour remains a high priority because of its immediate effect on the increasingly educated and vocal population in wealthy developed countries.

2. Straightforward regulatory tools are urgently required.

3. Characterisation of sources using odour threshold measurements, combined with dispersion models and odour concentration standards seem to be the most straightforward and practicable regulatory approach.

4. In the Netherlands this approach has been used since 1984 and has met with acceptance, in spite of the existing inaccuracies of olfactometry.

5. Improvement of repeatability and reproduceability of olfactometry is an imperative pre–requisite for regulation based on these measurements. Inter laboratory differences should be less than a factor 2–3

6. Standardisation and quality assurance structures are required to achieve the desired performance and traceabilty of olfactometry.

7. Recent experiences with advanced, computer controlled olfactometers indicate that the performance required as a pre–requisite for odour regulation on the basis of source characterisation using olfactometry can be achieved on a routine basis.

8. The developments in the Netherlands of the past fifteen years can serve as a predictive model for the developments that could take place in the U.S. and some European countries as a result of continuing and increasing complaints about odour nuisance as an indicator of general environmental annoyance.

References

[1] Feenstra, J.F., Brouwer, J., *Analysis of environmental complaint telephone lines in the Netherlands (preliminary study)*, in Dutch, Instituut voor Milieuvraagstukken, 84/10, Amsterdam, 1984.

[2] Anon., *Brochure Veehouderij en Hinderwet.*, Ministerie van Landbouw en Visserij, Staatsuitgeverij, 's Gravenhage, 1976.

[3] Köster, E.P. et.al., *Population panels in odour control: the development of a direct method for judging annoyance caused by odours*, Proceedings of International Symposium 'Characterisation and control of odiferous pollutants in process industry', Louvain–la–Neuve, April 1984, pp.99–126.

[4] anon., *Standaardisatie Olfaktometers*, Publication series on Air number 49, Ministry VROM, 's Gravenhage, September 1985, ISBN 90 346 0657 0.

[5] anon., *Geurnormering*, Publication series on Air number 11, Ministry VROM, 's Gravenhage, 1983, ISBN 90 346 0169 2.

[6] Wijnen, H.L.J.M.,*Guideline for olfactometric measurements in the Netherlands: Comparison with Western European Guidelines*, In: Odour Prevention and control of organic sludge and livestock farming, ed. Nielsen,V.C., Voorburg, J.H., L'Hermite, P., Elseviers Applied Science Publishers, London, 1986.

[7] Anon., *Indicatief Meerjarenplan Lucht 1985–1989*, Staatsuitgeverij, 's Gravenhage, 1984.

[8] anon., *Ringonderzoek van Olfaktometers in Nederland*, Publication series on Air number 80, Ministry VROM, 's Gravenhage, 1989, ISBN90 346 2036 0.

[9] Suro, R., *Grass roots groups show power battling pollution close to home*, The New York Times, New York, Sunday July 2 1989.

1 pig unit	equals	1	fattening pig
1 pig unit	equals	1	veal calf
1 pig unit	equals	1.5	sows
1 pig unit	equals	15	chickens (layers), wet manure storage
1 pig unit	equals	30	chickens (layers), dry manure storage
1 pig unit	equals	100	chicks (young broilers)

Table 1: Conversion of livestock into *pig units* for odour zoning purposes

Methodology

Method	Forced choice, 1 out of 2 (or more)
number of dilution steps	minimally 5
Order of dilution steps	increasing odour intensity
Difference in concentration from step to step	between $\sqrt{2}$ and 3 with 2 as 'common' value
Number of repetitions of dilution series	at least 2
Time allowed for stimulus eavluation	At least 10 seconds, 20 at most
Interstimulus time	At least 30 seconds

Panel selection

Number of panelists	Provisionally 8, but remains to be defined

Instrumentation

Dilution range	At least up to 25,000 dilutions
Actual dilution	Within 5% constant for duration of one setting
Repeatabilty of actual dilution	Within 10% of setpoint
Stabilisation time between settings	Sufficient to avoid memory effects
Air flow from sniffing port	At least 20 $l \cdot min^{-1}$
Shape of sniffing ports	Conical (angle no more than 7 degrees), with diameter between 4 cm and 7 cm

Calibration

Technique	Calibration of actual dilution factors using tracer gas
Frequency	As often as necessary to avoid significant drift (at 95% reliability, Student-t test)
Reporting	The calibration status has to be indicated in a clearly visible manner on or near the olfactometer
Air flow from odour ports	Difference between ports no more than 5%

Odour room

Odour free environment	Recirculated air via activated carbon filters, recycling 10–20 times an hour; refreshment ratio of air: 10%
Temperature	21–23 °C
Relative humidity	40–60%
Seating	Comfortable
Static electricity buildup	To be avoided

Table 2: Principal characteristics of the draft of the Dutch Preliminary Standard on Olfactometry (NVN Olfactometrie)

	Odour threshold values		
	$^{10}\log\mu g \cdot m^{-3}$	$\mu g \cdot m^{-3}$	ppb
Number of observations	14	14	14
Minimum	1.838	69	23
Maximum	2.708	511	169
Geometric mean	2.265	184	61
standarddeviation	0.208	n.a.	n.a.

Table 3: Odour threshold values for n-butanol, determined with the **olfaktomat** olfactometer according to standard NVN2820 on traceable reference material at various concentration levels, unknown to the operator. Preliminary data, January - May 1990.

Category I	Domestic neighbourhoods and other nuisance prone objects such as hospitals
Category II	Scatterred clusters of domestic, non agricultural dwellings, day recreation areas
Category III	Isolated domestic dwellings in otherwise agricultural area
Category IV	Agricultural housing only

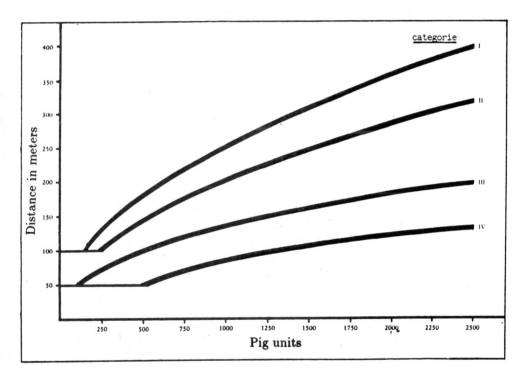

Figure 1: Guideline for zoning around stables, minimal distance to dwellings determined by number of *pig units*, 1976.

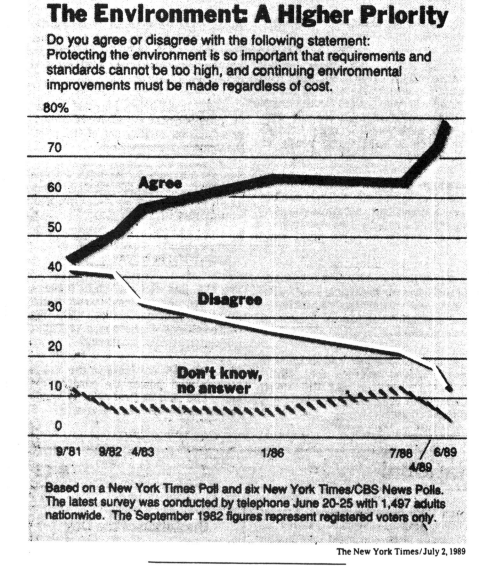

The Environment: A Higher Priority

Do you agree or disagree with the following statement:
Protecting the environment is so important that requirements and standards cannot be too high, and continuing environmental improvements must be made regardless of cost.

Based on a New York Times Poll and six New York Times/CBS News Polls. The latest survey was conducted by telephone June 20-25 with 1,497 adults nationwide. The September 1982 figures represent registered voters only.

The New York Times/July 2, 1989

Figure 2: The trend of the priority that the general public in the U.S. gives to a clean environment, from a New York Times poll

ACTIVITIES OF THE VDI COMMISSION ON AIR POLLUTION PREVENTION TO STANDARDISE THE METHODS OF ODOR ASSESSEMENT

M. Hangartner ,
Institute of Hygiene and Applied Ergonomics
Federal Institute of Technology Zurich, Switzerland

G. Winneke,
Inst. of Environmental Hygiene, University of Düsseldorf
W.Germany

M. Paduch
Association of German Engineers, Düsseldorf, W.Germany

Administrative odor regulation strategies in several European
countries are reviewed, and West Germany technical guidelines
concernig different aspects of odor measurement and annoyance
assessment are listed. Results from several national and
international ring-tests on odor threshold determination based
on VDI Guideline 3881 parts 1 to 3 still exhibit an
unsatisfactory variability among the laboratories, with factors
varying between 1.5 and 4 repeatability) and 5 or more
(reproducibility). Odor regulation based on a risk-assessment
approach requires valid exposure-response functions. Results
from a field study, relating odor frequencies based on actual
monitoring by human observers to odor annoyance responses from
social surveys, show a linear function on annual odor frequency
between 0 and 30%. Odor regulations based on annual odor
frequencies seem to be both a valid and appropriate strategy.

1. INTRODUCTION

Legislative regulations on odor abatement are rare in Europe. First attempts to cope with the problem were made in Switzerland[1], West Germany[2,3] and the Netherlands[4].

The relevant administrative recommendations in West Germany have no binding character for the nation and vary in region and approach: there are limitations of odor emissions or of the frequency of odor events in ambient air per year and also minimum requirements of the odor abatement efficiency of waste gas purification plants. Special agreements were found for livestock farming of pigs and hens; minimum distances to residential areas were set on the basis of practical experiments in the field[5,6]

The Netherlands prefer the "nuisance approach", i.e. an existing nuisance with every odor event is assumed. Therefore, one aims at reducing the frequency of odor events per year. The assessment method applied is dispersion calculation, either based on field inspection to determine the emission flow from the source or based on emission concentrations of the source to determine the residual concentration level in ambient air at a certain distance and the pertinent frequency per year (4).

In view of the fact that there is a want of harmonisation in measurement methods in the European Community, a working group of the Commission of European Communities established recommendations on odor threshold measurements to have a common basis for the determination of the key quantity in odor assessment[7].

2. VDI GUIDELINES ON ODOR ASSESSMENT AND ABATEMENT

The Association of German Engineers has about 100'000 registered members and is a non-profit organisation. There are 17 departments in the offices of VDI working on different technical problems. One of them is the Commission on Air Pollution Prevention. This commission is divided into 5 subcommissions which are dealing with the following topics:

> I   Emission Control
> II   Environmental Meteorology
> III Effect of Air Pollutants on Man, Animals, Plants and Materials
> IV Measurement Techniques
> V   Abatement Techniques

External and internal specialists in science, economics and administration are meeting regularly to establish guidelines on an honorary basis and on their own responsibility, but on behalf of VDI. The guidelines are used in the preparatory stage of West German legislation. As accepted engineering standards, these guidelines provide information on the state of the art of air pollution control. They are updated continuously.

VDI guidelines are first published as drafts (green papers), which are submitted for public objection by an announcement in the Federal Gazette and in the technical press. This guarantees that the frequently diverse opinions of the parties concerned are taken into consideration before the final version (white paper) is issued.

A few guidelines describe measurements of odor properties. Four guidelines on odor threshold determination were already published as well as a draft guideline on odor intensity measurements. In spite of a few disadvantages,it was decided to use a 7-point category scale. Category scaling is simple and can be used by practicioners. For scaling of hedonic tone the bipolar 9-point scale of Peryam Pilgrim is proposed (guideline in preparation). The assessment of ambient odors is approached by field inspection with the help of human observers and dispersion modeling. Assessment of odor annoyance is made with special opinion polling techniques.

The numbers and titles of the pertinent guidelines are listed in table 1.

VDI Guidelines on various methods of waste gas purification for odor abatement are listed in table 2. At present, biological methods are the most promising and progressing methods for many organic and some inorganic substances.

3. VARIATIONS IN ODOR THRESHOLDS

In order to check the effect of guideline VDI 3881 Part 1"Olfactometry - Odor threshold Determination - Fundamentals", which covers the measuring procedure such as instruments, presentation of stimuli, indication of response, panel and calculation, several ring-tests were carried out with the regard to the remaining variation in threshold values [8-12].

It turned out that there is a significant difference in threshold levels with regard to the age of the panelist, the questioning technique (forced choice vs. yes/no reply) and the type of olfactometer used. The most important results are shown in table 3.

It is obvious from the spread of the results that efforts must be made to reduce the variability of the measurement steps (see table 4) still admissible in accordance with guideline VDI 3881 parts 1 to 3, all the more since there are laboratories which achieve a repeatability of about a factor 2 (quotient of limits of 95 % confidence interval) when they follow the sequence of measurement steps familiar to them. At present, odor threshold determinations based on guidelines VDI 3881 parts 1 to 3 produce short-term repeatabilities of 1.5 to 4 (hydrogen sulfide) and long-term repeatabilities (period of several weeks between two measurements) of a factor of 5 and may even reach 100.

A more specific presentation of the results is given in figure 1. The overall median of all the panel thresholds was determined, irrespective of the odorant (basic line of figure 1) and indicated in odor units per cubic meter (OU/m3). A rank order of laboratories was found based on the mean value of the absolute differences of the panels' four threshold medians from the overall median. The dots are the absolute differences of the panel thresholds per odorant to the overall median. The marked spread per panel represents the above-mentioned mean values of absolute differences and the standard

51

deviation. All distances may be read in logarithmic units (left) and in numerical units (right).

Based on these mean differences, a rank order between the laboratories could be established, which is independent of the odorants.

Figure 1 shows that the stability of results within the individual laboratories is generally up to a factor of 2. It seems that there are systematic errors in the individual laboratories. These systematic errors can be eliminated by subtraction of the mean median. Then the variation in terms of quotient of 95% confidence interval decreases to a factor of 2, which is in the range of repeatability of a good working laboratory.

The conclusion out of these results was that it is possible to introduce a correction factor based on several odorants, which involves the need of regular ring-tests.

4. RELATIONSHIP BETWEEN EXPOSURE AND ANNOYANCE

Both emission as well as immission standards are necessary for effective air pollution control. Response to sensory events in the environment, e.g. noise or odors, are typically discussed under the heading of annoyance. However, whereas noise annoyance has succcessfully been related to the physical attributes of sound, this has not yet been achieved for odor annoyance in a convincing manner.

The establishment of clearcut exposure response functions is a prerequisite for the regulation of environmental odors within a risk assessment approach. Results are reported here from a field study relating frequency of annual odor events to the degree of annoyance of the exposed population.

4.1 Exposure Assessment

Odor events were observed, recorded and calculated as the "annual odor frequency" in the vicinity of a chemical plant using a systematic stratified monitoring program based on guideline VDI 3940. Covering a twelve-months period,each of 50 monitoring locations of a spatial matrix surrounding the plant was visited 26 times by observers selected from a basic observer panel following a systematic plan of rotating permutations. An observation was coded as an "odor hour" if the cumulated odor duration exceeded 10% of the monitoring time of 10 minutes duration. The annual odor frequency is definded as the percentage of "odor hours" as compared with 8760 annual hours.

4.2 Annoyance Assessment

The degree of annoyance was measured with the questionnaire approach, with the recommendations of guideline VDI 3883 Part 1 being taken into account. Face-to-face interviews were conducted by trained interviewers following the random-route procedure. Twelve locations out of 50 monitoring locations were selected for annoyance assessment in such a way that the range of observed odor frequencies was fully covered. Between 20 and 50 interviews were conducted in each of the chosen locations.

The mean value of an annoyance score in an area was calculated from the judgement of at least 20 persons on an 11-point self-rating thermometer[13] . Figure 2 shows the results of this field study arount a chemical plant in the Rhine-Ruhr area of the Federal Republic of Germany [14]. The frequencies of the annual odor events are represented on the x-axis. These points represent different areas at varying distances from the plant. The mean annoyance score is shown on the y-axis.

Linearity of the exposure-response relationship occurred at an annual odor frequency of between 0 and 30%. No further increase in annoyance was observed at frequencies exceeding 30%. So the conclusion is that exposure-response relationships may be established for environmental odors with odor frequency as the exposure variable. Additional efforts will be necessary to find out whether the odor flow of emissions may be used for planning purposes to accurately predict the odor frequency in the vicinity of an odor source with dispersion calculation. It will, furthermore, be necessary to check whether and to what extent the results of the present study may be generalised across different sources, as odor emissions differ widely in terms of their hedonic tone.

5. CONCLUSIONS

The present lack of rigid legislative prescriptions on how to deal with an odor situation in ambient air leaves the chance to proceed in two important directions:

1. Improve the standardization of the complete method of odor threshold determination to achieve better results with regard to repeatability and reproducibility and consequently get more stable values for measurement quantities depending on the odor concentration, such as efficiency of waste gas purification in terms of odor, odor intensity, hedonic tone, prognosis of odor concentration in ambient air based on emission measurements etc. Demand a quality proof of any measuring laboratory by means of its repeatability with defined standard odorants.

2. Investigate into the relationship between exposure and response around various odor emitters with different types of odors to find out which odor property or combination of odor properties is decisive in the generation of odor annoyance; check whether such findings are transferable to other plants, odors and situations.

6. REFERENCES

1. M. Hangartner,"Odour measurement and assessement of odour annoyance in Switzerland" In:"Volatile emissions from livestock farming and sewage operations" V.C.Nielsen, J.H.Voorburg and P.L`Hermite (eds.) Elsevier Applied Science Publishers, London and New York (1988)

2. First General Ordonnance for the Federal Immission Control Act GMBl. G 3191 A Bd. No. 7 dated 28 Feb. 1986 (Federal Republic of Germany, in German)

3. Performance of the Federal Immission Control Act - Best available technology to abate odorous emissions with biofiltration) RdErl. d. MB dated 21 July, 1983, Nds. MBl. No. 39/1983, p 749 (in German)

4. H. Wijnen,"Air quality standards on odours in the Netherlands" In: VDI Berichte 561, Düsseldorf, VDI-Verlag GmbH, pp 365-386.

5. VDI 3471 Emission Control - Livestock Farming - Pigs (in German, June 1986), Berlin, Beuth Verlag GmbH.

6. VDI 3472 Emission Control - Livestock Farming - Hens (in German, June 1986), Berlin, Beuth Verlag GmbH.

7. M. Hangartner, J. Hartung, M. Paduch, B.F. Pain, J.H. Voorburg,"Improved Recommendations on Olfactometric Measurements."
    Environmental Technology Letters, Vol 10, pp 231-236 (1989)

8. V. Thiele, "Olfactometric measurements of hydrogen sulfide - Results of a VDI ring-test" Staub-Reinhaltung der Luft 42 (1982) No.1, pp.11-15 (in German)

9   V. Thiele, "Verification and Standardization of olfactometric measurements" Staub-Reinhaltung der Luft 45 (1985) No. 5, pp.200-203 (in German).

10. H. Bahnmüller, "Olfactometry of dibutyl amine, methyl acrylate, isoamyl alcohol and a spray thinner for car varnish - Results on a VDI interlaboratory test" Staub -Reinhaltung der Luft 44 (1984) No. 7/8 pp.352-358 (in German)

11. H.W.O Dollnik et al. "Olfactometry with hydrogen sulfide, n-butanol, isoamyl alcohol, propionic acid and dibutyl amine" Staub - Reinhaltung der Luft 48 (1988) No. 9, pp. 325-331 (in German)

12. M. Hangartner, "Standardization on Olfactometry with respect to Odor Pollution Control" Paper 87-75A.1. 80th Annual Meeting of APCA, New York, June, 1987.

13. M. Hangartner, "Assessment of Odor Annoyance in the Community.Paper 87-75B.3. 80th Annual Meeting of APCA, New York, June, 1987.

14. G.Winneke, M. Paduch, "Measurement and evaluation of odors in air quality control" Proceedings of 8th World Clean Air Congress, 11-15 September 1989, The Hague Holland.

Published:

VDI 3881 Part 1:	Olfactometry - Odor Threshold Determination Fundamentals (German/English, May 1987)
VDI 3881 Part 2:	Olfactometry - Odor Threshold Determination Sampling (German/English, Jan. 1987)
VDI 3881 Part 1:	Olfactometry - Odor Threshold Determination Olfactometers with Gas Jet Dilution (German/English, May 1987)
VDI 3881 Part 4:	Olfactometry - Odor Threshold Determination Application and Performance Characteristics

In preparation:

VDI 3882 Part 1:	Olfactometry-Determination of Odor Intensity
VDI 3882 Part 2:	Olfactometry-Determination of Odor Hedonics
VDI 3883 Part 1:	Effect and Valuation of Odors - Annoyance Assessement by Opinion Polling Technique
VDI 3883 Part 2:	Effect and Valuation of Odors - Annoyance Assessement by Population Panels
VDI 3940	Determination of Odor Immissions by Field Inspection
VDI 3781 Part 3:	Determination of Stack Heights for Odor Emissions
VDI 3782 Part 4:	Dispersion of Odorant

Table 1: VDI Guidelines On Odor Measurement and Assessment

VDI 2442	Waste Gas Cleaning by Thermal Combustion (German/English, June 1987)
VDI 2443	Waste Gas Cleaning by Oxidative Scrubbing (German/English, Jan. 1980)
VDI 3478	Biological Waste Air Purification-Bioscrubbers (German/English, July 1985)
VDI 3677	Fabric Filters (German, July 1980)
VDI 3679	Wet Precipitators (German, May 1980)
Drafts	
VDI 3476	Catalytic Methods of Waste Gas Purification (German Jan. 1988)
VDI 3477	Biological Waste Gas/Waste Air Purification Biofilters (German, Feb. 1989)
VDI 3674	Waste Gas Purification by Adsorption - Surface Reaction and Heterogeneous Catalysis (German, June 1981)
VDI 3675	Waste Gas Purification by Absorption (German, May 1981)

Table 2: VDI Guidelines on odor abatement methods

Ring-test	Odorant	Number of labs	Threshold concentration ppb	Ratio P_{16}/P_{84} n
1984	hydrogen sulfide	15	1.3	6
1984		13	1.8	3
1988		20	1.2	14
	-FCT	5	0.2	4.8
	-yes/no	15	2.1	6.5
	-ipt	7	2.3	5.3
	-other	13	0.9	17
1988	n-butanol	18	127	11
	-FCT	4	33	19
	-yes/no	14	180	5.3
	-ipt	5	287	3.7
	-other	13	87	12
1988	isoamyl alcohol	20	33	31
	-FCT	5	5	8
	-yes/no	15	56	19
	-ipt	7	94	9
	-other	13	18	32
1988	propionic acid	19	36	23
	-FCT	5	12	13
	-yes/no	14	48	22
	-ipt	5	102	9
	-other	14	21	22

Table 3: Results of various olfactometric round robin tests (8-11) in ppb.

ipt: most used commercial olfactometer in West Germany
FCT: forced choice technique

S A M P L I N G

Static sampling (Various bag materials)	Dynamic sampling (Various duct materials)
with without sample conditioning	with without sample conditioning

(Dust removal, predilution, heating)

O L F A C T O M E T E R S

Commercially available olfactometers	Modified commercially available olfactometers	Individually designed olfactometers

T Y P E O F D I L U T I O N A I R

Synthetic Air	Conditioned ambient air

P R E S E N T A T I O N O F T H E S M E L L S A M P L E

Method of limits		Constant method	
Dependent on the breathing cycle	Independent of the breath-ing cycle	Dependent on the breathing cycle	Independent of the breath-ing cycle

Questioning per measuring series/ per set of measurements	Questioning in turns with con-stant dilution

A N S W E R S T O T H E S T I M U L U S

Yes/No (1 sniffing port)	Yes/No with reference air (2 sniffing ports)	Forced choice (2 or 3 sniffing ports)

E V A L U A T I O N O F T H E R E S U L T

Geometric mean value	Characteristic curve of the odor threshold	Probit analysis

Table 4: Optional items of the complete olfactometric method

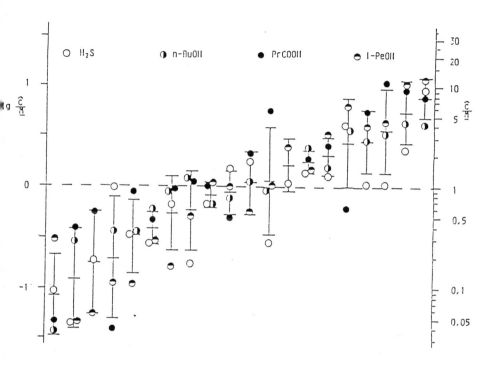

Figure 1: Odorant-independent rank order of the participants in the ring-test 1985 (11). Deviation of laboratory odor threshold of from substances from overall threshold (median).
H2S: Hydrogen sulfide, n-BuOH n-butanol, PrCOOH propionic acid, I-PeOH isoamylalcohol

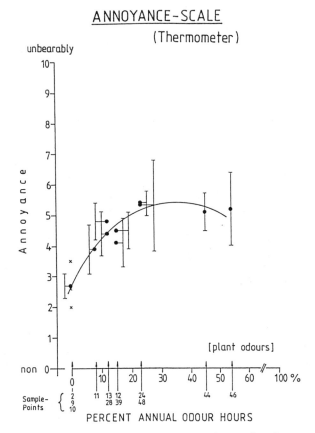

Figure 2: Exposure - annoyance relationship in the surroundings of a chemical plant

Odour Management and
Regulations in Denmark

Kasper Rovsing Olsen
DK-TEKNIK
Danish Boiler Owners' Association

In 1985 the Danish Environmental Protection Agency published
"Guidelines for Abatement of Odour Pollution". Hereby
regulations for measurement, assessment, prevention, and
emission of odour were promulgated.

According to these guidelines, odour samples should be
collected during the period of maximum emission under normal operating conditions. The sampling time should be 1 - 5
min and the odour concentration should be measured within 24
hours with a dynamic olfactometer using a panel of at least
6 observers.

In order to minimize the variations in measured odour
thresholds due to the use of different olfactometers and
panels, the results are divided by a sensitivity factor
which is calculated from the odour thresholds for hydrogen
sulfide and butanol, respectively. The ground level odour
concentration is then calculated by a standard procedure.
The Danish Guidelines prescribe that the calculated ground
level concentration should not exceed 5-10 OU/m^3, depending
on the location of the source.

For area sources, the same regulations are applied, but no
defined method has been established in the measurement and
assessment of emission and odour nuisance. At DK-TEKNIK
assessments are made by a combination of odour observations by
neighbouring and non residents, upwind sampling and headspace odour analysis in laboratory and in-situ. To complete,
continuous monitoring of local meterorological conditions
during observation period is performed

Applicable results have been achieved.

1. Introduction

The Danish Society has become more and more aware of odour nuisance since the mid 1970's. In order to facilitate the odour management the Danish Enviromental Protection Agency published "Guidelines for Abatement of Odour Pollution" /1/. Hereby regulations for measurement, assessment, prevention and emission of odour were promulgated.

2. Measurement

According to the guidelines, odour samples should be collected during the period of maximum emissions under normal operating conditions. The sampling equipment used should be made of e.g. teflon tubing and TEDLAR^r bags. The sampling period should be 1-5 min.
The guidelines do not recommend a specific number of samples to be collected in order to obtain a reliable estimate of the odour emission. Usually, but evidently dependent on the process, two or three samples are collected under the above mentioned conditions. The resulting odour concentration is determined as the arithmetic average of the measured odour concentrations.

Odour concentration is determined by sensoric analysis within 24 hours using panellists. By definition the odour concentration expressed in OU (odour units)/m^3 is the number of dilutions needed to attain the odour threshold. This should be measured by dynamic dilution using a forced or semi-forced choice method. Further more the dilution air to the panellists should be of approximately 20 l/min and the exposure time should be short to avoid adaptation.

The odour threshold is the odorant's concentration which just initiates perception of that odour. It is defined as the concentration of an odour sample at which 50% of the testing panel is unable to distinguish it from odourless air.

3. Sensitivity factor

In order to minimize the variations in measured odour thresholds due to the use of different olfactometers and panels, the guidelines introduce a sensitivity factor,

$$p = \left(\frac{0,0006}{C_s} \times \frac{0,05}{C_b} \right)^{0,5}$$

where C_s and C_b are the measured odour thresholds in ppm for hydrogen sulfide and n-butanol respectively.

In practice this means that the odour thresholds for n-butanol and H$_2$S are determined every time a test is run. A test consists of no more than 7 samples with unknown odour

62

concentration and two samples with known content, (usually 59 ppm H_2S and 260 ppm C_4H_9OH).

The measured odour concentration, C, determined in a sample is divided with the sensitivity factor P to attain the odour concentration C_{50}, which is the concentration used when the groundlevel concentration is calculated. At DK-TEKNIK the usual sensitivity factor is of the order of 2,0 - 2,5.

4. Panellists

The panellists are from the age of 18 to 50 and should be able to reproduce odour threshold determination for the hydrogen sulphide and n-butanol samples within $\pm$ 30%.

In practice 6 panellists are used for each series of tests, which normally takes about three hours and the panellists are not used further the same day.

It is important that the panelists have no preliminary knowledge of the origin of the samples that are to be tested, in order to avoid the influence of e.g the name of a plant with a "bad" reputation.

5. Ground Level Odour Concentration

The ground level odour concentration is calculated by a standard procedure, (based on formulas from LEE & STERN, 1973) from knowledge of stack height, odour emission rate, temperature, flow and the surrounding topography including buildings.

The Danish Guidelines prescribe that the calculated ground level concentration should not exceed 5-10 OU/m^3 as 99% percentile value, depending on the location of the source (residential area, urban area). The averaging time is 1 min.

6. Area Sources

For odour emission form area sources such as landfill or sewage treatment plants the same regulations are applied, but no defined method has been established in the measurement and assessment of emission and odour nuisance.

At DK-TEKNIK (Danish Boilers' Association) we have made sucessful assessments by handling this problem in one or more of the following ways:

-   Odour observations by neighbouring residents in different wind sectors around the potential source usually over 3 months.

-   Odour observation by non-residents in the vicinity.

- Sampling at different altitudes from the source, both up-wind and down-wind

- Headspace odour analysis of water samples in laboratory and in situ.

- Continuous monitoring af local meteorological conditions during observation period.

6.1 Odour Observations

Neighbouring residents in different wind sectors and distance from the source are invited to participate in the odour observation. In the selection of neighbouring residents for odour observation it is important to be aware of their association with the source.

It is also very important to be aware of the influence of the source in the surrounding society. In small towns a large factory can have an important influence and the neighbours can adapt to or accept the odour that by non residents is perceived as malodour.

The selected observers are asked to make observations three times per day, where they characterise the percieved odour on a scale from 0-5, where the numbers refer to the following:
0   no odour
1   odour, not annoying
2   odour, a little annoying
3   odour, annoying
4   odour, very annoying
5   odour, extremely annoying

These observations are complemented with observations by non residents e.g. observers from DK-TEKNIK or the local environmental protection agency.

6.2 Measurements

In order to complement the observations and to quantify them, odour sampling is performed on the site.

At DK-TEKNIK we use different techniques.

6.2.1 Upwind Sampling

Odour sampling is made at different altitudes up-wind and down-wind from the source. This method is not very adequate, because it is necessary to attain the boundary layer and thereby be able to define the emission area.

At DK-TEKNIK we usually make simultaneous measurements of
wind velocity and air sampling at the same altitude.

An idealized wind velocity and odour concentration profile
is shown in figure 1.

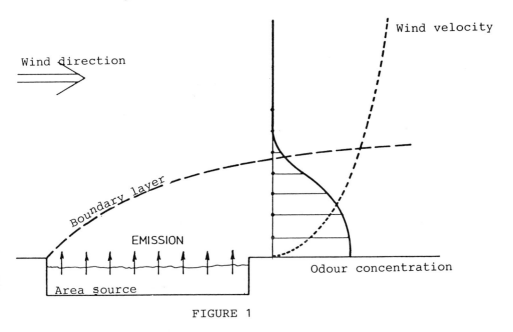

FIGURE 1

By normal wind velocities a boundary layer of about 13 meters
has been found.

Despite the uncertainty (a factor af about 2) from the
acquired results one obtains an estimate of the emission
and thereby the possible impact on the surroundings.

6.2.2 Headspace Analysis

By leading 1 l/min of clean air into the headspace of a
flask containing 1 l of the odourous substance, 10 l of
"odour contaminated air" is sampled in a TEDLAR[r] Bag.

The odour concentration is determined and the total
emission is calculated as

$$Q = k \left(\frac{V_s \times (C - C_0)}{P} \right)^{0,5} \times A, \qquad OU/s$$

where
- Q is the odour emission from the area source
- k is a constant 1,6 for non aerated reservoirs
        4,4 for aerated reservoirs

- V_S is the average wind velocity at 1 m above surface
- C is the odour concentration determined in the sample
- C_O is the odour concentration of the headspace of a clean water sample
- P is the sensitivity factor

This is a result of a series of experiments made by K. Bo-holt, DK-TEKNIK /2/.

6.2.3 Headspace Sampling in Situ

DK-TEKNIK has developed a method that enables collecting samples of the headspace in-situ.

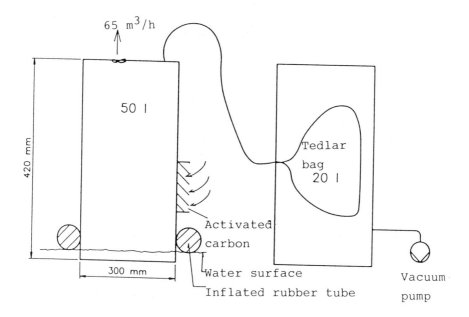

FIGURE 2

Ambient air is passed through activated carbon whereafter it aerates the known surface of the source within the vessel. Air is sampled through the teflon tubing inserted into the Tedlar bag.

The total emission from the reservoir is calculated as:

$$Q = \left(\frac{R \times C}{a \times P} \right) \times A$$

where,

R is the flowrate of clean air into the vessel in m^3/s
C is the measured odour concentration in ou/m^3
a is the area limited by the vessel in m^2
P is the sensitivity factor
A is the area of the reservoir in m^2

At the same time as these in-situ head space samples are collected up-wind and down-wind sampling is performed. The head space analysis is used in an effort to minimize the number of samplings and to minimize the uncertainty which can easily be of about a factor 2. It can be difficult to be certain that the maximum emission occurs when these measurements are made, because it is influenced by a number of parameters, eg pH and temperature. Therefore it can be necessary to complete the results by more subjective observations.

6.3 Results Of The Assessment

At the end of the observation period the observations are compared with the results of the measurements and the meteorological conditions during the period. In addition a comparison can be made with results of calculations of ground level concentrations.

This method of assessment when the source is complex is proven to be applicable.

The above mentioned method has shown that e.g. in the neighbourhood of a sugar factory at least 20 OU/m^3 (one minute average) is acceptable. This is very important because it could cause enormous investments in odour reducing equipment if the authorities only permit 5-10 OU/m^3.

In order to obtain more accurate and reliable results, further experiences with head space odour analysis are necessary.

The Danish regulations are to be revised in the near future (1990) in order to incorporate the experience that has been obtained since it was promulgated.

References

1. Danish Environment Protection Agency, " Guidelines for Abatement of Odour Pollution". 4/1985, Miljøstyrelsen, 1985 (In Danish)

2. K. Boholt, "Determination of Odour Emission From Area Sources", Miljø & Teknologi 2, 1988 (in Danish).

A REGULATORY APPROACH FOR CONTROLLING ODORS IN THE TERPENE CHEMICAL MANUFACTURING INDUSTRY

Wayne E. Tutt
Bio-Environmental Services Division
Department of Health, Welfare, and
Bio-Environmental Services
Jacksonville, Florida

Located in Jacksonville, Florida are two of the largest terpene-based synthetic organic chemical manufacturing plants in the world. These plants fractionally distill crude sulfate turpentine (CST) into its principal pure components, and through specialized organic chemistry, transform many of these chemicals into flavoring and fragrance compounds. An analysis of citizen odor complaints indicated that the two terpene chemical plants, along with two kraft pulp mills and a municipally-owned sewage treatment plant, were the leading causes of objectionable odors in the City.

During 1987, the terpene chemical manufacturers, together with a third resin chemical producer, met with Jacksonville's Bio-Environmental Services Division, and formed a voluntary, cooperative group to study the odor problem from the chemical plants, and identify cost-effective solutions. This group, called the Chemical Odor Task Force, but informally known as Tri-Chem, identified the major odor producing areas within the chemical manufacturing process as follows: process equipment, including reactors and distillation columns; storage tanks; loading and unloading of liquid materials; spills and accidental releases; transfer vessel (tote tank) cleaning; fugitive leaks; and wastewater collection and treatment.

Each of the three plants developed its own odor abatement plan, designed around its specific needs. The two terpene manufacturers each identified more than fifty discrete odor abatement tasks in the various areas of plant operation mentioned above. This paper will discuss the major odor abatement tasks, the results to-date, and the future plans for development of a Best Management Practices for Odorous Substances rule, for chemical manufacturing.

A REGULATORY APPROACH FOR CONTROLLING ODORS IN THE TERPENE CHEMICAL MANUFACTURING INDUSTRY

INTRODUCTION

Jacksonville, Florida, located near the pine forests and palmetto scrub of Northeast Florida and Southeast Georgia, has been a commercial center for turpentine processing and the Naval Stores industry for more than 100 years. This connection began with the production of gum turpentine, and later of steam-distilled wood turpentine, and finally with the production of sulfate turpentine, a by-product of the kraft pulping industry.

Sulfate turpentine is produced by the condensation of vapors liberated during the process of cooking wood chips in a caustic solution at elevated temperature and pressure. During this process, known as digestion, reduced sulfur compounds are produced, and those of lesser volatility, principally dimethyl sulfide and dimethyl disulfide are condensed and recovered as a malodorous contaminate in the turpentine.

Located in Jacksonville, Florida are two of the largest terpene-based synthetic organic chemical manufacturing plants in the world. These plants fractionally distill crude sulfate turpentine (CST) into its principal pure components, and through specialized organic chemistry, transform many of these chemicals into flavoring and fragrance compounds.

An analysis of citizen odor complaints indicated that the two terpene chemical plants, along with two kraft pulp mills and a municipally-owned sewage treatment plant, were the leading causes of objectionable odors in the City. The wastewater streams from both chemical plants are treated at the City's Buckman Sewage Treatment Plant. The objectionable odors from the chemical plants are caused by the volatilization and escape of reduced sulfur gases.

Unlike the situation in kraft pulp mills, where most reduced sulfurs are emitted from a select number of process vents under elevated temperature and height, the chemical plant emissions are predominantly low level, fugitive emissions at atmospheric temperature and pressure. Although the mass emissions from the chemical plants are much less than from kraft mills, and they fumigate a smaller geographic area, the chemical plants do not benefit from good atmospheric dispersion. Also, the scattered and unconfined nature of the odor emissions makes odor control and odor regulation difficult.

During 1987, the terpene chemical manufacturers, together with a third resin chemical producer, met with Jacksonville's Bio-Environmental Services Division, and formed a voluntary, cooperative group to study the odor problem from the chemical plants, and identify cost-effective solutions. This group, called the Chemical Odor Task Force, but informally known as Tri-Chem, identified the major odor producing areas within the chemical manufacturing process as follows:

1. Process equipment, including reactors and distillation columns.
2. Storage tanks.
3. Loading and unloading of liquid materials.
4. Spills and accidental releases.
5. Transfer vessel (tote tank) cleaning.
6. Fugitive leaks.
7. Wastewater collection and treatment.

Each of the three plants developed its own odor abatement plan, designed around its specific needs.[1] The two terpene manufacturers each identified more than fifty discrete odor abatement tasks in the various areas of plant operation mentioned above. This paper will discuss the major odor abatement tasks, the results to-date, and the future plans for development of a Best Management Practices for Odorous Substances rule, for chemical manufacturing.

REGULATIONS AIMED AT ODOR REDUCTION

In 1984, the City Council of the Consolidated City of Jacksonville adopted a comprehensive ordinance to address environmental problems in Duval County, including odors. Known as the Omnibus Ordinance, this bill delegated administrative rule-making authority to a nine-member Board, the Environmental Protection Board. The ordinance directed the Board to develop three types of rules to address odors:

1. Odor emission limits.
2. Ambient odor standards.
3. Incineration/control technology standards.

These rules were required to specify the methodology for arriving at the standards, and the specific analytical procedures by which a violation is established.

Bio-Environmental Services Division (BESD) is a local pollution control program, and a part of the City Health Department. In matters of rulemaking, the BESD staff conducts workshops and does the technical development, whereas the Board does the actual rule adoption. In responding to the direction of the Council, BESD actively participated in the development of a state of Florida rule limiting Total Reduced Sulfur (TRS) emissions from kraft pulp mills. This rule was adopted by reference as a local rule in 1985.

In September 1988, the Board adopted a standard limiting the maximum ground level concentration of Aggregate Reduced Sulfur (ARS) to 55 parts per billion for any consecutive three minutes. Measured exceedances, standing alone, cannot be used for direct enforcement. Rather, the standard represents a benchmark to determine the adequacy of source-specific odor abatement rules. Ambient standards alone are difficult to enforce, due to evidentiary problems. Emission limits and control technology or management practice rules may or may not reduce ambient odorant concentrations to below levels which cause community annoyance. An ambient standard based upon the annoyance threshold for ARS provides a linkage between easily enforced source-specific rules and citizen annoyance.[2]

BESD still lacked a specific, enforceable rule for controlling odors from the terpene chemical plants. Work began on a Best Management Practices (BMP) rule for terpene chemical manufacturers, using information generated by the Chemical Odor Task Force.

ODOR ABATEMENT TASKS AT THE TERPENE CHEMICAL PLANTS

Turpentine is a mixture of unsaturated, bicyclic, monoterpene hydrocarbons, $C_{10}H_{16}$. The principal components are alpha-pinene and beta-pinene. The term terpenes has come to include various acidic, aldehydic, ketonic, and other derivatives. Examples of terpene and terpene derivative products include pine oil, linalool, geraniol, citral, citronellol, menthol, and others. These products are used as fragrances in soaps and detergents, in the manufacture of disinfectants and cleaners, in the syntheses of Vitamins A and E, as well as many other products.

The raw chemical feedstock for terpene chemical plants is crude sulfate turpentine, or CST. CST from southern mills contains from one-half to two and one-half percent TRS as a contaminant.[3] These malodorous compounds must be removed from the terpenes, in order not to impart an unpleasant and unacceptable aroma to the fragrance products.

CST arrives at the Jacksonville plants primarily by railroad tank car, with a lesser amount by tank truck. The liquid CST is unloaded, and pumped into large storage tanks. From the storage tanks, the CST is first passed through a stripping column, where the more volatile TRS compounds are removed. The product from the stripping column, Topped Sulfate Turpentine (TST), is fractionally distilled into its major components, chiefly alpha and beta pinene, which are then further desulfurized in an absorption column. Various contaminated wastewater streams are discharged into the wastewater collection and pretreatment system, where significant potential exists for further volatilization and release to the atmosphere.

Both terpene manufacturing facilities in Jacksonville, SCM Glidco Organics and Union Camp Corporation, have identified the areas of significant potential for odor release, and have undertaken many tasks to reduce the release of TRS to the atmosphere.[4] The following narrative describes the odor problems, and odor abatement tasks undertaken by the manufacturers in the various major chemical plant process areas.

LOADING AND UNLOADING OF LIQUID MATERIALS

Because of the extremely low odor threshold of reduced sulfur compounds, even a small leak or spill of CST onto the ground can cause an objectionable odor in surrounding neighborhoods. At the beginning of the odor task force, one terpene manufacturing plant had a dirt bed and wooden cross ties under the railcar unloading area. When a quantity of CST was spilled, there was no efficient means to clean it up. The liquid would rapidly soak into the soil, and into the wooden cross ties. Whenever a rainfall occurred, these porous materials acted like a wick, drawing up the oils from the ground and causing odors. To prevent this problem, the entire rail bed was dug up and replaced, concrete cross ties were installed, and the entire area around and between the rails was paved. The truck unloading areas were also paved, and a new concrete under-drain system was installed. Spills, when they do occur, can now be treated with hypochlorite, then quickly flushed into the drains, which lead to foul water holding tanks. Portable hypochlorite wash systems are kept available in the railcar and tank truck areas. These small portable tanks, actually plastic garden sprayers, are available for quickly neutralizing the odor of CST spills.

Railcars and trucks are bottom unloaded, with liquid material then pumped to storage. Even as small a quantity as several pints of CST left in a hose can drain out when the discharge line is disconnected and laid on the ground, causing odors. To prevent this problem, one facility has installed dry, self-sealing disconnect valves on the tanks and the hoses. These small spills have been all but eliminated, with the exception of an occasional broken or clogged fitting.

In addition to liquid leaks, the CST unloading operation can cause malodorous vapors to escape from the domes of the trucks or railcars. Prior to unloading a railcar, the tank must be gauged to determine water content. In addition, the hatch covers must be left slightly ajar to allow air to enter the car as the CST is pumped. A gentle breeze can aspirate fumes from the car, creating an odor problem. One company has modified its dome hatch covers to allow gauging the car by opening a small cap, rather than the entire hatch cover. Conservation vents were added to allow air to enter without having a large opening that permits the escape of fumes. For cars that could not be permanently modified, several portable hatch covers were fabricated, for use on differently configured railcar hatches. The other company opted to install fume collection hoods in the unloading areas. Made of flexible vinyl fabric, they are draped over the open hatches, and the collected fumes are ducted to a small hypochlorite fume scrubber, followed by a composted sludge bio-filter.

STORAGE TANKS

After being unloaded, CST is pumped to large storage tanks. These tanks are subject to the same type of working and breathing losses as any other petroleum or VOC storage tank. One terpene plant had collected displaced CST tank headspace gases with a noncondensable gas (NCG) collection system, which delivered the fumes to either a power boiler or a dedicated fume incinerator for combustion. The second plant used 55 gallon drums of activated charcoal for fume adsorption. This plant elected to fabricate composted sludge bio-filters to replace the activated charcoal drums. The composted sludge is generated at a local sewage treatment plant, where it is also used to remove TRS gases from sewage. The compost bio-filters proved unsuccessful in handling the odors from such concentrated TRS streams, and have subsequently been eliminated. Both facilities now incinerate the CST storage tank headspace gases in power boilers or incinerators.

PROCESS EQUIPMENT, INCLUDING REACTORS AND DISTILLATION COLUMNS

The distillation columns used in the terpene chemical industry are operated under a vacuum. This is necessary because without operating the columns under a vacuum, such high temperatures as would otherwise be necessary to boil the liquid terpenes at atmospheric pressure would cause unwanted chemical reactions among the various compounds. Traditionally, the vacuums in the columns were created by use of multistage steam eductors, each stage being followed by a tube and shell condenser. The foul condensate from the overhead condensers was drained through a barometric leg into a hotwell. This large volume of foul, hot condensate was discharged into the wastewater collection system. TRS compounds were volatilized from the hot, foul condensate in the hotwells. In addition, the hot condensate elevated the temperature of the wastewater downstream in the collection system. This added to the TRS concentration in the waste stream, more readily volatilized TRS already present in the wastewater, and hindered any potential for secondary waste pretreatment.

Initially, one plant covered and sealed the hotwells, ducting the vapors into an NCG system. Eventually, both facilities replaced the steam jet systems with mechanical vacuum pumps on all columns handling reduced sulfur compounds. This eliminated the hotwells as an odor source, and lowered the overall wastewater volume, temperature, and biochemical oxygen demand (BOD).

72

Noncondensables from the columns, containing TRS compounds, are collected via noncondensable gas (NCG) systems, and are burned.

In addition to fractional distillation columns, each facility has chemical reactors which emit malodorous gases. Pyrolysis units perform thermal cracking, and produce noncondensable gases such as carbon dioxide and methane, which leave the units saturated with unreacted terpenes. These noncondensable gases are now returned from the decant and oil tanks back into the pyrolysis furnace for incineration. Citral reactors also produce noncondensable gases which leave the unit saturated with terpenes. The NCG from the condenser, water tank, decant tank, and oil tank are now piped to the superheat furnace for incineration.

A number of steam driven reciprocating piston pumps were in use in the terpene plants for transporting turpentine, blend oil, and other liquids. These pumps used packing to achieve a seal, and were subject to frequent steam and hot water leaks. Depending on the material being pumped, a leak could cause a significant odor problem. All such steam pumps used for pumping malodorous liquids have been replaced with electrically driven positive displacement pumps with mechanical seals.

One facility had an odor problem due to emissions from an stratification vessel in the geraniol process. This sweet, fruity odor, although not as offensive as TRS odors, did cause complaints from the close vicinity of the plant. This problem was a result of too small a condenser on the geraniol reactor. A larger condenser was installed, with residual fumes not condensed routed to a scrubber for final removal of organics.

TRANSFER VESSEL (TOTE TANK) CLEANING

Because of the wide variety (hundreds) of intermediate and final products produced from turpentine, very few pieces of production equipment are dedicated to one process. Most operations are batch reactions, with each reactor or column capable of running any of several processes, depending upon market demands. This variety makes dedicated pumps and pipelines impractical in many plant operations. As a result, intermediate and final products are transported from place to place in stainless steel wagons, and in smaller portable tanks. In order to avoid cross-contamination of these flavoring and fragrance chemicals, whose end use requires high purity and aroma reproductibility, it is necessary to clean the tanks between uses.

Traditionally, these tanks have been steam cleaned, vaporizing residual organic liquids and occasionally contributing to odors off the plant site. One manufacturer has converted an existing building to use as a tank cleaning facility. All vessels are steamed out within this enclosed building, with fumes evacuated through a hood, and then treated in a wet scrubber.

The second manufacturer has limited tank cleaning to daylight hours, and has installed a tank cleaning station using an automatic, rotating sprayer, employing timed high pressure hot water. Far less organic material is now vaporized and released into the atmosphere.

WASTEWATER COLLECTION AND TREATMENT

The wastewater collection and treatment areas had the highest odor potential within the chemical manufacturing facilities, and the solutions to the wastewater odor problem have been more divergent than in other process areas.

One facility has an aerated lagoon, which has a retention time of about 30 days. It is important to remove as much of the terpenes, oils, and sulfur compounds as possible from the wastewater, prior to its discharge into the lagoon. To reduce the organic load to the aeration lagoon, a number of steps were taken, as follows:

1. Better oil separation at intermediate sumps.
2. Better oil separation by replacing the final clarifier just ahead of the lagoon.
3. Better aromatics removal by installation of a larger chiller.
4. Segregation of additional foul water streams for chemical oxidation prior to the lagoon.
5. Elimination of foul condensate from steam jet vacuum systems.

An old Dorr-Oliver clarifier immediately ahead of the lagoon has been replaced with an API oil/water separator for better oil removal efficiency. The new oil separator is completely enclosed to contain odors.

The aeration lagoon had previously suffered from poor aeration. Four additional 25 horsepower AIRE O_2 aerators were installed. The new aerators provided sub-surface aeration, and also agitated the water so as to prevent sedimentation which had previously occurred on the bottom of the lagoon. The impact of the increased aeration is being evaluated; however, some pond odors persist.

A number of intermediate collection sumps formerly used rope skimmers to recover oil prior to discharge. These skimmers employ a continuous loop of one-half inch plastic tubing which passes across the liquid surface and through the skimmer. These sumps were open to the atmosphere, contributing to plant odors. The rope skimmers have been replaced with more efficient belt skimmers, reducing the organic load at the treatment area. The skimmers and sumps have been fully enclosed, eliminating the sumps as odor sources.

Mixing of aromatics wastewater with wastewater containing plant slop oil causes formation of emulsions. These emulsions are hard to separate, and increase the organic load at the wastewater treatment area, contributing to odors. Oil separators employing pH adjustment have been installed to facilitate alcohol/oil separations.

Wastewater generated in different process areas was formerly sent to sumps via open U-drains. Malodorous wastewater streams created an odor problem, especially when the U-drains became clogged. Several streams were removed from open drains and hard piped to sumps. Improved housekeeping has eliminated clogging as a problem in other drains.

The other facility does not use a lagoon, but rather has three large tanks for sedimentation and oil recovery. The first of these tanks was a very old concrete clarifier, which was not equipped for efficient oil recovery. It has been replaced with a new oil separator. These three tanks are covered, but headspace gases were previously untreated, or scrubbed with wet scrubbers. Sewage sludge compost bio-filters were installed, and although not satisfactory for treating headspace gases from CST storage tanks, they have proven very successful in handling wastewater tank gases.

This facility has begun a complete replacement of existing underground drain lines with above-ground force mains from strategic collection areas down to the treatment areas. This will eliminate future soil contamination problems, which eventually contribute to odors in plant storm drains.[5]

Several foul water streams occur throughout the terpene manufacturing process, which can contribute to odors in the wastewater treatment areas if not pretreated. This was a significant odor reduction task for both facilities.

The primary foul water streams include - the water arriving in the incoming CST tanks, foul condensates from steam jet vacuum systems, and foul water dissolved in CST and fractionally distilled in the vacuum columns. As discussed earlier, the foul hot condensate from the steam jet vacuums systems was eliminated by installation of mechanical vacuum pumps.

The water phase in the CST is saturated with reduced sulfurs, and is extremely malodorous. It is not uncommon to find several feet of water in the bottom of a railroad tank car. This water is pumped along with the CST into storage tanks of approximately a million gallon capacity. In the past, this water was periodically drained into the sewers; or, in the case of one plant, into the aerated wastewater lagoon, where TRS vapors were stripped and discharged to atmosphere. Now, both facilities drain the water from CST storage tanks into foul water holding tanks, where it is pretreated with hypochlorite solution prior to discharge. The pretreatment with hypochlorite oxidizes the TRS compounds.

The water dissolved in the CST and removed during fractional distillation is small in quantity, only several gallons per day. It is very malodorous, and is likewise segregated, stored, and hypochlorite treated. Vapors from the hypochlorite treatment tank are controlled by incineration. The wastewater streams from both plants discharge to city sewers, and are received and treated at a regional sewage treatment plant. In the case of the facility without the aerated lagoon, tests showed that hypochlorite treatment of foul water streamed reduced the TRS going to city sewers from 270 pounds a day to less than 10 pounds a day.

SPILLS AND ACCIDENTAL RELEASES

Improvements to prevent odors due to spills or releases in the CST railcar unloading area were discussed previously. Several other changes were made in other plant areas as follows:

1. Product and CST sampling stations at various plant locations required flushing liquid lines prior to sampling. The flushed materials went into wastewater drains or onto the ground, causing odors. Flushing losses have been eliminated by installing circulation loops containing three-way sampling valves.

2. Several storage tanks surrounded by retaining dikes had earthen floors, from which spills could not easily be cleaned up. Concrete floors have been installed. In addition, some porous concrete block

dike walls have been sealed with a liquid tight fiberglass coating, to further protect against liquid spills.

3. Employee attention to detail and proper training is necessary in order to complete the task of odor prevention. An odor abatement manual has been written, with operating procedures periodically reviewed for improvement. Checklists for use by operators are part of the manual.

CONCLUSION

The odor abatement tasks described above focus on reducing the discharge of reduced sulfur compounds. In so doing, other organic odorants will also be reduced. The tasks described above were initially proposed and completed as a voluntary effort on the part of the terpene chemical manufacturers to eliminate objectionable odors.

Amendments to Jacksonville's Omnibus Environmental Ordinance in March 1988 provided that verified citizen complaints from five (5) or more different household within ninety (90) days constitute an objectionable odor; and are, in and of themselves, grounds for an administrative Citation. A Citation requires corrective action, and carries maximum penalties of $10,000.00 per day for each offense. The Citations are enforceable by use of a locally appointed administrative hearing officer, who issues findings of fact, conclusions of law, and makes a recommendation to the Environmental Protection Board. Also, Citations are enforceable by civil procedures in the Circuit Court.

This amendment was a very significant change in the way citizen complaints are used. Previously, complaints were used only as a threshold to identify community odor nuisances, but were not themselves grounds for enforcement action. The same ordinance amendments also contained a "safe harbor" provision, which reads as follows:

"However, if a person alleged to have caused an objectionable odor shows that an emission is made in compliance with odor emission standard, ambient odor standards, odor incineration standards, an odor compliance plan, or a consent order with respect to odor, such emission shall not be deemed an objectionable odor. Provided, however, that nothing contained herein shall be construed to prohibit abatement of or enforcement against objectionable odors from a source or sources not specifically regulated or not in compliance with the above mentioned standards, compliance plans, or consent orders."

These amendments brought about an equally significant change in the outlook of regulated industries toward odor regulations. As of this writing, a total of sixteen odor Citations have been issued against ten different facilities (including one city sewage treatment plant). Previously, regulated sources were willing to formulate voluntary odor abatement plans, but were unwilling to actively participate in rule development.

After the amendments described above, regulated facilities developed an interest in having enforceable standards or enforceable compliance plans. The limited "safe harbor" afforded by specific regulations was much preferred over the more subjective nuisance approach. Also, regulated facilities could know with greater certainty and clarity whether they were in compliance or not. Both terpene chemical manufacturers in Jacksonville have updated and amended their voluntary odor abatement plans, adding several sources not previously covered. Both plans have been reviewed by BESD and found acceptable, and have been adopted by the Environmental Protection Board as enforceable compliance plans. Failure to comply with a compliance plan is punishable by a civil fine of up to 500 dollars for each offense, for each day. This fine can be levied directly by the Health Department Director. These compliance plans cover the period of time necessary to complete the odor abatement tasks covered by the plan, and expire on December 31, 1989.

Meanwhile, the BESD has contracted with a consulting engineering firm to review all odor abatement actions identified by the voluntary odor plans, and develop a recommended rule, titled "Best Management Practices for Odorous Substances". The contractor is directed to consider other information as well, including United States Environmental Protection Agency (EPA) New Source Performance Standards for the synthetic organic chemical manufacturing industry (SOCMI), and leak detection protocols developed for the petroleum refining industry. The contractor must identify the odor control options for the various process areas discussed above, evaluate their effectiveness, define their costs, and define the cost to BESD to implement the rule. Based upon a consideration of the likely environmental impacts and costs, the contractor must define Best Management Practice for the control of odorous substances in the terpene chemical manufacturing industry. The final report must contain recommended regulatory language, including specific procedures and instruments for determining compliance. Both regulated companies are actively cooperating in the rule development.

The final technical report from the contractor is scheduled for completion by December 31, 1989. Following the completion of the contracted study, one or more technical workshops will be held, to receive comments from regulated sources and the general public. Following the workshops, a Hearing will be conducted by the Environmental Protection Board, and barring any unforeseen problems, the BMP rule should be adopted in early spring of 1990.

REFERENCES

1. W. E. Tutt, K. K. Mehta, et al, "Final Report - Chemical Odor Task Force", Jacksonville, Florida, 1987

2. W. E. Tutt, "Setting an Ambient Odor Standard", AWMA Annual Meeting, Anaheim, California, 1989

3. J. Drew, J. Russell, and H. W. Bajak, ed., Sulfate Turpentine Recovery, Pulp Chemicals Association, New York, 1971

4. K. K. Mehta, "Odor Abatement at Terpene Chemical Plants", Florida Section APCA, San Destin, Florida, 1988

5. "Evaluation of Wastewater Generation, Collection, and Treatment - SCM Glidco Organics", Evnironmental Science and Engineering, Inc., Gainesville, Florida, 1988

II. Odor Controls and Environmental Systems

COST EFFECTIVE ODOR CONTROL TECHNOLOGIES
IN THE PROCESS INDUSTRIES. WHAT IS THE
BEST APPROACH

Thomas J. McEnhill
Vice President
Randers Engineering, Incorporated

ABSTRACT

The topic will cover an evaluation of the many types of odor
control systems available to the process industries to control
the varied odorants generated from industrial and municipal
facilities. This "overview" of applicable control technologies
will provide a systematic approach to defining the odorants,how
to capture them, and how to eliminate them in the most cost
effective manner. A total cost approach (capital investment
and operating cost) will be discussed in determining the best
odor control system for your application.

Introduction

The control of odors from an industrial process plant or a municipal waste treatment facility is one of the more perplexing and difficult, technical problems facing an owner or a design engineer today.

This paper titled, "What is the best approach for selection of a cost effective odor control technology in the process industries?", is targeted towards solving odor problems for industrial manufacturers or muncipalities with wastewater treatment operations. The objective of this paper is to provide the generator of odors a systematic approach, from an engineering design perspective, to define the composition and quantity of the odor involved, to determine how to capture the odor and then how to eliminate them in the most cost effective manner. A total cost approach must be considered in determining the best odor control technology available since both equipment operating costs and capital investment costs will impact the decision on the equipment system to be recommended and selected.

The most difficult phase in solving an odor control problem is to accurately determine the quality, quantity, and make-up of the odor generating streams. Typically, when an odor becomes a problem, it has been identified by plant personnel within the plant's boundaries or from individuals from the surrounding community based on their sensitivities to smell. At this point, the degree to which the odor offends an individual is clearly a subjective evaluation. Since the design engineer must complete the necessary mass and energy balance calculations to size and select the most cost effective odor system, he must first have a sound design basis. This usually requires good analytical information. With odors typically being identified and measured in subjective units, the real challenge for the design engineer is to define the constituents in the odor stream so that an odor control system can be designed to effectively lower the odor concentrations to levels below the odor threshold limits or odor regulatory levels.

Discussion

The management of chemical, process and waste treatment facilities has a responsibility to operate their plant in an environmentally safe manner and a responsibility to be a good neighbor in their communities. This management responsibility requires that the company minimize and eliminate, if possible, the emission of odors to the environment. If odors are present, the company must systematically determine a control scheme that they can implement to solve the odor control problem. They must also be able to afford to install and efficiently operate the odor control equipment system. The necessary steps that must be taken on from an engineering perspective to solve the odor control problem are shown below. To solve an odor control problem, it is important that the owner or engineer follow this defined, systematic approach to obtain a cost effective solution.

1. Determine the source(s) of the odorant.

2. Define and determine the composition of odor being generated.

3. Change, modify or improve the operating steps in the process to minimize emission of the odorant.

4. Determine if the odorant concentrations, operating volumetric discharge rate, and temperature of the emission stream can be reduced to minimize overall emissions of odorants.

5. Determine how to capture the odor generating fumes.

6. Evaluate the viable odor control technologies available. Select the control technologies that have distinct advantages for the process system involved.

7. Complete a preliminary design to size the appropriate equipment systems.

8. Based on the preliminary design, determine the capital investment and operating costs involved for the viable technologies being considered. Select the most cost effective technology.

9. Review the proposed control scheme and preliminary design documents with the regulatory agencies and public sector personnel involved.

10. Finalize the design documents.

11. Construct odor control system.

It is critical to the success of the design and the selection of equipment for any odor control project that these design tasks be completed diligently and thoroughly. No step in this approach should be by-passed without being thoroughly evaluated by management and by the engineering design professional. There is usually a time pressure to solve odor control problems. The temptation to obtain a quick solution, although possible, must be resisted and the design process should not be compromised since the project's long-term success is dependent upon properly and systematically carrying out these design steps.

Odor Source Determination

It is very important that all potential sources for odor are identified within the plant. Many times the source of the emitter is obvious, but usually there are secondary sources contributing to the odor problem. The design engineer and the plant management must familiarize themselves with the operating procedures of all the existing equipment systems in the plant which could be contributing to the odor problem. Typically, the design and management team should review the process flow diagrams, piping and instrumentation drawings and plant layout drawings to determine the process venting streams. A complete plant field survey should be undertaken to verify and identify all process emission sources. The plant survey should also include a review of housekeeping, maintenance practices and general plant conditions. Very often odorants are being emitted from fugitive sources from trenches, dike areas, pump seals and process leaks. Fugitive emissions can be a major source of odor. Concurrent with the field survey, the process design engineer can prepare mass and energy balances on the individual equipment systems that are generating vent streams. These calculations are usually based on the process chemistry and process operating conditions. With the use of stoichiometry and the physical property data of the materials involved, the engineer can usually quantify the amount of emissions involved. Upon completion of the field survey, analytical testing should be completed on the suspected vent stacks to determine the components in the vent stream. After receipt of the analytical data results from the field testing of individual vent stacks, the engineer can verify the actual composition of the vent gas stream and by comparison with his theoretical calculations can quantify the amount of each constituent being emitted.

Odor Identification

Odors generally are made up from a complex mixture of constituents. The key to an analytical analysis is to pinpoint the components which are contributing to the odor so that the design engineer can have a sound basis to use for design. Usually, there are a few odorants which are responsible for the characteristic odor emanating from the plant. These odor producers must be defined by analytical lab testing of the vent gas streams or by the process engineer taking current plant process data and, with the use of stoichiometry, determine the expected emitters.

Lab testing of the odor components are typically completed using both analytical methods or sensory methods. Where there are only one or two components to the odor stream, a suitable analytical method is available. If more than one component is present in the gas stream, identification becomes more difficult. From an engineering design stand point, the quantitation information obtained from analytical testing method is preferred. Test methods such as using gaschromatography, mass spectrometry, infra-red adsorption and infra-red spectrophotometry clearly give excellent design basis information. Where these test methods are not possible, sensory measurements via "odor panels" can be helpful to the design engineers. The sensory testing methods tend to provide a more qualitative basis for the design engineer.

Process Improvements

During the process review and field survey, a determined effort should be made to change, modify or improve the processing steps. These improvements can often reduce the concentration or amounts of emissions leaving each part of the operation. There needs to be a constant awareness by every one involved in the plant that if the odorants are contained and not emitted, then they will not create odor problems.

Simple process changes or modifications at the early stages of an odor control project that reduce odor emissions are typically the least expensive and most cost effective control measure to pursue. These changes can sometimes solve entire plant odor problems. Some typical process changes include:

*   Modifying the process chemistry such that the odor generating material is reduced, replaced or charged at a lower flow rate. These modifications will reduce the concentration of the odor producing components in the venting system.

*   Reduce the operating vent gas volumetric flow rate.

*   Reduce the operating temperature of the process equipment or reduce the temperature of the vent gas stream. The vapor pressure or "evaporation rate of liquids" increase with temperature such that the lower the vent stream temperature the lower the concentration of odor producers.

*   Enclose the operation which generates the odor such that fugitive emissions do not exist. Generally, enclosing an operation also reduces the operating volumetric flow rate required, and thus reduces the size of any control equipment that may be required.

Odor Fume Capture

The design engineer must investigate methods to assure that fumes are efficiently captured and properly routed to the control device. The design principles defined in the Industrial Ventilation Handbook published by the American Conference of Governmental Industrial Hygienists should be strictly adhered to. The principles shown in this manual will provide optimum ways and methods to capture fumes via hoods and enclosures such that the volume of air required is minimized. A thorough understanding of ventilation principles is important since installation costs and operating costs are directly related to the volume of gas to be handled.

Odor Control Technologies

The selection of a specific odor control technology is usually made based on two main factors: 1) its effectiveness at reducing odor levels below a predetermined odor threshold limit and 2) its total cost impact to the owner. The primary odor control technologies are listed below:

*   Vent Gas Condensation

*   Vent Gas Collection/Dispersion

*   Odor Modification

*   Adsorption and Oxidation Reaction

* Adsorption

* Incineration

Vent Gas Condensation

Typically, odors are carried to the atmosphere in a gas stream which is vented from a vessel or process in the form of a vapor or mist. These vapors can usually be condensed by direct contact with a solvent, usually water, or by indirect contact in a shell and tube condenser. In a direct contact condenser, the vapor is intimately contacted with the solvent providing excellent mass and heat transfer. But, disposing of the condensate from a direct contact condenser is usually expensive, therefore, a shell and tube condenser (non-contact condenser) is generally preferred. The coolant selection is dependent on the vapor pressure of the gas stream, but is usually cooling tower water, chilled water, a brine solution or direct expansion refrigerant. The colder the coolant that is used, the better removal of the vapor from the air stream.

Vent Gas Collection/Dispersion

It is important to any odor control technology implemented that all the odor producing streams are captured and collected into a common ductwork system and routed to the control device. The use of dispersion through tall stacks is a relatively inexpensive method of odor control; whereby odors are not oxidized but dispersed at such a high level that dilution via wind and stack discharge velocity reduce odors below these threshold levels.

This method is usually effective for handling odors that are marginal at meeting odor threshold limits.

Odor Modification

Odor modification is a procedure where an additive is introduced into the gas stream and mixed with the odiferous gases to produce a mixture of gases which is far less objectionable than the original odor. Introducing an "odor modifier" changes the perception and intensity of the odor for the receiver. This "masking" is a less expensive method odor control which will require "use testing" of different agents to find the best modifier. This method does not eliminate or oxidize the odor producers and may not specifically meet the air quality standards established, but it is clearly an inexpensive option that can be considered. It should be noted that "odor modifiers" should not be used to "mask" hazardous or toxic substances.

Adsorption/Oxidation Reaction

Control of odors in a gas adsorber is accomplished by intimately contacting the contaminated gas stream with a solvent in which the gas stream and its components have a high solubility.

The odorant is transferred from the gas phase to the liquid phase. After contacting the liquid, the uncontaminated gas phase is then released to the atmosphere and the liquid phase can be removed or chemically treated (oxidized). Oxidation in the absorber can be carried out by selectively adding the proper reagent to the recirculation liquid to react with the odor producing components in the gas stream. Adsorption and gas phase oxidation reactions are usually carried out in countercurrent, cocurrent or crossflow packed columns (called scrubbers) where a liquid and gas pass over a pre-selected packing material or where a gas is contacted with a vapor or mist within a column. The packing provides a torrentous path for the gas to follow and where it has intimate contact with the liquid. The reaction or adsorption process takes place as the two materials contact each other either in the packing or by contact with the spray mist.

Adsorption

Adsorption is the process by which molecules from a gas or liquid stream attach themselves on the surface of a solid. The process is based on the attractive forces between the solid surface and the adsorbed molecules. An effective adsorbant is generally a highly porous solids with large internal areas. Activated carbon is currently the most important adsorbant odor control. Carbon is very effective in removing most organic odor producing substances.

The odor bearing gas stream is passed through a bed of carbon where the odor bearing component is captured, and an uncontaminated gas stream is released to the atmosphere. The collected material on the carbon may be stripped off for recovery or disposal. Adsorbers are typically set-up with parallel columns such that one column is always "in operation" while the other is being regenerated.

Incineration

Incineration of odor bearing gas streams comes the closest to complete removal and oxidation of odors, but typically with high installation and operating costs. Incinerators are high temperature oxidation devices where odorous gases are converted primarily to carbon dioxide and water. Incinerators are either thermal incinerators or catalytic incinerators. Catalytic incinerators can be operated at much lower temperatures with the use of a properly selected catalyst. Both incinerators can be provided with heat recovery equipment to reduce operating costs. Typically, a thermal incinerator operates at 1,200-1,800°F and a catalytic incinerator operates at 600-1,000°F. The length of reaction time at the elevated temperature in a turbulent environment is critical to assure complete oxidation of the odor bearing compounds in these set-ups.

Preliminary Design

After a thorough review of the process and the odorants involved, the viable control alternatives for reducing the odors to below required threshold level should be selected. A preliminary design and the selection of the equipment should be completed for each alternative. This is necessary so that a budget capital cost estimate (±25% accuracy) can be determined and operating costs verified.

Based on this preliminary design and cost estimating, the most cost effective control technology can be determined.

The typical installed costs and operating costs for most systems can vary substantial depending on each application. Generally, the volume of gas to be handled, and the materials of construction have the most dramatic impact on operating and installation costs. Figures 1 and 2 give a general comparison of operating and installed costs for each type system. The costs are based on some typical, recent applications and give a general comparison of costs for each type system.

Clearly, not each control scheme will eliminate odors to the odor threshold level required. But, once the viable control schemes are selected and a preliminary design completed, cost effective decision can be made based on the cost information that is generated.

Design Review

After the preferred technology has been selected, the proposed scheme and preliminary design document should be reviewed, in detail, with the appropriate personnel from the environmental regulatory agencies, the plant personnel and the community at large. The specific reasons for selecting the proposed equipment systems and the expected performance should be openly reviewed. It is important to keep all parties informed on the details for the project so they can have their inputs to the project design. The scope of the project should be clearly defined at this point, such that any environmental or installation permits can be applied for and obtained.

Final Design/Construction

Once all involved parties have committed to the project, all the final construction quality design drawings and specifications can be completed. The equipment can be ordered, construction completed and the equipment brought on-line.

Summary

The best approach to solving an odor control problem is to address the problem in a systematic fashion as we have described such that all aspects of the problem have been clearly defined, and that there has been sound basis for design established. If steps in the approach are by-passed, important components of the design may have been overlooked and a more costly or less effective control system may have been inadvertently selected.

The most cost effective control technology to use for odor control system is different for each application. Many times there is only a single choice for a control technology. Often times, a number of control technologies are available. The selection of the best system to use is usually based on a total installed cost and operating cost of the system. However, in the final selection the engineer should also evaluate the need for future expansion of the odor control system and the flexibility of the odor control equipment to changes in process loads and process conditions.

Design Steps to Solve Odor Control Problems

1. Source Determination

2. Odor Composition

3. Process Improvements

4. Odor Containment/Fume Capture

5. Odor Control Technology Evaluation

6. Preliminary Design/Cost Evaluation/Equipment Selection

7. Review Design with Regulatory Agencies/Community

8. Finalize Design/Construct

9. Start-up/Testing

Primary Odor Control Technologies

1. Condensation

2. Odor Modification

3. Dispersion

4. Gas Absorption/Oxidation

5. Gas Adsorption

6. Thermal Incineration

7. Catalytic Incineration

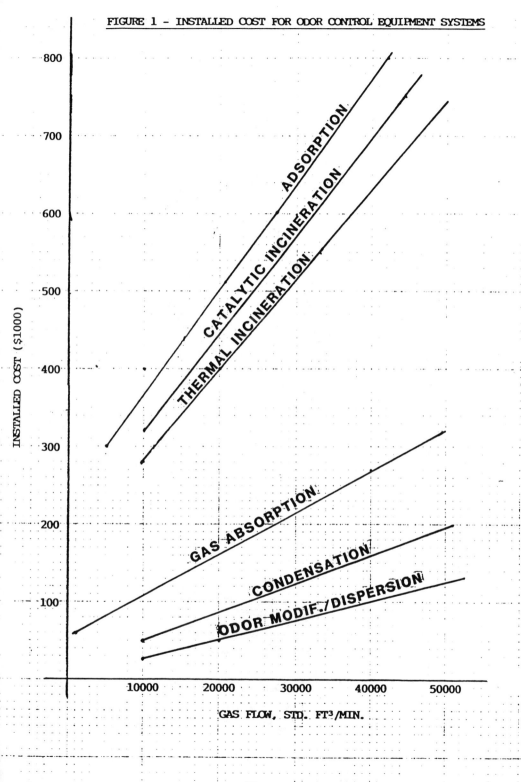

FIGURE 1 - INSTALLED COST FOR ODOR CONTROL EQUIPMENT SYSTEMS

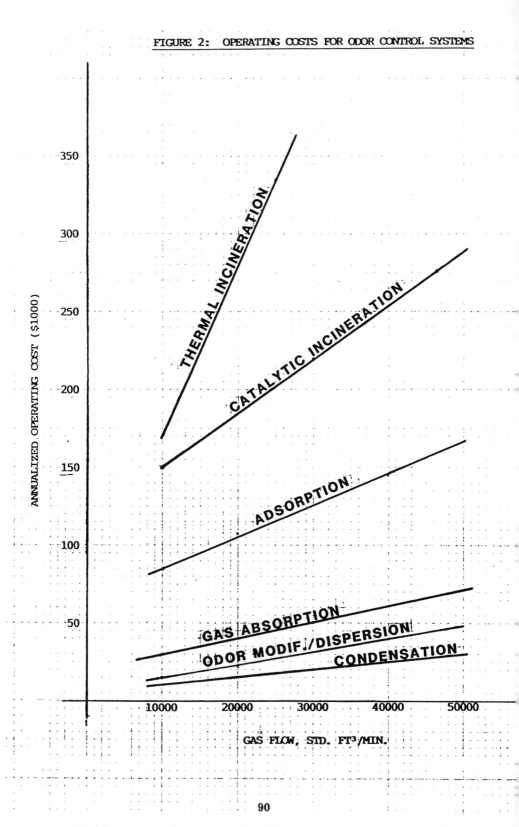

FIGURE 2:   OPERATING COSTS FOR ODOR CONTROL SYSTEMS

USE OF BIOFILTRATION FOR ODOR CONTROL

Ned Ostojic, Richard A. Duffee, Martha A. O'Brien
Odor Science & Engineering, Inc.
57 Fishfry Street
Hartford, CT   06120

The last several years have witnessed a rapid growth in the interest in biofiltration as an odor control technique. In many instances biofiltration offers a cost effective alternative to conventional add-on odor control technologies such as wet and dry scrubbing and it can often be used to supplement these techniques.

This paper reviews the principle of biofiltration, factors affecting its performance and the design practices in the US and abroad. Further progress in development of the technique is hampered by a lack of quantitative data on its odor removal efficiency. This paper presents several case histories describing applications of biofiltration in control of odors. odor removal efficiencies were measured by means of dynamic olfactometry, using panels of screened odor observers. Biofiltration was found to be particularly effective in removal of odors related to municipal wastewater sludge.

INTRODUCTION

Biofiltration is a relatively recent development in the area of odor control. It offers a welcome alternative and a supplement to the other techniques used in this field such as wet scrubbing, dry scrubbing and incineration. In biofiltration, odorous gas is brought in contact with a biologically active substrate. The odorous compounds from the gas are first adsorbed on the surface of the substrate or absorbed in the water which is typically present in the substrate. Subsequently, the compounds are biodegraded by a variety of microbiological processes.

FACTORS AFFECTING PERFORMANCE

Rate of Odorant Transfer

The removal of odor by biofiltration occurs in several steps. The rates of odorant transfer vary from step to step. The mass transfer rates have often been analyzed in terms of their inverse, i.e. the resistance to mass transfer. Insofar as the steps are independent and consecutive, the resistances from the individual steps are additive. The first of these resistances is known as the "gas phase" resistance which pertains to the transfer of the odorant from bulk gas to the surface of the substrate. This step is also present in two other techniques used for odor removal: wet scrubbing (gas - liquid contacting such as absorption) and dry scrubbing (gas - solid contacting such as adsorption). These techniques have been studied extensively and several empirical relationships have been established for the mass transfer rates involved. The driving potential between the bulk gas and the solid surface is generally described by:

$$\frac{dy}{dt} = \frac{k_g a}{R} (x-x^0) \tag{1}$$

where y = dimensionless solid phase concentration,
t = time,
k_g = gas phase mass transfer coefficient,
a = external surface area of sorbent particles per unit packed volume,
x = dimensionless bulk gas concentration,
x^0 = gas concentration in equilibrium with the outer surface of the solid, and
R = partition ratio as defined by Vermuelen (1984).

The most applicable to biofiltration is the relationship for the mass transfer coefficient in gas-solid contact in packed bed operations, developed by Wilke and Hougen (Hougen, 1945) and modified by Vermuelen et al.(Vermuelen, 1984):

$$k_g\bar{a} = \frac{10.9\ (1-E)}{d_p}\left(\frac{F}{S}\right)\left[\frac{D_g}{d_p\ (F/S)}\right]^{0.51}\left(\frac{D_g P_g}{\mu}\right)^{0.16} \tag{2}$$

where E   = void fraction,
       d_p = particle diameter,
       F   = volumetric gas flow rate,
       S   = superficial cross sectional area,
       D_g = diffusivity in gas phase,
       P_g = partial pressure, and
       μ = velocity,
       all in consistent units.

Dimensional analysis of expression (2) shows that the mass transfer coefficient increases with a 0.49 power of the superficial gas velocity (F/S). For a given depth of substrate, the rate of mass transfer increases with gas velocity. However, the residence time of the gas within the substrate decreases linearly with the gas velocity. Thus, although the transfer rate increases with an increase in gas velocity, the time available for the transfer decreases even faster.

In the simple case of physical adsorption, the overall rate of the process is affected by the partition ratio (R in expression (1)). This quantity is related to the equilibrium between the gas and solid phases, the adsorption isotherm. As the concentration of the odorant on the surface of the adsorbent increases, the adsorption slows down. In the case of biofiltration, however, the concentration of the adsorbed odorant is constantly being reduced by the microbiological activity. Most of the factors affecting the rate of the microbiological activity have been identified. However, the interrelationship is quite complex and highly specific even for relatively simple systems and only limited number   of parameter correlations can be found in literature. Thus for foreseeable future, biofiltration will remain an empirical technique.

In general, humidity and temperature are the critical factors affecting the process rate irrespective of the type of the microbiological system. In some systems, pH of the substrate is very important.

<u>Resistance to gas flow</u>

The resistance to gas flow through a biofiltration substrate can be approximated by the corresponding relationship developed for adsorption (CEP, 1952) which was modified by substitution of the values for air at $25^{\circ}C$ and barometric pressure in the appropriate variables:

$$\Delta P = 2.66 \cdot 10^{-5}\ \frac{(1-E)^2}{E^3}\ \frac{U}{d_p^2} + 1.02 \cdot 10^{-6}\ \frac{(1-E)}{E^3}\ \frac{U}{d_p} \tag{3}$$

where ΔP = pressure drop through the sustrate (inch W.C. per
                inch of substrate), and
       U = gas velocity (ft/min).

The factors affecting the resistance are gas velocity (U), void volume (E) and particle size (d_p). The effect of the void volume is very pronounced. For example, reduction of the void volume from 0.45 to 0.30 (a 33% reduction), increases the pressure drop through a one inch deep bed from 0.10 in W.C. to 0.51 in. W.C. (a fivefold increase). In biofiltration practice with soil beds having low void space, the gas velocity must be reduced drastically to compensate for the above effect. The data on void space for various materials are limited. Effective void space can be calculated, however, from bulk density, composition and specific gravity of the main constituents of a biofiltration substrate. Table 1 lists some of these properties for some materials. It should be realized that increased moisture in the substrate (such as from rain or condensation), effectively reduces the void space and could substantially increase the pressure drop through the substrate. Particulate in the gas stream could have the same effect.

DESIGN PARAMETERS AND PRACTICE

Bed Size

Two types of biofiltration processes have evolved with respect to bed size: in-soil filters and biotowers. For in-soil filters the substrate is applied on land surface and the gas is distributed below the substrate, typically by means of extensive manifolds. In biotowers the substrate is confined within vessels similar to the activated carbon beds used in dry scrubbing for odor control. The in-soil filters occupy a much larger space than do biotowers of comparable capacity. Therefore the choice in part depends on the availability of sufficient land area for construction of an in-soil filter. European practice has emphasized use of biotowers, while in-soil filters are more popular in the USA.

A design parameter used in selecting the size of the biofilter is the gas loading, typically expressed in terms of ft^3/ft^2min or m^3/m^2-hr. This parameter, which can be interpreted as superficial velocity (in ft/min or m/hr), is closely related to the size of the bed and to the resistance to gas flow. Superficial velocity is a reasonable measure of actual velocity only when the void space is quite large such as in packed towers used in wet scrubbing. Not surprisingly, the gas loadings in packed towers significantly exceed those used in biofiltration. Table 2 compares the gas loading rates used in biofiltration with those used in wet and dry scrubbing. The gas loadings in a packed tower are more than ten times as high as the highest loadings used in biofiltration. The range of loadings used in activated carbon adsorption and similar processes, partially overlaps the range of loadings used in biofiltration. The highest loadings used in adsorption (about 100 ft/min) are still almost four times as high as those used in commercially offered biotowers.

Bed Height

Most biofilters utilize bed depth of about 3 ft (1 m). That height has typically been sufficient to assure adequate odor removal. In addition to providing sufficient contact time, bed depth also serves to equalize the resistance to air flow and thus provide a more uniform distribution of gas through the bed. Biotowers are sometimes supplied with several consecutive substrate beds typically in stacked configuration. That configuration makes it possible to use modular unit design.

Uniformity of Gas Flow

Uniformity of gas flow is one of the most important factors affecting biofilter performance. Flow channeling through low resistance sections of a bed results in lower residence time in those sections of the bed. This in turn may reduce odor removal efficiency. Just as importantly, reduced flow in some sections of the bed may result in oxygen deficiency in those sections which could upset the microbiological activity. Anaerobic conditions, which may develop, could give rise to odors which could seriously impair biofilter performance.

Uniform air flow is just as important in wet and dry scrubbing. Those techniques, however, offer much better control of substrate uniformity which in turn makes it easier to control uniformity of gas flow. Substrates used in biofiltration are much less homogeneous than those used in dry scrubbing. In addition they are more likely to change in the course of the biofiltration process, in part due to fluctuation of moisture levels.

Gas flow distribution in earth filters is most commonly achieved by means of a manifold system emloying perforated pipes. Use of cinder blocks appears to offer a promising alternative (Eitner ?). In biotowers, as in wet and dry scrubbing, the gas is supplied into a portion of the vessel which forms a plenum. The flow resistance caused by the substrate pressurizes the plenum which results in an even distribution of the gas through the substrate.

Bacteria Culture

Many biofiltration substrates such as compost, peat and soil, naturally support wide variety of bacteria. Once such substrate is brought in contact with the odorous gas, the natural equilibrium of the nutrients in the substrate is altered. In a process of natural biofilter optimization the bacteria adjust to the change, with some strains becoming predominant.

Bacterial strains known to be especially effective in metabolizing certain odorants can be introduced into the substrate to optimize its performance. This is especially promising in the cases when the odors are caused by only a small number of compounds or classes of compounds. Thus Thiobacillus Thiaparus has been found most effective in oxidation of reduced sulfur compounds (Atlas 1981, Grant 1981). In many applications, however, the composition of odorous gas streams is quite complex and the best effectiveness would most likely require a presence of a wide spectrum of bacterial strains. It should be realized that a bacteriological transformation of many compounds, especially the more complex ones, involves many steps which may require various strains.

Substrate Conditioning

Moisture in the biofilter substrate has been identified as one of the most important parameters affecting biofilter performance. In many situations, proper moisture level (range here or above) cannot be maintained solely by controlling the humidity in the gas stream. Many biofilters are equipped with a sprinkler or a spray mechanism to maintain the moisture in the proper range.

In some cases, nutrient addition was found to be beneficial for biofilter performance (Don, 1984). Adjustment of pH was found to be very important in some cases where bacteriological processes result in pH change in the substrate (Eitner, 1984).

Other Factors

Gas conditioning - humidifying, cooling or filtration - may be required to maintain proper operating conditions within a biofilter. Ideally gas temperature would be in the $59^{\circ}F-95^{\circ}F$ range with high humidity.

The in-soil filters should be designed for proper drainage and collection of the leachate and condensate for treatment. Inadequate drainage could lead to development of septic conditions.

Any residual odor from an in-soil filter is emitted at ground level from an area much larger than a typical point source. Point source emissions typically occur at some height above the ground level. Dispersion of any residual odors may be significantly affected by the type of source and may affect the choice of the odor control technique. Residual odors may be "secondary" odors caused by the substrate itself, such as occasionally observed "compost" odors when compost is used as the primary biofilter substrate.

EXAMPLES OF BIOFILTER APPLICATIONS

All odor data presented in the following examples were obtained using the Odor Science & Engineering, Inc.'s forced choice dynamic olfactometer and panels of screened odor observers. The method employs an ascending series of odor concentrations, doubling at each step, in full conformance with the recommendations of ASTM Standard Practice for Determination of Odor and Taste Thresholds by Forced Choice Ascending Concentrations Series of Limits (ASTM E679-79)

60,000 ACFM Biofilter at a Composting Facility

The composting facility in Akron, Ohio, is designed to process 70 dry tons of municipal sludge per day, using a modified Fairfield process. The facility uses biofiltration for control of the odors from the dewatering, mixing and some process areas. The combined exhaust is transported to the biofilter in two 48 inches wide circular ducts.

The biofilter is of an in-soil filter type and occupies an area of 50,000 ft^2. The total area is divided into two equal rectangular sections each 250 ft by 100 ft. The two air ducts deliver the air into a centrally located 13 ft wide and 250 ft long aisle which divides the biofilter in the two sections. The two ducts are closed at the end, thus forming two plenums from which the air is delivered into the two biofilter fields. A total of 30 perforated pipes 100 ft long and 12 inches in diameter, are used to distribute the air into each half of the biofilter. Sixty stainless steel butterfly valves, located at the entrances to the pipes, provide individual flow control for each pipe. The pipes are positioned directly on top of a layer of woodchips placed on an 8 inch thick slab of reinforced concrete which serves as a base for the filter. The slab is inclined for drainage of the condensate and leachate which are transported to the adjacent wastewater treatment plant.

A four ft high bed of compost, produced on site, is used as the biofiltration substrate. At a loading rate of 1.2 ft^3/ft-min, the pressure drop through the substrate was 4 inches W.C. The compost is placed directly on top of the distribution pipes which may inhibit lateral distribution of the gas. The main reason for omitting a flow distributing layer (such as gravel) at the bottom of the bed was the ease of bed replacement.

At the time of the biofilter evaluation, the substrate was in place for approximately four weeks. There was no vegetation cover on top of the bed.

Odor samples were collected from a flux chamber positioned on top of the biofilter. Odor concentrations at the inlet to the biofilter were measured using the samples from the supply plenums.

When the composting process was operating in a positive mode (i.e. with the air passing upwards through the composting reactor), odor concentrations at the inlet to the biofilter ranged from 281 dilutions to threshold (D/T) to 758 D/T. Odor concentrations in the samples collected at various points on the surface of the biofilter ranged form 3 D/T to 26 D/T. It is important to note that a significant change in odor character took place as a result of biofiltration. The dominate character of the inlet was that of municipal sludge, in large part attributed to the reduced sulfur compounds. The outlet odor character was entirely that of compost.

When the composting process was operated under negative mode (downwards flow through the reactor), the odor levels from the composting process increased. The biofilter was operated at maximum loading of 2.4 ft^3/ft^2-min. The odor character at the inlet to the biofilter contained a noticeable "compost" component. With an inlet odor concentration of 890 D/T, the odor at the outlet of the biofilter decreased to 300 D/T under these conditions.

The capital cost for the above biofilter is approximately $200,000, according to Burgess and Niple Limited of Akron, Ohio, designers of the system. The cost of replacing the biofilter substrate, assuming no recovery of the distribution pipes is about $43,000.

The filter has now been operating for approximately one and half years without a need to replace the substrate. Thick shrubbery was allowed to grow on top of the bed, which was judged beneficial.

12,000 ACFM System for a Wastewater Treatment Plant/Composting Operation

A municipal wastewater treatment plant in New England utilizes composting for treatment of 7 dry tons of sludge per day. The combined 12,000 ACFM exhaust from the screens, sludge holding and blending tanks and from the composting process operated in the negative mode, is treated in a biofilter for odor control.

The biofilter is located within the composting building within an 8 ft tall concrete rectangular enclosure 16 ft by 200 ft (area 3,200 ft^2). The bottom layer of the biofilter consists of a one foot deep layer of crushed stone about 2 inches in size. The air is distributed into that layer by means of "half pipes", i.e. horizontally positioned pipes whose bottom halves have been removed. The stone layer promotes lateral distribution of the air. Positioned above the stone layer is a one foot deep layer of wood chips, followed by a three feet deep layer of compost where biofiltration takes place.

Smoke tests have revealed significant channeling, i.e. sections of widely different flow rates through the biofilter substrate. Sections of high flow could also be recognized by closely spaced small holes on the surface after the top of the biofilter had been wetted.

The total flow rate of air in the two biofilter supply ducts was measured at 12,000 ACFM. The actual quantity of air delivered to the bed was in all likelihood less because of some leaks in the distribution piping. With a total filter surface area of 3,200 ft2, a 12,000 ACFM flow rate translates into a superficial vertical velocity of 3.75 ft/min. The channeling increased this velocity significantly in some sections of the bed. Observations using smoke puffs indicated that vertical velocity at some locations could have approached 30 ft/min.

With the dewatering and composting processes not in operation, the odor concentration at the inlet to the biofilter ranged from 222 D/T to 383 D/T. With all systems in full operation, the odor level at the inlet to the biofilter increased to 653 D/T. Odor concentrations in the samples collected at various locations on the surface of the biofilter, ranged from 18 D/T to 103 D/T. The biofilter reduced ammonia concentration from 22 ppm at the inlet to 12 ppm at the outlet. The inlet and outlet values for trimethylamine were 30 ppm and 12 ppm and for dimethyl sulfide 2 ppm and 0.3 ppm respectively.

It is interesting to note that the odor removal performance of the bed in the high flow regions was higher than in the low flow regions. This in part may reflect a sampling error. The air leaving the biofilter had lower odor concentration than the rest of the air within the composting building. Therefore, at low flow rates from the biofilter, the sampling enclosure may not have been effectively purged with the biofilter air, in spite of the efforts to allow sufficient time prior to sampling and to sample at a low rate.

3,000 ACFM Biofilter at Wastewater Treatment Plant

A municipal wastewater treatment plant in Bristol, Connecticut, employed wet scrubbing for control of odors from sludge blending tanks. The scrubber utilized sodium hydroxide solution as the scrubbing liquid. The scrubber's odor removal efficiency was in the range of 37% to 40%. The odors in the scrubber exhaust were variable, ranging from 3,750 D/T to 15,000 D/T. Since oxidizing scrubbing liquids could not be employed in the scrubber because of material of construction limitations, biofiltration was considered for removal of the residual odors.

Biofiltration was tested on a pilot scale using plant's compost as the substrate. After two weeks of operation, the pilot unit reduced the inlet odor concentration from 15,000 D/T to 310 D/T (98% reduction).

Based on the results of the pilot tests, a full scale biofilter was constructed. The biofilter was in a form of a rectangle 40 ft x 52 ft. The area for the filter was excavated to a depth of two feet, providing a drainage slope with an incline of 1%. The area was covered with a plastic membrane to retain the leachate and condensate for return to the treatment plant. The air distribution system employed a system of pipes, 8 inches in diameter, equipped with a pair of 1 inch holes at six inch intervals. The pipe system was imbedded into a two ft deep bed of crushed stone which filled up the excavated area back to the level of the surrounding terrain. A three ft deep bed of compost was placed above the crushed stone as a biofiltration substrate.

The performance of the full scale bed was comparable to the pilot scale unit. An inlet odor concentration of 9,700 D/T was reduced to 190 D/T and 37 D/T at two locations on the bed surface. The character of the odor emitted from the biofilter was described as "earthy" and "musty" which is attributable to the compost substrate. The "sludge" character was completely absent. The cost of installing the biofilter was estimated at $25,000 by Keyes Associates of Wethersfield, Connecticut, designers of the installation.

COMPARISON WITH OTHER ODOR CONTROL TECHNIQUES

Table 3 lists some advantages and some disadvantages of biofiltration. The technique is the best suited for treatment of gas streams of rather constant composition, "ambient" temperature and high humidity. It is especially advantageous in those instances when successful odor removal would otherwise require use of several stages of treatment. Odors from some sources in municipal wastewater treatment plants, for example, fall into that category. These odors are often caused by a mixture of low molecular weight reduced sulfur compounds and various other compounds. The highly odorous reduced sulfur compounds are relatively easily oxidizable into less odorous forms by techniques such as wet scrubbing. However, wet scrubbing with oxidative reactions is sometimes less effective with other types of odorants. High levels of odor control have therefore sometimes required the use of activated carbon adsorption to "polish" the odors remaining after wet scrubbing. Biofiltration was shown to be capable of achieving high odor removal efficiency without a need for additional treatment.

The capital cost of an in-soil type biofilter is typically lower than the cost for wet and dry scrubbing equipment of comparable capacity. For instance, the $200,000 installed cost of the 60,000 ACFM biofilter in Akron, described above, would compare with an estimated $300,000 for a single stage wet scrubbing system and about $400,000 for an activated carbon adsorption system.

Theoretically, an in-soil type biofilter could continue to operate indefinitely without a need to replace the substrate. In that case, the operating cost would consist entirely of the energy required for overcoming the resistance to air flow. That cost is roughly comparable with the corresponding power requirements in wet and dry scrubbing.

Should replacement of the substrate prove to be necessary, the cost for a 60,000 ACFM in-soil system in the above example from Akron is estimated at $43,000, including replacement of the distribution piping. Corresponding cost for off-site reactivation of activated carbon in a comparable system would be about $100,000. The cost for the chemicals used in wet scrubbing depends on the concentration of the odorants. A $50,000 annual expenditure for scrubbing chemicals could easily be exceeded in the case of a 60,000 ACFM wet scrubbing system.

Based on the above discussion, both the capital and operating costs for in-soil biofiltration compare favorably with the wet and dry scrubbing alternatives.

As stated previously, biotowers operate at lower gas loadings than wet and even dry scrubbers. A biotower vessel is thus correspondingly larger than either a wet or a dry scrubber of comparable capacity. Consequently, the capital cost for a biotower installation is likely to be higher than the cost for wet or dry scrubbing. Any requirements for gas conditioning would further increase the cost differential. It appears that the biotowers would find application primarily in special circumstances when other techniques fail to provide sufficient odor removal efficiency and sufficient land area is not available for in-soil biofiltration. "Custom" strains of bacteria for specialized applications are more easily introduced and monitored in biotowers than in the in-soil installations.

REFERENCES

1. Atlas, R.M. and R. Bartha, (1981), "Microbial Ecology: Fundamentals and Applications", Addison Wesley Publ. Company, Reading, MA, 1981, USA.

2. C.E.P., "Fluid Flow through Packed Columns", Chem. Eng. Progr., 48 (2): 89(1952).

3. Don, J.A., and L. Feenstra, (1984), "Odour Abatement through Biofiltration", a paper presented ata a Symposium in Louvain-La-Neuve, April 1984, Belgium.

4. Dragt, A.J., and S.P.P. Ottengraf, (1985), "Biofiltration A New Technology in Air Pollution Control Experiences in the Netherlands", File 4-4508-23-10, Eindhoven University of Technology, may 1985, The Netherlands.

5. Eitner, D., (1984) "Untersuchungen über Einsatz und Leistungsfähigkeit von Kompostfilteranlagen zur biologischen Abluftreinigung im Bereich von Kläranlagen unter besonderer Berücksichtigung der Standzeit (Investigations of the the Use and Ability of Compost Filters for the Biological Waste Gas Purification with Special Emphasis on the operation Time Aspects)', GWA, Band 71, RWTH Aachem Sept. 1984, West Germany.

6. Grant, W.D., and P.E. Long, (1981), "Environmental Microbiology", Halsted Press, NY, 1981, USA.

7. Haug, R.T., "Compost Engineering, Principles and Practice", Ann Arbor Science Co., Ann Arbor, MI, (1980), pp 223.

8. Karney, P.T., (1985), "Odor Control, Sludge is the Solution. The Jacksonville Experience", a paper presented at Annual Meeting of the Florida Section of the Air Pollution Control Association at St. Augustine, Florida, Sept. 1985, USA.

9. VDI (1984), VDI-3477, "Biologische Abluftreinigung - Biofilter (Biologial Waste Air Purificaiton - Biofilters)", VDI-Handbuch Reinhaltung der Luft, Band 6, Dec. 1984, West Germany.

10. Vermuelen, T., M. Douglas LeVan, N.K. Hiester and G. Klein, "Adsorption and Ion Exchange", chapter in "Perry's Chemical Engineer's Handbook", 6th ed. McGraw-Hill Book Co., New York. 1984, pp 16-22.

11. Wilke, Hougen, Trans. Am. Inst. Chem. Eng., 41: 445 (1945).

TABLE I.
BULK DENSITIES AND VOID SPACE FOR SOME MATERIALS

Material	Bulk Density (lb/ft^2)	Void Space[a]	Source
Activated Carbon	22-31	0.5	(MSA, 1973)
Compost (60-80% solids)	30-45	0.4-0.65	(Haug, 1980)
Earth, dry, loose	76	0.27[b]	(Perry, 1984)
Earth, dry, packed	95	0.08[b]	(Perry, 1984)

(a) volume occupied by gas as a fraction of total volume (gas, solid and liquid), not including internal pore volume

(b) calculated assuming 15% moisture, 20% organics @ 62 lbs/ft^3, 65% inorganics @ 155 lbs/ft^3.

TABLE II.
SAMPLES OF GAS LOADING RATES

Contactor Type	Gas Loading Rate $\frac{ft^3}{ft^2/min}$	$\frac{m^3}{m^2/hr}$	Residence time for a 3 ft deep bed (sec)	Source
Packed tower wet scrubber	300-500	6,600-9,100	0.5-0.36[a]	
Activated Carbon Adsorber (high flow)	100	1,800	1.8	
Activated Carbon Adsorber (medium flow)	50	900	3.6	
Activated Carbon Adsorber (low flow)	25	450	7.2	
Biotower	54	1,000	3.3	(Karney, 1985)
Biotower	44	800	4.1	(Eitner, 1984)
Biotower	27	500	6.7	(Dragt, 1985)
Biofilter, commercial	16-27	300-500	11-6.7	
Biofilter	<22.5	<410	>8	(VDI, 1984)
In-soil Biofilter	<12	<220	>15	(VDI, 1984)

(a) bed depths in packed towers used for odor control are typically in the range of 6-10 ft.

102

TABLE III.
SOME ADVANTAGES AND DISADVANTAGES OF BIOFILTRATION

ADVANTAGES	DISADVANTAGES
- high odor removal possible	- bacteria cultures sensitive to changes in inlet conditions (temperature, composition) (danger of bed sterilization)
- low capital and operating cost for in-soil biofilters	
- low maintenance (primarily for in-soil biofilters)	- may require gas conditioning
	- space requirements (for in-soil filters)
- versatility in performance (e.g. in comparison with wet scrubbing which offers limited types of reactions)	- potential deterioration of the uniformity of flow distribution

METHODOLOGY FOR DETERMINING THE ODOR BUFFER DISTANCE FOR SANITARY LANDFILLS

Ralph Kummler, Ph.D., and Dukman Song
Department of Chemical Engineering
Wayne State University
Detroit, Michigan

Thomas Wackerman, George Kandler and Joe O'Brien
Applied Science & Technology, Inc.
Ann Arbor, Michigan

Peter Warner, Ph.D.
Air Pollution Control Division
Wayne County Health Department
Detroit, Michigan

Chuck Hersey
Southeast Michigan Council of Governments
Detroit, Michigan

A cooperative research program, funded by the City of Ann Arbor and the State of Michigan through the Southeast Michigan Council of Governments, was designed to create and test a methodology to determine an acceptable odor buffer distance for sanitary landfills in the State of Michigan. The overall project was managed by ASTI and was completed by comparing 1990 proposed operations to acceptable 1984 operations with a 1200 foot isolation distance and to current operations to predict odor units at adjacent receptors as close as 600 feet from operations.

ASTI made flux measurements at the open face, the temporary cover, the final cover, vents, and the composting zones of the Ann Arbor Sanitary Landfill. Gas velocities and in-ground concentrations were determined to allow a quantification of the total and methane gas flow rates.

The Wayne County Air Pollution Control Division performed ASTM odor panel measurements to determine the odor intensity in odor units at the corresponding sites.

Wayne State University used the flux and odor panel measurements in the Industrial Source Complex Terrain Model to determine the hourly averaged highest and second highest odor levels at 175 receptors placed at the property boundary and all nearby residential locations.

Using measured values for velocity, subsurface CH_4 concentration and source odor intensity, it was determined that a buffer distance of 600 feet provided at least a factor of five protection below the odor threshold for all receptors, and provided dilution protection equal to historic operations with a 1200 foot isolation distance.

1 INTRODUCTION

Michigan's Solid Waste Management Act(1978 P.A. 641) requires county plans for solid waste management. Permits issued pursuant to Act 641 by the Michigan Department of Natural Resources(MDNR) for sanitary landfills must be consistent with the law, regulations and the approved county plans. As required by the regulations(Michigan Solid Waste Management Regulations), new sanitary landfills, or expansions to existing sanitary landfills, must maintain a minimum horizontal isolation distance. This isolation distance is to be established for each specific landfill taking into account various factors including the control of odors. The minimum distance is 300 feet from any existing domicile, and 100 feet from the property line. However, greater isolation distances may be required if other regulations apply. Although each county plan can individually address air quality issues, no other requirements for operational or design control of odors are included in these solid waste management rules.

In reviewing sanitary landfill permit applications, the MDNR has frequently applied one other state regulation to control odors(as required by Part 3, 12(b)(iv) of the Act 641 regulations). Air Pollution Control Commission Rule 901, issued under Michigan's Air Pollution Act(1965 P.A. 348) specifies that air pollution (including odors) cannot cause unreasonable interference with the comfortable enjoyment of life and property.[1] Based on this regulation, the MDNR has required various sanitary landfills to increase the isolation distances from the minimum 300 feet in order to increase odor dilution.

Although no general horizontal isolation distance has been established by the MDNR, sanitary landfill facilities have typically been required to provide up to 1,500 feet of isolation from adjacent domiciles.

Historically, a tool developed to quantify trade-offs between control and isolation distance, and to evaluate whether the ambient air is acceptable is the dispersion model. The USEPA has developed, validated, and employed such models for two decades. Their purpose is to use the source strength, source velocity, area and height, with typical meteorology, to calculate representative ambient concentrations at designated receptors.

However, odors have long been difficult to quantify. The perception of odor, or the odor intensity, follows a Weber-Fechner law like most physiological responses. Thus, the sensation of odor is expected to be proportional to the logarithm of concentration, not to the concentration itself.[2,3] This is a crucial issue when attempting to implement trade-offs between isolation distance and source strength. Moreover, in complex situations such as landfills with their myriad chemical composition, it is not easy to identify the chemical constituent which is causing the odor problem. One cannot quantify the source strength in absolute terms without identifying the chemical species involved.[4] Hence the usual air dispersion techniques must be modified, at least philosophically, to recognize these difficulties or the interpretation and use must be made differently than they would be for the normal application.

The objective of this study is to evaluate reductions in the off-site migration of odors resulting from operational and design changes in new or proposed expansions to sanitary landfills in Michigan. The Ann Arbor Sanitary Landfill was used as a test case.

2 THEORETICAL MODELING BACKGROUND

2.1 Brief Synopsis of Gaussian Dispersion Modeling

Given a quantified source strength, dispersion modeling will allow a quantitative prediction of receptor concentrations, and assuming the use of real past meteorology, will allow sufficient opportunity to find practical maximum conditions for design alternative case comparisons. We also assume that inhomogeneity in chemical composition, geometry, time, cover, etc. is averaged over the period considered, relative source strengths by position or type are known, and worst case strengths are used.

Two Gaussian plume models approved by the USEPA were applicable to this study; the Real-Time Air-Quality-Simulation Model(RAM)[5,6] and the Industrial Source Complex(ISC) Dispersion Model. The Industrial Source Complex(ISC) Dispersion Model is a multiple source, multiple receptor Gaussian plume model including Briggs plume rise and stack tip down wash, building wake effects, and complex terrain adjustments. Concentrations are subject to hourly averaged meteorology and hence, the ISC model is designed for short term use. Its use is described by Bowers, et al,[7,8] and it is generally believed to be an improvement on RAM. ISC was the model chosen for this study.

2.2 Odor Modeling Strategy

In conventional dispersion modeling,[9] we predict an absolute concentration which can be compared, for example, with the National Ambient Air Quality Standards for criteria pollutants. However, if the concentration of a pollutant at the source as well as the source strength are known, then dispersion models can also be used to predict the dilution factor as a function of location. For odor modeling purposes, the source strength can be taken to be the product of the measured odor units($o.u.$) times the measured gas production velocity times the cross sectional area of the source. The threshold for source odor measurement by a trained panel is 25 odor units, and the threshold for odor detection in the community (i.e. at receptors) is defined to be 1 $o.u.$

Let
Q = area source strength($o.u./\sec$)
υ = surface velocity($m/\sec$)
A = cross sectional area (m^2)
C = concentration in odor units($o.u./m^3$)
χ = ground-level concentration ($o.u./m^3$)

Then,

$$Q = \upsilon A C \qquad (1)$$

The dispersion model predicts a concentration, χ, which is proportional to Q. The dilution factor is defined to be:

$$D = C/\chi \qquad (2)$$

The dilution factor is independent of the actual source strength. Thus, any consistent set of units can be used for the input concentration.

We also postulate that we are not dealing with chemicals which synergistically reinforce odors. Thus, if we research all cases where the sum of the odor concentrations is ≤ 1 odor unit, and find none, then we can also assume that the effect of each source separately is ≤ 1 odor unit. Thus, the recommended procedure is as follows:

(1) measure the odor unit levels at the sources.
(2) measure the source velocities and areas.
(3) run the dispersion model.
(4) evaluate whether the odor levels are ≤ 1 odor unit at or beyond the property boundary.
(5) if odor levels exceed 1 odor unit, modify the source strength or geometry to reduce the concentrations until the odor levels are below 1 odor unit.

3 METHODOLOGY

3.1 CH_4 Flux

We measured the CH_4 flux rates from surface sources and vents using an isolation flux chamber, which was designed and operated according to USEPA guidelines,[10] at the open face, the temporary cover, the final cover, vents(direct measurements) and the composting zones of the Ann Arbor Sanitary Landfill.[11]

Total hydrocarbons as methane were measured using a Century 128 Organic Vapor Analyzer(OVA). The OVA utilizes a Flame Ionization Detector(FID) to monitor the presence of organic vapors, and it has a detection limit of 0.2 ppm as methane.[11]

3.2 Total Gas Flux and Surface Velocity

Based on the results of flux rate calculations for surface sources using USEPA guidelines[10], and the subsurface monitoring conducted during the site survey, surface velocities were estimated for use in the dispersion modeling. Surface velocity estimates were calculated at standard condition(STP, 1 atm and 25 °C).

Let
v_{avg} = average surface velocity for a specific source at STP($ft/$min)
Q_{tot} = total flow rate of source surface(SCFM)
A_C = inside cross sectional area of flux chamber(1.40 ft^2)
Q_{CH_4} = flow rate of methane at source surface(SCFM)
C_S = measured concentration below surface, total hydrocarbons as methane($ppmv$)

$$v_{avg} = Q_{tot} / A_C \qquad (3)$$

$$Q_{tot} = Q_{CH_4} / C_s \qquad (4)$$

This method was used to estimate average surface velocities of total gases at the Ann Arbor Sanitary Landfill, and to compare the results to predictions of surface velocities based on gas generation rate models.[10,11]

3.3 Odor Panel Measurements

The response of odor panels using a forced choice method was employed to estimate dilutions to threshold from selected sources at the landfill. Those dilutions to threshold define the value of odor units.[2,3,12] The odor intensity samples were collected and analyzed according to the principles of the American Society for Testing and Materials(ASTM) method number D-1391,[11,13] by the Wayne County Air Pollution Control Division. We used a five-member odor panel.

3.4 Summary of Sampling Results

The experimental values for the odor units, surface velocity, and CH_4 concentrations are presented in Table I. The statistical lower detection limit of the odor panel method is listed as 25 odor units($o.u.$). This value is provided for all samples where the odor threshold is less than 25 dilutions of the original sample.

Table I. Odor Intensity Sample Results[11]

Location	Odor Units ($o.u.$)	Average Velocity (m/sec)	CH_4 Concentrations ($ppmv$) Average Flux Chamber	Subsurface
Open Face	900	$5.0 \cdot 10^{-5}$	5.4	70
Temporary Cover	≤ 25	$2.0 \cdot 10^{-5}$	28.3	1,000
Final Cover	≤ 25	$1.3 \cdot 10^{-8}$	7.4	380,000
Compost	≤ 25	$1.5 \cdot 10^{-6}$	1.4	600
Vent	8,000	$1.2 \cdot 10^{-1}$	380,000	380,000

4 MODELING APPROACH

4.1 Modeling Scenarios and Input Data

Because actual case studies involve complicated geometry and source functions, the modeling was performed in two parts. First a hypothetical square area as a "landfill" was evaluated. Second the site specific geometry of the Ann Arbor site was used.

Standard preprocessed meteorological data from the Detroit Metropolitan Airport, for a 31 day period in January or October of 1976, were employed for all scenarios.

The highest and second highest concentrations observed are presented, in odor units, directly comparable to the measurements made by Wayne County.

4.1.1 Hypothetical Landfill Study

We used a simple circular geometry to illustrate principles rather than final specific concentrations. Thus, the polar coordinate option with 16 compass points of the ISC model was chosen to surround a square landfill zone.

4.1.2 Ann Arbor Case Study

The hypothetical model demonstrated the need to consider actual geometry for regulatory decision making. Therefore, using the real geometry of the Ann Arbor Sanitary Landfill and the complex terrain option of the ISC model, we predicted the highest and second highest concentrations at 175 receptors from 162 grid cells, 3 vents, and the open face treated as a point source. The open face was modeled as a point source to avoid ascribing a false or numerical dispersion on a grid of 100 meters: the open face area is small compared to a grid cell.

We used three operational scenarios ; the configuration of the landfill as sampled in 1988, the historic or baseline configuration in 1984, and the proposed configuration in 1990. For all three scenarios the relative source strengths remained constant. Each grid location was assigned a source type based on historic or proposed operations(temporary cover, final cover, or compost pile), or was considered background, for each of the three scenarios. Four point sources(3 vents and the open face) were included in 1988 and 1990 scenarios.

4.2 Discussion of Modeling Results

4.2.1 Hypothetical Landfill Study

With the prevailing winds from the southeast, we would expect the maximum concentrations to occur in the northeast sector, along compass directions 360° , 22.5° , and 45° as shown in Figures 1, 2, and 3. The maximum projected concentrations were all well below 0.1 $o.u.$[11]

4.2.2 Ann Arbor Case Study

The maximum odor levels occur along the northern and eastern boundaries and at residences on the northern boundary. There is some seasonal variation, but in no scenario do odor levels become higher than 0.2 $o.u.$, a level five times below the odor threshold.[11] Because the baseline 1984 scenario involves a landfill with no history of odor complaints, this represents a qualitative confirmation of our methodology and the ISC predictions.

It was determined that a buffer distance of 600 feet provided at least a factor of five protection below the odor threshold for all receptors, and provided dilution protection equal to historic operations with 1200 foot isolation distance.

5 CONCLUSION

This study is a unique approach to odor prediction at sanitary landfills using surface gas velocity measurements, odor panel measurements, and dispersion modeling. Odor units and surface velocities were used to determine source strengths. Source strength, actual site topography, and geometry were used as inputs to the ISC Dispersion Model.

The results for various hypothetical modeling scenarios indicate that the adequate horizontal isolation distance for odor control is site specific. Although source strengths vary with operational practices, the most important elements in determining an appropriate isolation distance appear to be the surface area and volume of the landfill and local topography.

Because of a lack of site-specific research, this method must include some basic assumptions about odor generation and control at typical landfills. However, because of the Weber-Fechner relationship of odors, approximations based on available information may be acceptable for conducting comparisons to the odor threshold. Modeling will indicate whether remediation is needed to achieve a one odor unit value or less at receptors. The one odor unit strategy avoids the nonlinearity problem of the Weber-Fechner effect and protects the public interest.

For effective odor reduction, horizontal isolation distance must provide at least 10^3 dilutions[14] from the original concentration. Although the number of dilutions per foot of isolation distance is dependent on meteorological conditions, topography, and surface velocity, the Ann Arbor case study indicates that a 10^3 dilution can be achieved at as little as 600 feet under certain operating and site conditions. The procedure proposed herein was found to be a good method for estimating the required isolation distances on a site specific basis.

REFERENCES

1. C. Hersey, "Project Description and Scope of Work for an Analysis of Air Quality Problems at Sanitary Landfills," SEMCOG, Jan. (1988).
2. W. Summer, "Thresholds of Olfaction," in Odour Pollution of Air-Causes and Control, Leonard Hill, London, P.18 (1971).
3. W. Summer, "Panels and Evaluation," in Odour Pollution of Air-Causes and Control, Leonard Hill, London, P.86 (1971).
4. A. Dravnieks, "Odor Character Profiling," JAPCA, Vol. 33, No. 8, P.775, Aug. (1983).
5. D.B. Turner and J.H. Novak, "User's Guide for RAM-Data Preparation and Listings," EPA-600/8-78-016b, Volume II, Nov. (1978).
6. R.H. Kummler, B. Cho, G. Roginski, R. Sinha, and A. Greenberg, "A Comparative Validation of RAM and Modified SAI Models for Short Term SO_2 Concentrations in Detroit," JAPCA, Vol. 29, No. 7, P.720 (1979).
7. J.F. Bowers, J.R. Bjorklund, and C.S. Cheny, "Industrial Source Complex (ISC) Dispersion Model User's Guide," EPA-450/4-79-030, Volume I, Dec. (1979).
8. J.F. Bowers, J.R. Bjorklund, and C.S. Cheny, "Industrial Source Complex (ISC) Dispersion Model User's Guide," EPA-450/4-79-031, Volume II, Dec. (1979).
9. D.B. Turner, "Workbook of Atmospheric Dispersion Estimates," Public Health Service Publication, No. 999-AP26 (1969).
10. EPA, "Air Emissions from Municipal Solid Waste Landfills-Background Information for Proposed Standards and Guidelines, Preliminary Draft," Emission Standard Division, March (1988).
11. T. Wackerman, G. Kandler, J. O'Brien, R. Kummler, D. Song, and P. Warner, "Odor Prediction and Control Study-Ann Arbor Sanitary Landfill," Applied Science & Technology, Inc., Volume 1, March (1989).
12. P.J. Young and A. Parker, "Origin and Control of Landfill Odors," Chemistry and Industry, May 7 (1988).
13. S. Calvert and H.M. Englund, "Odor Sampling and Analysis," in Handbook of Air Pollution Technology, John Wiley & Sons, Inc., P.847 (1984).
14. J.M. Baker, C.S. Peters, R. Perry, and C.P.V. Knight, "Odor Control in Solid Waste Management," Effluent and Water Treatment Journal, April (1983).

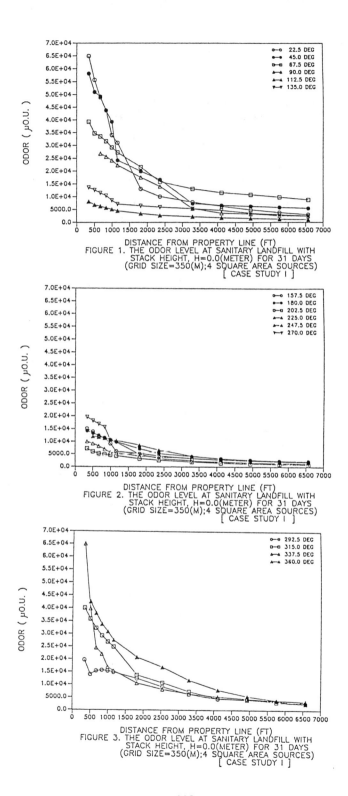

DISTANCE FROM PROPERTY LINE (FT)
FIGURE 1. THE ODOR LEVEL AT SANITARY LANDFILL WITH
STACK HEIGHT, H=0.0(METER) FOR 31 DAYS
(GRID SIZE=350(M);4 SQUARE AREA SOURCES)
[CASE STUDY I]

DISTANCE FROM PROPERTY LINE (FT)
FIGURE 2. THE ODOR LEVEL AT SANITARY LANDFILL WITH
STACK HEIGHT, H=0.0(METER) FOR 31 DAYS
(GRID SIZE=350(M);4 SQUARE AREA SOURCES)
[CASE STUDY I]

DISTANCE FROM PROPERTY LINE (FT)
FIGURE 3. THE ODOR LEVEL AT SANITARY LANDFILL WITH
STACK HEIGHT, H=0.0(METER) FOR 31 DAYS
(GRID SIZE=350(M);4 SQUARE AREA SOURCES)
[CASE STUDY I]

112

WASTES GENERATED FROM THE REMOVAL OF SULFIDE ODORS

Amos Turk, Harry Karamitsos, Khizar Mahmood,
Jahangir Mozaffari, and Rivka Loewi
Department of Chemistry
The City College of the City University of New York
New York, NY   10031

and

Anthony N. Borowiec
Vice President
Metcalf & Eddy, Inc.

This study describes laboratory methods for testing and regenerating exhausted caustic-impregnated or ammonia-injected activated carbons used for removal of sulfide odors in wastewater treatment plants (WWTPs). Partially exhausted carbons are subjected to accelerated exhaustion in the laboratory to help predict the properties of the fully exhausted carbons. Tests carried out at WWTPs in New York City show that the exhausted carbons do not exhibit the E.P.A. "characteristics of hazardous wastes," but that the leachates from the caustic-soaking regeneration do have the characteristics of "corrosivity" and perhaps of "reactivity." A method of treatment is suggested.

INTRODUCTION

Malodorous reduced sulfur compounds can be removed from air streams by treatment with granular absorbents, by liquid scrubbing, or by incineration. The odor thresholds of these compounds are so low that even highly odorous streams may carry too little combustible matter for economical incineration. Liquid scrubbing with caustic solutions can remove acidic reduced sulfur compounds, but is not effective for neutral odorants, such as hydrocarbons or fatty oils, or for basic ones, such as amines, that may be present in the same air streams. Activated carbon is a practical medium for removal of many organic odorants, but its capacity for light gases such as hydrogen sulfide and methyl mercaptan is low. The capacity of activated carbon for removal of sulfide odors can be greatly enhanced, however, by impregnating it with NaOH (_) or KOH, or by injecting a small concentration of ammonia gas into the air stream preceding the carbon (_). Furthermore, the exhausted caustic-impregnated or ammonia-injected carbon can be regenerated in situ by a series of soakings with water and caustic solutions (_). The acceptable number of such treatments is limited by the buildup of a non-regenerable "heel," so that the carbon must eventually be discarded as a solid waste. The leachates from the caustic washings, too, must be discarded. Since the use of these carbon systems for control of sulfide odors is fairly widespread, especially in sewage treatment plants, it is important to establish the degree to which such wastes may be characterized as hazardous.

Subpart C of 40 Code of Federal Regulations (CFR) Part 261 identifies four characteristics of hazardous wastes as (1) ignitability, (2) corrosivity, (3) reactivity, and (4) EP toxicity. This study examines these criteria as applied to exhausted carbons from the treatment of sulfide-containing odorous gas streams and to leachates from the regeneration of such carbons.

Experimental methods

Exhausted or partially exhausted carbons were obtained from the Owls Head and North River Water Pollution Control Plants in New York City. Samples included carbons that had been impregnated with NaOH or KOH, or that had been injected with ammonia. These samples were then tested for their degree of exhaustion and, if found to be exhausted, were characterized as hazardous or non-hazardous. Some of the exhausted samples were regenerated and subsequently examined. If the samples were not exhausted, they were subjected to accelerated exhaustion in the laboratory, regenerated, and characterized. Figure 1 outlines these sequences.

114

Analysis of Carbon Samples for Exhaustion

Breakthrough capacity for hydrogen sulfide: The procedure for determining the breakthrough capacity of GAC for H_2S (4, ___) was similar to that now being considered by ASTM Committee D-28 on Activated Carbon. A closely packed carbon bed lies in a vertical glass cylinder of 2.54 cm (1 in) inside diameter and 22.86 cm (9 in) in height. The challenge contains 1% (10,000 ppm) of H_2S by volume at 80% relative humidity. The effluent stream is monitored by a lead acetate paper tape system until the H_2S concentration reaches 50 ppm. The amount of H_2S adsorbed, expressed as g of H_2S per cm^3 of carbon, is the "breakthrough capacity," which thus represents a drop from an undetectable level ("100%" efficiency) down to 99.5% efficiency [10,000-50)/10,000]. The capacity of a new carbon freshly impregnated with NaOH or KOH ranges from about 0.12 to 0.16 g H_2S/cm^3, and for a new carbon used with the ammonia process is about three times these levels. Capacities that are much lower than these values characterize exhausted or partially exhausted carbons.

Sulfur content of the carbon. The initial rapid reaction of H_2S with caustic (NaOH or KOH) carbon is neutralization to form Na_2S or K_2S. These salts then react with atmospheric oxygen to form polysulfides, elemental sulfur, and more highly oxidized compounds such as sulfites and sulfates. The predominant product, however, is elemental sulfur. When virgin carbon is used with the ammonia process, elemental sulfur is the exclusive product. The elemental sulfur content of the exhausted carbon is therefore a good index of the amount of hydrogen sulfide that the carbon has removed from the air stream, as distinct from sulfur compounds on the carbon from other sources. The determination is made by Soxhlet extraction of a 40g sample of the exhausted carbon with carbon bisulfide, followed by ambient temperature evaporation of the solvent. The residue contains elemental sulfur plus any organic matter extracted by the solvent. The latter is selectively removed by washing with a little ice-cold ethyl ether, in which sulfur is only sparingly soluble. The dried sulfur, as rhombic or monoclinic crystals, is weighed and the hydrogen sulfide removed is calculated as (34/32) x wt of sulfur/cm^3 of carbon.

Organic content of the carbon. Hydrogen sulfide is an odorous air stream is usually accompanied by other reduced sulfur compounds (_) as well as by various other organic gases and vapors. It is therefore possible for the non-H_2S components of an air stream to be primarily responsible for the exhaustion of the carbon, even when the hydrogen sulfide is a major odor source. The components of an exhausted caustic-impregnated carbon may be grouped into five categories: the carbon itself, moisture, sulfur, inorganic salts, and organic matter. Since organic matter, especially the less volatile portion, is not easily removed from the carbon, one approach to its assay is to measure everything else and calculate the organic content by difference. The carbon content of the sample can be estimated by comparing its apparent density (_) with that of the original dry, unimpregnated carbon substrate. Then the proportion of carbon is simply the ratio of the density of the substrate to that of the exhausted sample. If the former is not known, they all the adsorbed matter must be stripped from the exhausted carbon by superheated steam (__), essentially reactivating it, and calculating the carbon proportion from the weight loss. Moisture is determined by xylene distillation (__), and sulfur by the extraction procedure described above. Inorganic salts may be removed by washing with cold water (all the sodium or potassium salts that may be formed from the caustic carbons are soluble), but calculation of the resulting weight loss must take into account the change in water content produced by the washing. An unimpregnated carbon used with ammonia injection does not accumulate inorganic salts, so the latter procedure is unnecessary.

If it is important to identify the organic components on the carbon, the oily condensate from the stream reactivation could be separated or extracted with a suitable solvent and analyzed by any of the following methods, in increasing order of cost: (a) composite infrared (ir) scan (identifying the organic compounds by class, such as hydrocarbon, aldehyde, amine, etc., and "fingerprinting" for comparison with other samples); (b) combined gas chromatograph (gc) and ir analysis (adding information on number and light and heavy organic components, with possible identification of some of them); (c) combine gc and mass spectrometry (ms) analysis (identification of many and possibly most, components).

Accelerated Exhaustion of Non-Exhausted Carbons

A caustic-impregnated or ammonia-injected carbon that is exhausted by sulfide odorants may be regenerated in situ. A decision about the advisability of such regeneration should be based on its expected cost and effectiveness, and on whether or not the wastes ultimately generated will be classified as hazardous. If the decision is made before construction, it may also influence the design of the carbon vessels and ancillary equipment.

In order to may early predictions, a sample of the carbon can be exhausted in an accelerated laboratory procedure. The method should be rapid enough to be completed in a reasonable time, without allowing the carbon to be heated (by the adsorption, neutralization, a nd oxidation reactions) to a level approaching its kindling temperature. Also, the sample should be sufficient for all subsequent testing on it. We have found that the following conditions satisfy these requirements:

The test sample (either new impregnated or unimpregnated carbon, or a sample of partially exhausted carbon removed at any stage from its vessel during operation) is placed in a 3 inch (7.62 cm) i.d. vertical cylinder to a height of 6 in (15.24 cm). The challenge gas, a mixture of 25 mL/min of pure H_2S and 2 L/min of purified air at 100% R.H., is streamed into the bed to a 50 ppm H_2S breakthrough, just as in the test for the carbon's capacity. If the carbon is unimpregnated, a side stream of ammonia of about 10 mL/min is added. Under these conditions, the bed temperature rises to a maximum of about 150°, well below its kindling temperature. The exhausted sample, about 700 mL, is removed, allowed to cool, and homogenized before subsequent use.

Regeneration of Exhausted Carbons

The procedure for full-scale in situ regeneration of carbon exhausted by exposure to sulfidic odorants is specific to the manufacturer's instructions and to the particular carbon vessels and ancillary equipment and piping on hand. However, all such procedures have several steps and objectives in common.:

(a) The carbon is first washed with water to remove any water-soluble material. The action is slow because the water must penetrate the carbon granules. An added benefit is that any insoluble oils that may be on the exterior of the carbon (not adsorbed) are floated to the surface and can be skimmed off.

(b) Any acidic matter on the carbon (from the action of H_2S, CO_2, or other acidic gases) is neutralized to prepare for the next treatment, which is the critical part of the regeneration.

(c) The carbon is soaked in a concentrated caustic (NaOH or KOH) solution, which reacts slowly with elemental sulfur to yield soluble products. The caustic may also solubilize fatty esters by saponifying them. [note to lab: we should test this -- will be an interesting little project]

(d) The chemical actions are now complete, but the carbon is "blinded" by too much caustic solution on the exterior of the granules, and is dangerous to handle, so the excess caustic must be washed away while the interior impregnation is preserved. This objective is accomplished by a schedule of washings with less concentrated caustic solutions and with water. The carbon is finally air dried in place.

In the laboratory, these objectives may be realized as follows:

(a) Exhausted carbon from actual service for from the accelerated laboratory procedure is placed into a 3 inch (7.63 cm) i.d. cylinder as described earlier, to any desired height. The cylinder must be provided with suitable valves and tubes for holding liquid, draining, and top-skimming. The cylinder is filled with water to about 2-3 inches above the carbon layer, and any oily supernatant is skimmed off. The water is drained off and this step is repeat rapidly twice, after which a final 24-hour water soak is applied.

(b) If the original pH of the carbon being regenerated was less than 7, then cover the bed with dilute (about 2%) caustic solution. Let sit for 24 hours and drain.

(c) Cover the carbon with concentrated (about 45%, depending on manufacturer's recommendation) caustic solution, and let it sit for 24 hours. Drain the solution, but collect enough of it in a labeled bottle to save for further testing for possible hazard classification. This leachate is a strongly caustic solution which may contain sulfides, and should be recognized as a hazardous laboratory waste.

(d) Complete the schedule of dilute caustic and water washing as specified by the manufacturer or as will be used by the facility if regeneration is to be carried out on full scale. Dry the carbon with dry air overnight at a flow of about 12 L/min. Finally, remove the dry carbon from the cylinder and homogenize it.

General note: The scheduling of the procedure is critical, so it should be consistent from sample to sample, especially when comparative tests are made. For this reason, it is convenient for the long soaks to be standardized as 24-hour treatments.

Characterization of Exhausted Carbons

Ignitability. The EPA criterion (40 CFR Subpart C) characterizes a solid waste as ignitable if it "exhibits the characteristic of ignitability..." We have assumed that the most appropriate test for activated carbon is its ignition temperature. The ASTM procedure (___) involves a programmed heating in a quartz tube, and ignition is defined as the point at which the carbon temperature suddenly rises above

the temperature of the air entering the bed. When the
procedure is applied to exhausted carbon, adsorbed vapors
are desorbed during the heating and release odors into the
laboratory atmosphere. Appropriate hooding and venting are
therefore necessary.

Corrosivity. The EPA characteristic of a solid waste
is one "that exhibits the characteristic of corrosivity..."
and for an aqueous liquid is one that "has a pH less than or
equal to 2 or greater than or equal to 12.5..." We have
therefore assumed that an appropriate test for activated
carbon is the pH of its hot water leachate (__).

Reactivity. The relevant EPA characteristic for
activated carbon that has been exposed to reduced sulfur
compounds is that it "is a cyanide or sulfide bearing waste
which, when exposed to pH conditions between 2 and 12.5, can
generate toxic gases, vapors or fumes in a quantity
sufficient to present a danger to human health or the
environment." If the pH of the exhausted carbon is below 7,
any sulfide ions would long since have been converted to H_2S
and blow out of the carbon. Similarly, if the carbon is
alkaline, volatile amines would have been blown out. It is
therefore necessary only to alkalize an acidic carbon or
acidify an alkaline one, and test for gas evolution to
establish whether the carbon has the characteristic of
reactivity.

If the carbon is regenerated, the leachate is strongly
alkaline (pH near 14). Any elemental sulfur on the carbon
is thereby converted to a sulfide, and this possibility must
be tested for. A sample of the cooled leachate is carefully
acidified with a 5% solution of sulfuric acid in a flask
provided with a slow flow of nitrogen onto a test strip or
other analytical indicator to monitor for gas evolution.

EP Toxicity. The EPA criterion is: "A solid waste
exhibits the characteristic of EP toxicity if...the extract
of a representative sample of the waste contains any of the
contaminants listed in Table 1 at a concentration equal to
or greater than the respective value given in that Table."
The table referred to lists eight metals (As, Ba, Cd, Cr,
Pb, Hg, Se, and Ag) and six chlorinated hydrocarbon
pesticides, each with a specific maximum concentration.

For metals, a sample of the carbon may be extracted
with an appropriate aqueous solution or, more
conservatively, the carbon itself may be the sample. For
most facilities, the best choice is probably the use of a
commercial laboratory that routinely carries out elemental
analyses.

The chlorinated pesticides, if present, may be extracted from the carbon with an organic solvent such as carbon bisulfide. An alternative source is the condensate from the steam reactivation described above. Any pesticide components will be part of the non-aqueous phase containing other organic matter that is likely to be seen as a supernatant oily layer, which can be separated.

The simplest approach is then to test the oily desorbate for the presence of chlorine. If none is found, then chlorinated hydrocarbons must be absent or in lower concentration than the detection limit of the test. If chlorine is present, tests for the specific EP contaminants must be carried out. A fast, simple method is the Beilstein test ( ), carried out with a copper wire in a Bunsen burner flame. The detection limit for that test was found to be about 200 mg Cl/L of oil. Let

D_{oil} = detection limit of test in mg Cl/L of oil

D_{carbon} = detection limit of test in mg contaminant/L of carbon sample

then

$$D_{carbon} = D_{oil} \times \frac{\text{volume of oil desorbate, mL}}{\text{volume of C sample, mL}} \times \frac{\text{molar correction}}{\text{for contaminant}}$$

The molar corrections are obtained as follows:

Contaminant	Formula & Mol. Wt.	Weight of Cl Atoms	Molar Correction (Wt. of Molecule/ Wt. of Cl)
Endrin	$C_{12}H_8Cl_6O, 381$	6 x 35.45 = 213	1.8
Lindane	$C_6H_6Cl_6, 291$	6 x 35.45 = 213	1.4
Methoxychlor	$C_6H_{15}Cl_3O_2, 346$	3 x 35.45 = 106	3.3
Toxaphene	$C_{10}H_{10}Cl_8, 414$	8 x 35.45 = 284	1.5
2, 4-D	$C_8H_6Cl_2O_3, 221$	2 x 25.45 = 71	3.1
Silvex	$C_9H_7Cl_3O_3, 270$	3 x 35.45 = 106	2.5

Example. 1 mL of oil is desorbed from 100 mL of a carbon sample. The Beilstein test is negative. Can the carbon exhibit the characteristic of EP toxicity for Endrin? For Methoxychlor?

Answer. D_{oil} for the Beilstein test is 200 mg Cl/L. The molar correction for Endrin is 1.8. Then, for Endrin,

$$D_{carbon} = 200 \times \frac{1}{100} \times 1.8 = 3.6 \text{ mg Endrin/L}$$

This means that the concentration of Endrin in the carbon is less than 3.6 mg/L, but since the maximum permissible level is 0.02 mg/L, further testing is necessary.

For methoxychlor,

$$D_{carbon} = 200 \times \frac{1}{100} \times 3.3 = 6.6 \text{ mg methoxychlor/L}$$

Since the maximum permissible level for methoxychlor is 10.0 mg/L, no further testing is necessary.

Results

   Ignitability.   The ignition temperature of the fresh unimpregnated coal-based carbons is about 460°C (860°F), and of the fresh or exhausted caustic-impregnated carbons is about 260° (500°F).   By comparison, the ignition temperature of paraffin wax is about 475°F.   The exhausted carbon is moist, and not ignitable by friction or spontaneous change, and therefore does not have the characteristic of ignitability.

   Corrosivity.   The pH values of the hot-water leachates from the various carbons, whether exhausted in service or in the laboratory, were generally acidic, but no lower than 2.95.   One sample was slightly basic, with a pH of 8.25. None of these values falls into the range of wastes that "exhibit the characteristic of corrosivity."

   Reactivity.   None of the acidic carbons, when alkalized, released any amines or other toxic gases, and the basic carbon, on acidification, did not release any H_2S or other acidic gas.   However, the leachates from the caustic soaking solutions in the laboratory regeneration procedures were highly alkaline (pH 13 to 14) and did release H_2S acidification, and were therefore hazardous by virtue of being both corrosive and reactive.   The sulfide content of the leachate can be precipitated by addition of a solution of ferrous sulfate, $FeSO_4$, and the resulting FeS can be removed as a filter cake.   FeS does have the characteristic of reactivity (H_2S is evolved at pH 2), but it constitutes a small volume compared to the leachate and can be easily transported.

   EP toxicity.   Activated carbons used for air purification in municipal wastewater treatment plants do not address any gas phase releases from industrial sources, and are not exposed to liquid or solid wastes from any source. Nonvolatile substances, such as heavy metals or their compounds, therefore do not have access to the carbon beds.   For these reasons, the exhausted carbons would not be expected to constitute hazardous wastes.   This expectation was confirmed by the finding that name of the exhausted carbons contained any of the listed contaminants in concentrations equal to or greater than the specified maximum concentrations. The superheated steam desorbates yielded a highly malodorous oily layer whose infrared spectrum indicated a complex mixture consisting primarily of hydrocarbon and various oxygenated organics.

Conclusions

When sulfide odors, usually from H_2S, mercaptous, and other reduced sulfur compounds, are removed from air streams by granular media, the exhausted media ultimately become solid wastes. If the medium is activated carbon, the sulfur compounds are either physically adsorbed or converted to elemental sulfur or other, more highly oxidized, forms. In addition, odorous compounds other than sulfides accumulate on the granular bed. Exhausted carbons do not generally have the characteristics of ignitability, corrosivity, or reactivity, as defined by E.P.A. criteria, but these properties can be evaluated by standard laboratory procedures. The characteristic of "EP toxicity" can be assayed by testing the exhausted carbon itself, its aqueous or organic solvent extract, or its superheated steam desorbate. Carbon used for air purification in a wastewater treatment plant is not ordinarily exposed to industrial sources of such toxic contaminants.

The exhausted carbons tested in this study were not found to have any of the E.P.A.-legislated "characteristics of hazardous wastes." The leachates from the caustic-soaking method of in situ regeneration of exhausted carbon, however, are hazardous by virtue of their high pH values and also, possibly, because of their sulfide content. The latter can be removed as a solid FeS cake by precipitation with ferrous sulfate.

References

"Proposed Standard Test Method for the Determination of the Accelerated Hydrogen Sulfide Breakthrough Capacity of Granular Activated Carbon," Calgon Corp., Pittsburgh, PA, 1986.

ASTM D 3466-76 (1983) Standard Test Method for Ignition of Granular Activated Carbon.

"Standard Test Method for pH of Activated Carbon," ASTM D 3838 - 80.

D.L. Pavia, G.M. Lampan, and G.S. Kriz, Introduction to Organic Laboratory Techniques, 2nd ed., Saunders College Publ, Philadelphia. 1982 p. 402.

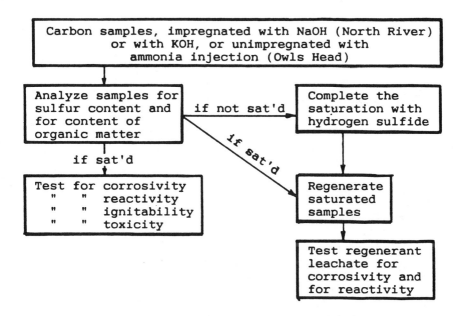

Figure 1   Testing sequence for exhausted carbons

REMOVAL OF ODORANT COMPOUNDS
BY PACKED TOWER SCRUBBING

Michael G. Ruby,
Victoria R. Stamper, and You-ran Wang

Envirometrics
4128 Burke Ave N
Seattle WA 98103

A pilot-scale wet scrubber has been constructed to make comparative tests of a new monolithic packing and a conventional random dumped packing. Both odor removal efficiency and pressure drop were measured at various gas and liquid flow rates. Hydrogen sulfide was removed from a air stream with a basic scrubbing liquor of sodium hydroxide and water at a pH of 12.5.

While the pressure drop through the monolithic packing was about one-tenth of that through an equal height bed of random dumped packing, the mass transfer coefficient was higher for the random dumped packing. Combining these two measurements, the pressure drop per transfer unit was found to be significantly lower for the monolithic packing. Thus, the monolithic packing will be more cost efficient over the life of the packing.

Introduction

One popular method for removing odors from exhaust gas streams is by packed tower scrubbing. The odorous component in the gas is absorbed into a liquid in the tower by intimately mixing the gas and a thin film of the liquid in the tower packing. The liquid is then separated from the gas stream, leaving only clean exhaust gas. The few small droplets that become airborne are removed in a droplet filter.

However many odors are objectionable at even very low concentrations. Additionally, many odors are complex mixtures, not just one chemical. This presents serious difficulties for packed tower scrubbing. At low concentrations the gradient across the gas-liquid interface will be relatively small, making absorption relatively slow. Any level of control can be achieved by lengthening the time the gas and liquid are in contact, but that means increasing the height of the tower and, consequently, the pressure drop across the tower. The complex chemistries of odors can also be overcome, but often at the cost of sequential towers with different scrubbing liquors. For example, ammonia requires an acid scrubbing solvent, hydrogen sulfide (H_2S) a very basic solvent, and dimethyl disulfide a neutral solvent.

When large volumes of air must be cleaned the fan power costs due to the high pressure drop across a tall tower or two towers may make the packed tower a very expensive solution. For this reason, manufacturers are constantly exploring packings with lower pressure drop. The amount of pressure drop through a packed tower is determined, primarily, by the amount of liquid supported by the air flow through the tower and, secondarily, by the frictional forces of the air flow through the tower packing.

Two main types of packings are available today: random dumped packing and structured packing. Structured packing has less void space than random dumped packings. Thus, less liquid is held up in the structured packing, resulting in lower pressure drops through the packed tower. However, structured packing can be expensive to manufacture and time consuming to install. Monolithic packing is a simpler structured packing. It is less expensive to manufacture, easy to install, and can be operated at a much lower pressure drop than random dumped packings. This paper reports a study of the pressure drop and mass transfer rates of a recently available monolithic packing and compares it to a widely used random dumped packing.

Theory

The mass transfer coefficient is a measure of the tendency of a minor gas constituent, in this case a strong odor, to transfer from the gas phase into the liquid phase. It is dependent on the operating conditions of the packed tower, the chemistry of the pollutant/scrubbing liquid combination, and the amount of surface area created by the tower packing. In order to calculate the value of the mass transfer coefficient, it is necessary to evaluate the driving force which causes the odor to move from the gas phase into the liquid phase. The driving force is a function of the concentrations of odor in both the gas stream and the scrubbing liquor. If there is a chemical in the scrubbing liquor that reacts with the odor, the concentration of the solute (i.e., odorous compound in solution) in the liquid phase is essentially zero.

In this study, H_2S was absorbed using a scrubbing liquor of sodium hydroxide (NaOH) and water. This system was chosen so previous studies reported in the literature[1] would provide a basis for comparison of the results. Future studies will involve other odors and other liquors.

125

The sodium hydroxide dissociates in water, forming a sodium ion (Na^+) and a hydroxide ion (OH^-). The H_2S then reacts with the dissociated hydroxide ion:

$$H_2S + OH^- \rightarrow HS^- + H_2O$$

Since the dissociation equilibrium drives to the right the concentration of solute in the liquid phase is essentially zero, and the mass transfer driving force approaches the product of inlet pollutant concentration and system pressure. Where the total system pressure remains constant, a log mean average driving force can be used:

$$\Delta P_{lm} = P \frac{y_i - y_o}{\ln(y_i/y_o)} \tag{1}$$

The solute transferred may be expressed in terms of the overall gas-phase mass transfer coefficient, K_gA, as:

$$N = AZK_gAP\Delta P_{lm} \tag{2}$$

When very low concentrations of solute are being removed from the gas phase, the amount of solute transferred from the gas phase is:

$$N = G_m(y_i - y_o) \tag{3}$$

Equations (1), (2), and (3) can be combined to give:

$$G_m = \frac{AZK_gAP}{\ln(y_i/y_o)} \tag{4}$$

The concentrations of odor in the gas and in the liquid change from the bottom to the top of the tower. A decrease in the odor concentration in the gas by $1/e$ is termed one transfer unit. This has a given physical height for a specific system. Thus as the height of a transfer unit decreases, the number of transfer units in a given tower increases, and the amount of mass transfer increases. With 6 transfer units, the odorous gas concentration will be reduced by better than 99.9%.

For an absorption system in which the pollutant gas is so soluble that there is no equilibrium vapor pressure of the pollutant over the scrubbing liquid due to the solute, the number of transfer units can be expressed as:

$$N_{TU} = \ln(y_i/y_o) \tag{5}$$

The height of a transfer unit can then be calculated as:

$$H_{TU} = \frac{Z}{N_{TU}} = \frac{Z}{\ln(y_i/y_o)} \tag{6}$$

Substitution of equation (6) into equation (4) and solving for the mass transfer coefficient gives:

$$K_gA = \frac{G_m}{APH_{TU}} \tag{7}$$

126

This equation was used to calculate the mass transfer coefficient in this paper.

Equipment and Materials

A wet scrubber pilot plant was built in order to test the different packings. The tower was designed for countercurrent flow with a packing depth of 3 feet and a diameter of 10 inches. The diameter of the tower is sufficient that wall effects are minimal. The gas flow was measured by a calibrated venturi. A rotameter was used to measure the liquid loading rate. A mist eliminator was used to collect any scrubbing liquor being carried out with the gas. Manometers were used to measure the pressure drop across the packed bed and the mist eliminator.

Two types of packings were tested: 1 inch Jaeger Tri-Packs* and a monolithic packing. The 1 inch Tri-Packs were selected as the most appropriate to the 10 inch diameter tower.

The scrubbing liquor was applied to the packing from a single nozzle at the center of the tower cross-section. Two different nozzles were used to give the best full cone spray at different liquid flow rates. The chemistry of the scrubbing liquor was maintained by a pH sensor and a metering feed pump, which added NaOH to the scrubbing liquid.

H_2S gas was injected into the ambient air stream upstream of the fan so it would be well-mixed with the inlet air. Samples of the polluted gas were collected in Tedlar bags. A Jerome 621 H_2S analyzer was used to measure the concentration of the H_2S in the gas upstream and downstream of the packed tower.

A diagram showing the arrangement of the wet scrubber pilot plant is provided in Figure 1.

Experimental Procedure

The first task was to measure the liquid holdup and pressure drop across the packed bed for each of the packings. These were measured at gas loading rates ranging from 1000 to 3200 lb/ft^2 hr and liquid loading rates varying from 2000 to 6000 lb/ft^2 hr. Water was used as the scrubbing liquor with approximately 1 to 2% NaOH added to bring the pH to about 12.5. Ambient air was used as the gas through the scrubber. Many of the higher flow rates measured for the monolithic packing could not be duplicated with random dumped packing because the pressure drop exceeded the capacity of our fan. The range was extended slightly by multi-linear regression on all the data collected for the random dumped packing. This provided an estimate of the pressure drop at higher gas loading rates. This regression analysis was also done for the data collected from the monolithic packing for comparison.

Next, each packing was tested for its efficiency in removing H_2S from ambient air. For each test, the inlet H_2S concentration was varied between 1 and 4 ppm. To determine the inlet and outlet concentrations, the air was sampled directly by filling up Tedlar bags through valves at points immediately downstream of the fan and between the packing bed and the mist eliminator. Before each reading of the bag with the H_2S analyzer, the bags

---

* Tri-Packs is a registered trade mark of Jaeger Products, Inc.

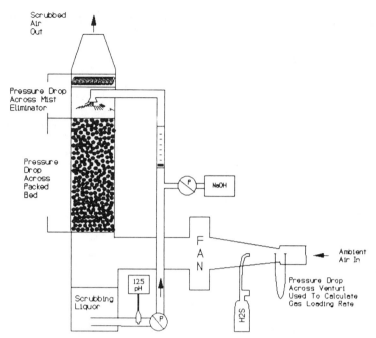

Figure 1. Diagram of wet scrubber pilot plant.

were agitated so that the H_2S would be well mixed with the air. To increase the accuracy of the measurements, three readings were taken from each bag.

Other measurements necessary for the calculation of the mass transfer coefficient were also recorded. The system pressure at the top of the tower was measured to provide the log mean estimate of the pressure through the packing. Also, the air temperature was recorded to convert the measured gas flow rate to standard conditions. The ambient pressure was assumed to be 1 atm for all test runs.

To keep the NaOH concentration constant in the scrubbing liquor, the pH was maintained at 12.5. Because the system was operating as a batch reactor rather than with constant overflow, after about three runs of 6 upstream and downstream samples each, the scrubbing liquor was flushed out and replaced with fresh water and NaOH. This kept the concentration of HS^- in the scrubbing liquid at sufficiently low levels.

Each packing was tested at the same liquid and gas loading rates. First, the gas loading rate was kept constant at 150 cfm (2240 lb/ft^2 hr), while the liquid loading rate was varied from 1.6 gpm (1475 lb/ft^2 hr) to 5.5 gpm (5070 lb/ft^2 hr). Second, the liquid loading rate was kept constant at 5.5 gpm (5070 lb/ft^2 hr), while the gas loading rate was varied from 150 cfm (1244 lb/ft^2 hr) to 370 cfm (3070 lb/ft^2 hr). Lastly, the liquid and gas loading rates were kept constant at 150 cfm and 5.5 gpm, while the pH of the scrubbing liquor was varied by controlling the amount of NaOH added to it.

Results and Discussion

The results of the pressure drop experiment found the pressure drop through the monolithic packing to be about one tenth of the pressure drop through the random dumped packing. The liquid holdup experiment verifies these results with the liquid holdup of the random dumped packing approximately five times the liquid holdup of monolithic packing. The pressure drop and amount of liquid hold up increases with both increasing gas flow and increasing liquid flow. This can be seen in Figures 2 and 3.

The pressure drop through the packed bed is caused by any restriction to the movement of gas. This will include the shape of the packing, the amount of liquid being held up in the packing, and the viscosity of the liquid. Thus, if liquid is being held up in the packing, that liquid is narrowing the free air passages and restricting air flow. A packed bed can be modeled as a series of parallel, small diameter tubes of equal length and diameter.[2] Because the monolithic packing is structured packing, the theoretical tubes are of a larger diameter than in the random dumped packing. There is both a lower wall friction loss and less wall for the liquid to be spread upon. The random dumped packing "tubes" are many and short, while with the monolithic packing, the tubes are continuous from the bottom to the top.

Not only does the increase in pressure drop increase the cost of operation, but it limits the useful range of the scrubber because of flooding. Flooding occurs when the liquid and gas phases invert. At this point the fan cannot overcome the large amount of liquid holdup in the packing, resulting in minimal flow and a large increase in pressure drop. Packed towers are usually operated at 60 to 75 percent of the flooding gas flow rate. From the Figure 3, it appears that flooding is occurring in the random dumped packing at gas flow rates greater than about 1800 lb/ft^2 hr. This is apparent by the sudden increase in liquid holdup at that point. Also in Figure 3, the liquid holdup in the monolithic packing is still increasing slowly with increasing gas loading rate, indicating the system is not flooding at any of these gas flow rates. Thus, the random dumped packing cannot be operated at as high a gas flow rate as the monolithic packing.

Another problem caused by increased liquid holdup is clogging of the packing by precipitates of the scrubbing liquor. When there is a lot of liquid held up in the packing, a residue can form on the packing which can clog the packing and increase the pressure drop. We observed this in the random dumped packing in our tower over the short time this study was underway. The intricate design of the packing made it more susceptible to clogging, and there was more opportunity for precipitation due to much liquid holdup.

Figure 4 shows the mass transfer coefficient variation as the liquid flow rate is increased at a constant gas flow rate of 2240 lb/ft^2 hr. The K_gA for the monolithic packing is about one-fourth of the K_gA through the random dumped packing. This translates into a removal efficiency of approximately 85% for the 3 feet of monolithic packing as compared to 99% for the random dumped packing. One would expect that the mass transfer rate would increase with increasing liquid flow rate, given the greater liquid available for the pollutant to be absorbed into. The random dumped packing does behave in this manner. However, for the monolithic packing, it appears there is an optimal liquid flow rate of 3500 lb/ft^2 hr at which maximum mass transfer occurs.

Figure 5 shows the mass transfer coefficient variation with increasing gas flow rate at a constant liquid loading rate of 5070 lb/ft^2 hr. It is apparent from the graph that the K_gA of the monolithic packing again is about one tenth of the K_gA of the random dumped packing. But note that the monolithic packing behaves different from our expectations. One would expect the mass transfer coefficient to increase with increasing gas flow rate up to a point of

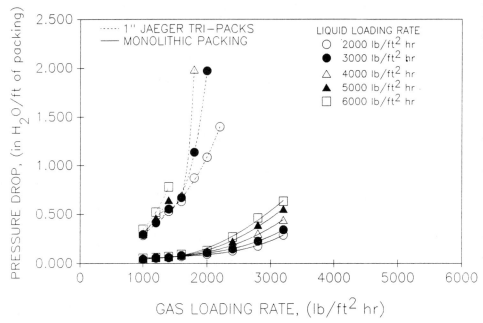

Figure 2. Comparison of pressure drop with random dumped packing and monolithic packing.

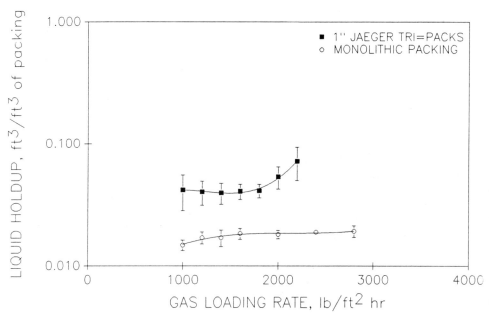

Figure 3. Comparison of liquid holdup for random dumped packing and monolithic packing at a liquid loading rate of 2000 lb/ft²hr.

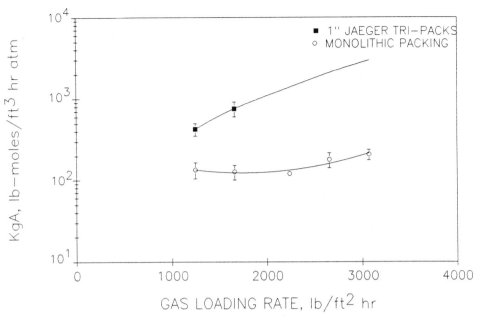

Figure 4. Comparison of mass transfer coefficient for random dumped packing and monolithic tower packing at a liquid loading rate of 5070 lb/ft^{2}hr and pH = 12.5.

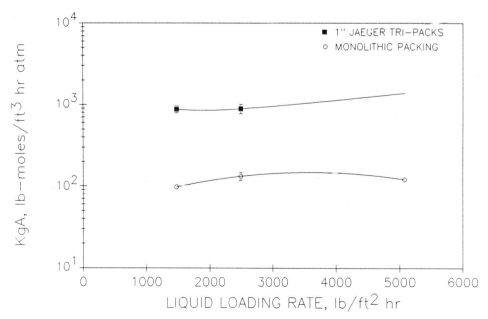

Figure 5. Comparison of mass transfer coefficient for random dumped packing and monolithic packing at a gas loading rate of 2240 lb/ft^{2}hr and pH = 12.5.

saturation. This was the behavior of the random dumped packing. However, the mass transfer through the monolithic packing decreased with increasing gas flow rate up to a point where it then began to increase.

One reason for the difference in behavior of the K_gA in the two packings may be the differences in the paths through the two packings. As stated above, the paths through the random dumped packing are short in length and small in diameter. So, when the polluted air is moving through the packed tower, it is constantly brought into contact with the wetted packing as it moves through the many different paths it takes to get to the top of the packed tower. The paths in monolithic packing, on the other hand, are nearly straight, large diameter paths from the bottom of the packed tower to the top. This feature accounts for the lower pressure drop but it also provides less contact between the wetted surface and the polluted gas.

If we calculate the Reynold's number for the monolithic packing, we see that laminar flow can be expected until the gas flow reaches about 1900 lb/ft^2 hr, and will not be fully turbulent for gas flow rates less than 2100 lb/ft^2 hr. During laminar flow, the bulk flow of the gas is in the same direction, in this case from bottom to top through the tower. Also, the greatest velocity is occurs in the center of the flow tube. The momentum of the air moving at the greatest velocity pulls the air moving at slower velocities toward the center of the tube. As the velocity of the gas increases, the momentum force pulling the slower velocity gas towards the center of the tube also increases. Thus, while in laminar flow, as the gas loading rate increases, less and less gas comes into contact with the sides of the flow tubes, allowing less mass transfer to occur. When the flow enters the transitional stage, the direction of micro-movement of the gas changes. It begins to move in different directions as the flow approaches turbulent flow. Thus, more and more contact occurs between the polluted gas and the wetted sides of the tubes. Once the flow is fully turbulent, the mass transfer coefficient through the monolithic packing begins to increase.

In determining the mass transfer rate through each packing, the pH was kept constant at 12.5. This value was chosen from information available in the literature for the removal of H_2S using NaOH and water. To verify that this value of pH would yield the highest rates of mass transfer, the mass transfer coefficient for varying values of pH were determined for each of the packings. From the graph of pH vs. K_gA, one can see that the mass transfer coefficient through each packing increased at the same rate as the pH increased. K_gA peaks at a pH of about 12.3, and then seems to remain fairly constant. Thus, using a pH of the scrubbing liquor of 12.5 yielded approximately the highest values of mass transfer coefficient for removal of H_2S using NaOH and water as the scrubbing liquor. This is shown in Figure 6.

All the test runs at pH = 12.5 were included in a multi-linear regression analysis for K_gA for the random dumped packing and the monolithic packing. The results are given in Table I. The L^2 term that appears in both the pressure drop and K_gA equations for the random dumped packing emphasizes the size differences in the theoretical flow tubes. The effect of the initial pollutant concentration is small, but larger than the L^2 term. One explanation for this effect may be a violation of the assumptions in equations (3) or (5) at higher pollutant concentrations.

Conclusions and Recommendations

The random dumped packing achieves high mass transfer at the cost of a higher pressure drop. This results in a lower value for the height of a transfer unit (H_{TU}) and thus a shorter tower. A tower packed with monolithic

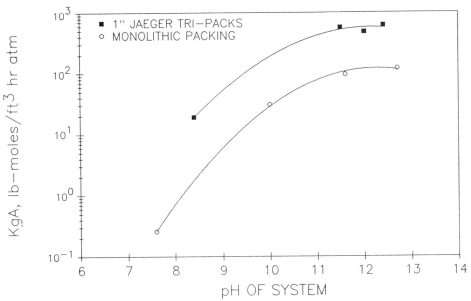

Figure 6. Comparison of mass transfer coefficient for random dumped packing and monolithic packing with change in liquid phase pH.

packing, however, has a low pressure drop, but the height of each transfer unit is greater, requiring a taller tower. Since total pressure drop is a function of the actual tower height for a given required removal, the proper comparison between the two packings is the pressure drop per transfer unit. This is the pressure drop that will occur at the tower height required for a specified mass transfer to occur.

This comparison is provided by the curves in Figure 7. We can see that not only is the pressure drop per transfer unit of the monolithic packing much less than that of the random dumped packing, but it also appears to be increasing at a slower rate than the random dumped packing at the higher gas loading rates. This tells us the monolithic packing will be the more cost efficient packing over the life of the scrubber.

Table I. Multilinear Regression Equations

Variable	Multi-linear Regression Equation	Correlation Coefficient
Monolithic Packing		
Pressure Drop =	$EXP[-2.00 + (5.03 \times 10^{-5})L + (4.44 \times 10^{-4})G]$	0.973
KgA =	$265.09 - (0.19)G + (5.17 \times 10^{-5})G^2 - (3.87 \times 10^{-3})ppm + (7.05 \times 10^{-3})L$	0.572
1" Jaeger Tri-Packs		
Pressure Drop =	$EXP[-1.51 + (1.19 \times 10^{-4})L + (7.11 \times 10^{-4})G - (9.91 \times 10^{-9})L^2]$	0.903
KgA =	$287.85 - (0.19)G + (5.36 \times 10^{-5})G^2 - (4.41 \times 10^{-3})ppm + (9.18 \times 10^{-7})L^2$	0.563

Note: L = liquid loading rate (lb/ft^2 hr), G = gas loading rate (lb/ft^2 hr), and ppm = inlet concentration of H$_2$S (ppm)

133

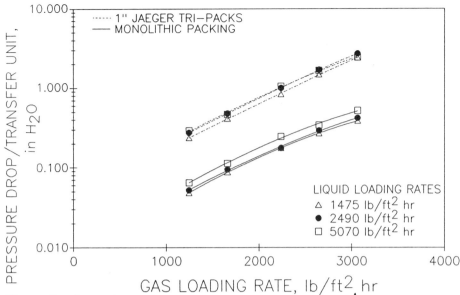

Figure 7. Comparison of pressure drop per transfer unit for random and monolithic tower packing.

The results presented here suggest several additional lines of inquiry. The K_gA may continue to increase at still higher gas loading rates. Depending on the increase of pressure drop the optimum design velocities may be still higher. The light weight of the monolithic packing suggests the construction of very large diameter chambers. It would be useful to compare the resulting designs against towers with random dumped packing using larger-sized packing, which has a lower pressure drop.

Nomenclature

A = cross-sectional area of tower, ft^2
ΔP_{lm}= log mean pressure driving force, atm
G_m = gas phase flow, lb-mole/hr
H_{TU}= height of a transfer unit, ft
K_gA= overall mass transfer coefficient, lb-mole/hr ft^3 atm
N = solute transferred, lb-mole/hr
N_{TU}= number of transfer units
P = system pressure, atm
y_i = mole fraction of solute in inlet gas
y_o = mole fraction of solute in outlet gas
Z = height of packed tower, ft

References

1. D.W. Van Krevelen, P.J. Hoftijzer, and C.J. Van Hooren, "Studies of Gas Absorption; IV. Simultaneous Gas Absorption and Chemical Reaction", Rec. Chim. Trav., **67**, 133-152 (1947).

2. G. Kozeny and J. Sitzber, cited in R.F. Strigle, <u>Random Packings and Packed Towers</u>. Gulf Publishing Co., Houston. 1987, pp. 1-2.

ODOR REMOVAL BY BIOFILTRATION

Hinrich L. Bohn
Department of Soil & Water Science
University of Arizona
Tucson, Arizona

Biofiltration utilizes sorption and microbial degradation by natural porous media such as soils and compost to remove odors from contaminated air. These media function like an activated carbon bed combined with a chemical scrubber or incinerator. Pollutants partition out on the surfaces of soil and compost as the air flows through soil and compost pores. Microorganisms oxidize the adsorbed organic gases to CO_2 and thereby renew the sorption capacity. Removal efficiencies are high for a wide range of organic gases. No fuel or chemicals are required because oxygen is the oxidant and the microorganisms catalyze the oxidation. Installation costs are low and operating costs for soil biofilters are virtually nil. Biofiltration of gases causes no soil or water pollution.

Of the senses, smell may have the strongest effect on human well-being. The nose, like our eyes and ears, responds logarithmically to intensity so to sense a significant odor reduction, odor concentrations have to be decreased by about 90%. This enables us to detect a very wide range of odor concentrations, but greatly increases the performance standards which odor control must satisfy.

Physicochemical methods of odor control either wash, adsorb, or oxidize, air pollutants. Biofiltration, in contrast, is a fixed-film biological process that combines washing, adsorption, and oxidation into a simultaneous process. Biofilters are beds of soil, compost or peat underlain by perforated pipes which allow contaminated air to flow through the bed, Figure 1. Pollutant gases adsorb on the surfaces of the media and are oxidized to CO_2 by the microbes. The oxidation regenerates the sorption capacity and requires no chemical oxidants or fuels. By combining washing, adsorption, microbial oxidation, silica and transition metal catalysis of inorganic oxidation, and acid neutralization, biofilters are effective for a very wide range of organic and inorganic gases.

Soils have been adsorbing and oxidizing organic gases and wastes for more than 109 years and have prevented organics from reaching the groundwater. Soil biofiltration of gases provides the same degree of environmental safety. The hazards associated with organic liquid disposal in soil are due to high loading rates, anaerobiasis, and subsurface burial. Gases in biofilters are sorbed at very low loading rates, under strongly aerobic conditions, and at the soil surface.

Soil Biofiltration

Soils are major sinks for air pollutants in nature. Their sorptive and oxidative properties can also treat contaminated air before the air is released to the atmosphere (Johnson, 1859; Pomeroy, 1957 and 1982; DuPont, 1964; Carlson and Leiser, 1966; Bohn, 1976; Bohn and Bohn, 1988). Soil beds indeed are the oldest and most widespread form of odor removal, but the beds are mostly small and unrecognized. Some 21 million soil beds in the USA alone remove odors from waste water leachlines (Reneau, Hagedorn, and Degen, 1989). These beds remove odors effectively and inexpensively; require no maintenance, fuel, or chemicals; and show no signs of diminished capacity with time.

From the standpoint of air pollutant sorption, soil can be considered as a mixture of activated carbon, silica, alumina, iron oxide, and lime. Soil has 40-50% porosity, 1 to 100 m^2/g surface area, silica and transition metal oxides to catalyze the oxidation of inorganic gases, and high acid neutralization and pH buffering capacities. Most importantly, soil contains a large amount of adsorptive humic matter and an enormous microbial population (10E9 bacteria, 10E7 actinomycetes, and 10E5 fungi per gram of soil) that oxidizes organic gases to CO_2.

The soil is at a steady state with the natural atmosphere. When contaminated air passes through the soil, the pollutants partition out on the soil surfaces and travel much more slowly than N_2, O_2, and CO_2 through the soil pores. The partition coefficients for odorous compounds are in the order of thousands, long enough to achieve 99% removal efficiency for odorous gases (Prokop and Bohn, 1985).

Compost and peat also sorb and degrade pollutant gases. Peat and compost beds are popular for odor control in northern Europe (Ottengraf, 1986), but must be kept at about 40% moisture content in order to maintain microbial activity and to prevent irreversible drying. In warmer and drier climates, peat and compost beds have dried out and their hydrophobicity prevents rewetting. Soils function over a wider range of bed moisture content and soil surfaces are more hydrophilic so soil beds wet and dry reversibly.

The sorption capacity of soils for gases is low. Biofilter beds remove pollutant gases effectively, however, because microbes continuously oxidize the sorbed gases and therefore continuously regenerate the sorption capacity. The effectiveness of soil biofiltration depends on the rate of gas oxidation and degradation rather than on the absorption capacity.

The degradation rate depends on the inherent biodegradability of the pollutants, the soil's ability to degrade, the sorptivity of the gases, the permeability of the soil, and the pollutant loading rate. Over a wide range of input air pollutant concentrations, the degradation rate is proportional to the input rate so the removal efficiency of the bed remains the same (Fischer, Homans and Bartke, 1984).

Deutsche Babcock & Wilcox marketed a compost-peat biofilter for a short time in the 1970s. It was unsuccessful because the gas residence times were too short and the beds dried out. In the 1980s more than 100 successful biofilters have been built in northern Europe for odor and VOC control (Ottengraf, 1986) and an equal number in Japan.

Messer Griesheim, a Hoechst subsidiary in West Germany, and BIOTON in the Netherlands market complex, completely contained peat biofilters for odor and VOC control. Research groups at the universities of Aachen and Eindhoven are developing peat and compost biofilters. Their emphasis is to reduce the land area requirements of biofilters and to investigate removal of slowly-biodegradable substances such as the halogenated hydrocarbons and aromatic (benzene ring) compounds such as xylene, toluene, toluol, and styrene (Fischer, Homans and Bartke, 1985; Ottengraf, 1986). Research on soil biofiltration is being carried out at the University of Arizona.

Input Gas Properties

The range of gases successfully treatable by biofilters is very wide, but some limitations exist:

1. The gas should be free of particulates. Solid particulates can clog the bed's pores. Aerosols of water or biodegradable liquids are permissible.

2. The range of gas temperatures is 50-140 F (10-60 C). The optimum is 97 F (37 C) but the temperature response curve is rather flat. The bed temperature takes on the temperature of the input gas so the ambient temperature is relatively immaterial.

3. The relative humidity of the input gas should be nearly 100% to maintain high microbial metabolism-degradation rates. At relative humidities of <95%, the air flow tends to dry out soil beds. The moisture content of soil beds can be maintained by irrigating, but in practice this has not been necessary because rainfall makes up the difference.

4. Halogenated compounds are ill-suited for biofiltration removal unless the residence time of the gas in the soil bed is long.

Applications of Soil Biofilters

Most of the biofilters operating are at waste treatment facilities. The economics of waste treatment require an inexpensive method of odor removal. In addition, the source of odor in waste treatment is usually anaerobic microbial degradation so the highly aerobic biofilter is readily recognized by waste treatment personnel as a feasible means of odor removal. Biofiltration achieves the degree of aerobiasis that trickling filters and activated sludge treatment of waste water strive to attain.

Waste Water Treatment

Pomeroy (1957) patented miocrobial treatment of odors from waste water treatment. In 1964 Carlson and Leiser (1966) designed fourteen soil biofilters to remove the odors at sewage pump stations in Mercer Island, Washington. The beds have required no maintenance since they were installed and people living on the adjoining lots are unaware of any odors. In 1971 the city of Coronado, California, installed a soil biofilter at its major sewage pump station (Pomeroy, 1982). In 1987 Coronado expanded the bed because of the increased sewage load and to replace activated carbon filters whose odor control was unreliable. At the same time the pump station and soil bed were made the center of an outdoor shopping and restaurant area. The area previously unusable because of malodor now generates considerable income for the city.

The cities of Tamarac, Florida, and Toenberg, Norway; A. Coors Co., Golden, Colorado; and A. J. Simplot Co., Hermiston, Oregon; are also successfully and inexpensively removing waste water treatment odors with soil biofilters.

These soil beds treat flows of up to about 1500 cfm (2500 m^3/hr). Soil beds to treat up to 120,000 cfm (200 000 m^3/hr) are in various states of design and completion for the cities of Springfield and Yarmouth, Massachusetts, and Madison, Wisconsin. The reasons for choosing soil biofilters are their effectiveness, simplicity, economy, and safety.

Composting

The city of Geneva, Switzerland, built a municipal waste composting plant on the French border in the early 1960s. The winds blew the odors back to Geneva so in 1964 a soil biofilter was installed to treat the 2200 cfm (3600 m^3/hr) air flow from the composting drum. A similar composting plant in Duisburg, West Germany, received odor complaints from as far away as 1.5 km because of the unreliability of a hypochlorite scrubber. In 1967 they began exhausting their contaminated air through a bed of mature compost. Now people live 100 m away without complaint (Bohn, 1976).

A large chicken farm in West, Texas, was forced to stop composting its manure because of odor complaints 1 mile away. The composting shed is now swept with 10,000 cfm (16 000 m^3/hr) of air which is then treated in a soil bed. The odor removal varies from 95 to 99% (Sweeten, 1988) because of the bed's fluctuating moisture content. Odor removal efficiency is lowest when the bed is very wet and very dry. Even at its worst, the odor control is satisfactory and the operators see no need to improve its performance.

Animal Rendering

The 600 cfm (1000 m^3/hr) process air from the cooker of an animal rendering plant in Tucson contains up to 200,000 odor units. A soil biofilter was installed in 1983 and its odor removal efficiency was measured 6 and 18 months later (Prokop and Bohn, 1985). The air leaving the soil bed contained 10-100 odor units and the average odor removal was 99.9%. Any large pores in the soil have since closed naturally and the odor now is undetectable when standing on the bed. This corresponds to probably 99.99% removal efficiency.

Industrial Odors and VOC

Other industries in North America have been slow to adopt biofiltration. Ebinger, Bohn and Puls (1987) and the S. C. Johnson Co. of Racine, Wisconsin, (Kampbell, et al. 1987) investigated propane and isobutane removal from contaminated air by soil biofilters. The beds remove about 90% of the hydrocarbons. As would be expected for a microbially-based process, the removal efficiency decreases at bed temperatures of <50 F (<10 C).

A fiberglass manufacturing plant in California is removing styrene from its plant exhaust with a soil biofilter. This is to end odor complaints and to comply with tighter regulations for VOC emissions in the Los Angeles Basin. Other hydrocarbons would be removed similarly. The sorption of hydrocarbons increases exponentially with carbon number (Bohn, Prososki, and Eckhardt, 1980) which offsets their somewhat slower biodegradability compared to analogous lighter-molecular weight compounds.

Odorous organic comounds characteristically contain oxide, hydroxyl, sulfhydryl, amino, nitroso, and phosphate functional groups. Functional groups greatly increase the molecule's sorptivity and usually also increase its biodegradability. The 99.9% removal efficiency of the Tucson rendering plant odors is indicative of the effectiveness of biofilters for alcohols, aldehydes, ketones, organic acids, and nitrogen- and sulfur-containing organic gases.

Cost

The major reason for avoiding odor control is its cost. Biofiltration costs much less than wet scrubbing, incineration, or activated carbon beds. The installation cost of monolayer soil biofilters ranges from $10/cfm ($6 per m^3/hr) for small beds down to $3/cfm ($2 per m^3/hr) for large beds, a fraction of the installation cost of other methods. These costs include soil modification and hauling in new soil, but do not include the collection systems nor the cost of the land. Since the land can be used simultaneously for parking, landscaping, equipment storage, etc., apportioning land costs is difficult. Multi-layer beds are more expensive to construct but require less land. The additional cost of a collection system is necessary for any odor control method, but can equal the cost of the bed itself.

The real savings of biofiltration are in operating costs. Biofilter beds require no fuel or chemical because oxygen is the oxidant and microbial metabolism is the catalyst. The lifetime of soil biofilters is indefinite and they require no maintenance.

The pay back period of biofiltration varies but is roughly 6 months compared to incineration and 12 months compared to activated carbon and chemical or wet scrubbing. Jaeger and Jager's (cited in Ottengraf, 1986) data summarizing the overall costs of treating the odors from waste water treatment, Table 1, indicate even faster pay back. Biofiltration is significantly cheaper than the alternatives for odor removal.

The data in Table 1 omit the secondary benefits obtained by effective odor control. Property values increase greatly when odor nuisances are removed. The buffer zone surrounding many odor sources becomes valuable and saleable real estate, greatly offsetting the cost of odor control. Removing odors also removes the stigma attached to many industries.

Limitations of Soil Biofiltration

Soil biofiltration requires much more space than incinerators, wet scrubbers, and activated carbon beds. Multi-layer beds and designs which allow parking and equipment storage on top of the biofilter greatly reduce the space requirement.

The real reasons for the slow acceptance of soil biofiltration are a lack of understanding about soil biochemistry and the misperception that the soil and groundwater will be contaminated:

1. Few people are aware of the soil's ability to adsorb and oxidize gases. The news media incorrectly view soils as pollutant sources, not realizing that the pollution comes from overloading, anaerobiasis, and subsurface

burial. The soil's role in nature is to remove pollutants from water and air and to safely recycle the chemical elements in pollutants back into their biogeochemical-environmental cycles.

2. Soil biofiltration appears at first clance to endanger groundwater quality by putting air pollutants into the surface soil. This fear arises from overloaded landfills or soils which have received great notoriety. These are liquid wastes at high loading rates in the lower vadose zone under anaerobic conditions--very poor conditions for sorption and degradation. Soil biofiltration in contrast treats gases at low loading rates, under strongly aerobic conditions, and at the surface where sorption capacities and degradation rates are high. Assuming 1000 ppm of pollutants in the input gas and a 40 hr work week, the annual loading rate is 0.04 lbs. of organic matter per square foot (0.2 kg/m^2). The annual input of natural organic wastes is 5-10 kg/m^2 and the degradation rate is directly proportional to the input rate. Compared to liquid and solid disposal, treatment of the gaseous state is much safer because gases are more rapidly biodegradable than liquids or solids and because biofilter beds are continuously supplied with excess oxygen.

3. In this high-tech age, people doubt the efficacy of apparently simple techniques. Biofiltration is actually an integrated washing-sorption-oxidation system which physicochemical methods can not emulate.

4, There is little economic incentive to promote soil biofilters. The medium is already available at the site or can be purchased for a few dollars per ton. Also, biofiltration has no repeat business because the beds are self-regenerating and need no replacement.

The major advantages of biofilters for odor removal are:

1. Complete odor removal is readily achievable.

2. Installation costs are low.

3. Operating costs are extremely low. Biofilters require no fuel or chemicals and soil biofilters last indefinitely.

4. Biofilters are effective for an extremely wide range of organic and inorganic gases.

5. Biofilters are nontoxic and nonflammable and create no secondary pollution. The major disadvantages of biofiltration for odor control are:

1. Soil biofilters require larger areas than other methods. Soils are less permeable to gas flow than activated carbon beds so larger area is required for the same backpressure. This is advantageous in that the longer residence time allows time for microbial degrdation of the pollutants.

2. The pollutants can not be recovered. The pollutants are adsorbed too tightly and degraded too rapidly.

3. Soil biofiltration is a form of land disposal and land disposal is sometimes misperceived as unsafe. The very low loading rates, strongly aerobic conditions, and very rapid biodegradation of the gases prevent soil and water pollution.

References

1. Bohn, H.L., "Soil and compost filters of malodorant gases," JAPCA 25:953 (1975).

2. Bohn, H.L., R.K. Bohn, "Soil beds weed out air pollutants," Chem. Engr. 95(6):73 (25 April 1988).

3. Bohn, H. L, G.K. Prososki, J.G. Eckhardt, "Hydrocarbon adsorption by soils as the stationary phase of gas-solid chromatography," J. Environ. Qual. 9:563 (1980.

4. Carlson, D.A, C.P. Leiser, "Soil beds for the control of sewage odors," J. Water Pollut. contr. Fed. 38:829 (1966).

5. Duncan, M., H.L. Bohn, M. Burr, "Pollutant removal from wood and coal flue gases by soil treatment," JAPCA 32:1175 (1982).

6. Dupont, G, "La desodorisation des gaz de fermentation des ordures menageres, dans les usines de compostage," Dept. Travaux Publics, Geneva, Switzerland (1964).

7. Ebinger, M.H., H.L. Bohn, R.W. Puls, "Propane removal from propane-air mixtures by soils," JAPCA 37:1486 (1988).

8. Fischer, K., W.J. Homans, D. Bartke, "Biologische Abluftreinigung," Umwelt Supplement 3/85:30 (1985).

9. Helmer, R., "Desodorisierung von geruchsbeladene Abluft in Bodenfiltern," Gesundheits-Ingenieur 95(I):21 (1974).

10. Johnson, S. W. "On some points of agricultural science," Am. J. Sci. Arts Ser. 2, 28:72 (1859).

11. Kampbell, D. H., J. T. Wilson, H. W. Read, T. T. Stocksdale, "Removal of volatilealiphatic hydrocarbons iin a soil bioreactor," JAPCA 37:1236-1240 (1987).

12. Ottengraf, S.P.P., J.J.P. Meesters, A.H.C. van den Oever, H.R. Rozema, "Biological elimination of volatile xenobiotic compounds in biofilters," Bioprocess Engr. 1:61 (1986).

13. Pomeroy, R.d., "Deodorizing gas streams by the use of microbial growths," U.S. Patent 2,793,096 (May 21, 1957).

14. Pomeroy, R.D., "Biological treatment of odorous air," J. Water Pollut. Contr. Fed. 54:1541 (1982).

15. Prokop, W.H., H.L. Bohn, "Soil bed system for control of rendering plant odors," JAPCA 35:1332 (1985).

16. Reneau, Jr., R.B., R.C. Hagedorn, M.J. Degen, "Fate and transport of biological and inorganic contaminants from on-site disposal of domestic wastewater," J. Environ. Qual. 18:135 (1989).

17. Sweeten, J.M., R.E. Childers, Jr., J.S. Cochran, "Odor control from poultry manure composting plant using a soil filter," Am. Soc. Ag. Engr. paper #88-4050 (1988).

Table 1. Total cost of odor removal at a waste water treatment plant in Heidelberg, West Germany, 1974.

	Deutschmark per 1000 m^3 of air	U.S. Dollars per million ft3 of air
Incineration	9.10	$130
Chlorine	4.20	60
Ozone	4.20	60
Activated Carbon, with		
Regeneration	1.50	20
Biofiltration	0.60	8

III. Odor Measurements, Modeling and Technology

VERIFICATION OF A PERCEIVED ODOR
PROBLEM IN A COMMUNITY

Dr. Ron Poustchi, P. Eng.
Louis Berger & Associates
East Orange, New Jersey

Dr. Alex Gnyp, P. Eng. and Dr. Carl St. Pierre, P. Eng.
Department of Environmental Engineering
University of Windsor
Windsor, Ontario

Of the various categories of air pollutants, odors are generally ranked as the major generators of public complaints to regulatory agencies in North American communities. It is estimated that more than 50% of the complaints related to air pollution deal with exposures to odors.[1,2] Surveys of citizens living in the neighborhoods of odorous stationary sources indicate that odors can cause mental and physiological stresses on humans. Receptors describe their perceptions of odorous environments in terms of tolerable; unpleasant; very unpleasant; terrible; and unbearable conditions. Typical human reactions are nausea; headache; loss of appetite; impaired breathing; and in some cases allergies.[3,4] On the basis of a recent study[5], it appears that more than half of the people in the 65 to 80 year age category have major olfactory impairments. Consequently many senior citizens can experience adverse health reactions without being aware of their exposure to odorous environments.

In recent years, odorous sources have received considerable attention from the public and regulatory agencies. In practice, the existence of an odor pollution problem must be established before any steps towards control can be attempted. Odor complaints are indicators that a potential odor problem may exist in the community[6]. In general, very few people register formal complaints with authorities with respect to any environmental problem[7]. A study of annoyance created by aircraft noise showed that the main characteristics distinguishing complainants from noncomplainants were related to education, value of their home and membership in organizations. On this basis, the number of complainants received by officials may reflect not so much the amount of discomfort experienced by the exposed population, but its social class composition and level of community organization[8,9].

However, studies have shown that persons who volunteer their opinions may tend to overstate their concern[10]. Some courts have indicated that the difficulty of relying on odor complaints to prove a nuisance lies in the possibility that the complainants do not represent the feelings of the community as a whole[11]. Social surveys can be used to estimate the true feelings of a community by including controls for bias. Field procedures involve asking questions whether odors have been perceived, whether they have caused annoyance and under what conditions[12]. Questionnaires are the most widely used method for collecting information about people's attitudes and behavior[13]. Questions about the backgrounds of the respondents are added to facilitate correlation of data. Additional questions about other forms of pollution are included, if the main interest in odors is to be de-emphasized to those being surveyed.

To establish if there is a recognizable odor problem in the community, records show that over several years many residents in the neighborhoods of odorous sources have complained regularly and vigorously about odor problems. In certain communities, as many as 40 complaints have been registered by some residents in one year[3]. It is important to appreciate that one or two complaints per year have been generated by many other residents in the same communities during the same period. Generally the number of different complainers is less than 15% of the number of documented residences. Although many complaints are validated by members of air pollution control agencies, municipal engineering or fire departments, no corrective actions are usually taken.

In fact, chronic complainers are treated with suspicion by odor emitting source owners and even some regulatory agency personnel.

The problem facing regulatory agencies is how

- to establish that there is a genuine odor problem in the community?

- to prove that spontaneous complainers are not just trouble makers?

Is it possible that chronic complainers are basically normal citizens responding to a definite community odor problem? To resolve this question it becomes necessary to determine whether people in the community under investigation are behaving normally, by comparing their attitudes towards commonly encountered odors with those of citizens in other neighborhoods or a control group. In addition, it is essential that they should be able to spontaneously identify any odors or odorous sources that are responsible for the alleged community problems.

To help regulatory agencies approach odor problems more objectively a public attitude survey was designed and implemented in a number of communities.

Development and Implementation of the Survey

Formulation of Questions

The survey was designed to expose individual attitudes towards commonly encountered odors on the basis of past experiences or prejudices. The ultimate goal was to determine if there were any odors or odorous sources in the neighborhood that would seriously bother the people. Questions were modified several times during preliminary field trials to accomplish the objective of the survey. The final version in Figure 1 represents the questions after a number of revisions. The questions were worded so that they could be understood by both young and old as well as people with limited educational backgrounds. During preliminary trials, changes were made after

- people complained that the questions were not clear enough

- participants had difficulty completing the questionnaire

- individuals found it difficult to record responses

- complaints were made that the questionnaire was too long.

Questions 1 to 5 were included to provide data that might relate responses to the backgrounds of the participants. Although questions 6 to 31 were formulated specifically to de-emphasize the objective of the survey, they helped to establish the hedonic ratings [pleasantness-unpleasantness] of various odors commonly encountered in any community.

The basic objective was to be achieved from answers to the question "What smells in the air seriously bother you? Please list" which was introduced as Question number 32.

Environmental Study

1. Age _____
2. Sex (circle one)   M   F
3. Address

_____

_____

4. Occupation _____
5. Health (check one)   Excellent ☐
                         Good   ☐
                         Average   ☐
                         Fair   ☐
                         Poor   ☐
   If problems, please list: _____

_____

    The following statements relate to various odors normally
present in the air. Please read carefully and then rate each one
individually based on <u>your personal reactions</u>. To complete the
question circle the arrow (↑) watch most closely matches how <u>you</u>
feel about it.

    Example: I like baseball.

  ↑_____↑_____↑_____(↑)_____↑
Not at all   O.K.   Very much

6. I like the smell of cut grass.
  ↑_____↑_____↑_____↑_____↑
Not at all   O.K.   Very much

7. I like the smell of car and truck fuels.
  ↑_____↑_____↑_____↑_____↑
Not at all   O.K.   Very much

8 through 31 classify in the same manner as 1 through 7:
    chicken outlets, sewers, vinegar, garbage etc.

2. What smells in the air seriously bother you? Please list:

_____

_____

_____

_____

3. This last question requires that you rank (in order of preference)
   5 odors. Write the number which indicates your preference, below
   each odor.

   1 - I like it best.
   5 - I like it least.

   Example: football   baseball   hockey   soccer   basketball
             (3)   (2)   (1)   (4)   (5)

   Please rank:

   May   Swimming Pool   Barbeque   Hamburger Joint   Garbage

The format of Question 33 is flexible and is designed to determine the citizens reactions to changes in the character of the odor in the community as a result of implementation of specific odor control techniques to an offending odorous source. The example in Figure 1 was formulated to assess the application of hypochlorite solutions to the control of odors from fast food restaurants.

The number of questions was limited to 33 in order to enable the participants to complete the questionnaire in 5 to 10 minutes.

Recording Responses

Every effort was made to simplify the recording of responses. Experience showed that individuals resisted quantifying their reactions numerically to any particular odor. In all the preliminary surveys, the young, old, educated and uneducated were prepared to respond in general terms using expressions such as "OK", "Not at All" and "Very Much". In order to simplify the field activities respondents were simply asked to circle the arrows that most closely represented their attitudes ranging from "Not At All" to "Very Much" as illustrated in Figure 1. For statistical purposes, a scale ranging from -10 to +10 was used to quantify the responses varying from "Not At All" to "Very Much".

Implementation of the Survey

The optimized version of the public attitude survey was presented to people in the neighborhoods surrounding a fast food restaurant; a municipal waste treatment plant; an automotive paint application facility (upwind and downwind); a foundry and an active municipal waste landfill site.

An indoor shopping mall in Windsor, Ontario was selected as a control area since it provided responses from different parts of the city as well as from out of town visitors.

In each community, residents were selected on a random basis. The responses were obtained voluntarily, with no compulsion. Care was taken to ensure that the presence of the survey team would not create any feelings of antagonism towards any existing or potential odor emitting sources by emphasizing the scientific nature of the investigation. The survey teams were made up of individuals who had no knowledge of the alleged odor problems in any of the communities under investigation.

In every neighborhood, many citizens refused to complete the survey because of apathy, defeatism or fear. Statements like "we have complained a number of times but nothing has happened", "this has been going on for many years", "I am very busy and have no time for that", were heard many times. In some apartment buildings close to odorous sources, the managers refused to cooperate. They claimed that they did not want tenants upset by such surveys. This was most common in areas where community action groups were organizing support against the suspected odorous sources.

TABLE I: Hedonic Ratings of Odors Expressed by People in Various Communities, The Control Group and Those Published by Dravnieks et al. (14,15)

Odor Description	Control Group	Fast Food Restaurant	Sewage Plant	Paint Upwind	Auto-Plant Downwind	Landfill Site	Foundry	Dravnieks et al.
cut grass	+2.6	+0.7	+3.7	+2.1	+2.2	+2.5	+2.8	-
car/truck fumes	-8.9	-9.1	-9.7	-8.7	-9.6	-9.3	-9.1	-
fried chicken outlets	-0.7	-1.3	+0.4	-1.1	-1.4	-0.8	-1.4	-
sewers	-9.6		-9.9	-10.0	-10.0	-9.9	-10.0	-9.2
vinegar	-3.9	-5.3	-4.0	-3.0	-4.0	-5.3	-2.9	-3.2
garbage	-9.8	-9.6	-10.0	-10.0	-10.0	-9.9	-9.8	-9.4
hospital	-4.7	-5.2	-5.7	-4.1	-6.4	-5.4	-4.2	-4.0
fruit market	+2.0		-1.0	+1.2	+1.1	+0.5	+1.8	
gasoline	-5.5	-3.8	-7.5	-5.0	-6.9	-5.6	-6.4	-2.9
hamburger restaurants	-0.9	-2.2	-2.1	-2.6	-1.2	-1.8	-2.3	
fresh popcorn	+4.7	+2.2	+4.2	+4.8	+5.0	+5.2	+3.9	+6.2
ammonia	-8.5	-8.2	-8.2	-7.6	-8.9	-8.6	-7.2	-6.2
paint	-5.9	-5.8	-6.1	-5.4	-8.4	-5.1	-5.5	-1.9
roses	+7.4	+6.0	+6.7	+7.3	+9.0	+7.3	+8.3	+7.7
beer	-1.6	-0.7	-3.5	-2.0	-2.7	-2.9	-4.3	-0.4
outside chinese restaurant	-0.7	-0.7	-2.0	-2.8	-2.5	-1.8	-2.7	
locker/dressing room	-8.4	-7.4	-8.9	-8.2	-9.0	-8.2	-9.0	
baking bread	+7.9	+6.0	+7.8	+7.4	+6.6	+7.7	+8.8	
wood fire	+6.0	+1.8	+4.2	+5.8	+2.3	+5.0	+6.5	+4.1
chocolate	+5.0	+3.2	+4.2	+4.2	+3.8	+3.5	+3.8	
cigarettes	-7.8	-6.5	-7.2	-6.5	-7.1	-6.3	-6.9	
carnival	-1.5	-2.4	-2.2	-2.9	-2.1	-3.0	-2.3	
peanuts	+2.8	+0.5	+2.2	+2.9	+2.9	+2.9	+3.0	
barbeque	+6.2	+6.0	+5.0	+5.8	+5.2	+5.7	+5.6	+5.9
leather jacket	+1.3	+0.1	-1.6	-0.2	-1.7	-0.7	-1.9	
coffee			+2.7	+3.3	+3.9	+4.2	+5.8	+5.8
Number of People	109	53	47	46	52	282	60	429

155

Results of the Survey

Analysis of Hedonic Ratings

Table I compares the hedonic ratings, of twenty-six commonly encountered odors, as expressed by residents in the communities surrounding the five selected odorous sources with those in a control area and similar study of 146 odor descriptors reported by Dravnieks et al. [14,15]. For comparison purposes, the Dravnieks data were multiplied by a factor of 2.5 to convert the ratings from a -4.0 to +4.0 scale to the -10 to +10 scale used in this investigation.

According to Table I, the overall community attitudes toward different odors are essentially identical in all the areas surveyed. It must be emphasized that in every neighborhood there are individuals whose appreciation of several odors would be distinctly different from the community average. A detailed analysis of Table I shows that, generally, people do not like the smell of gasoline; paint; cigarettes; locker/dressing rooms; ammonia; car/truck fumes; sewers; and garbage.

On the other hand, they do like the odors associated with baking bread; a barbecue; roses; wood fires; chocolate; and fresh popcorn. However, they are neutral with respect to the odors common to fried chicken outlets; hamburger establishments; and Chinese restaurants.

From the survey results of 649 people, it is possible to separate the 26 commonly encountered odors into the three categories pleasant; neutral; and unpleasant on the basis of the hedonic ratings falling in the ranges 10 to 5, 5 to -5 and -5 to -10, respectively. Table II shows the three classifications.

Table II.   Three catagories of odors based on their pleasantness, neutrality and unpleasantness.

Pleasant	Neutral	Unpleasant
roses	cut grass	garbage
barbecue	fried chicken outlet	sewers
baking bread	hamburger restaurant	car/truck fumes
	vinegar	ammonia
	beer	gasoline
	fruit markets	paint
	outside Chinese restaurant	cigarettes
	carnival	locker/dressing room
	peanuts	hospital
	coffee	
	fresh popcorn	
	leather jacket	
	chocolate	
	wood fire	

Table III summarizes the results of statistical t test for paired data. The good agreements between the ratings from the various communities, Dravnieks data and the control group illustrate the consistency of people's responses to common odors. Such a generality suggests that citizens reactions to odors in the neighborhoods of various odorous sources are no different from those in the control area. Consequently their spontaneous complaints can be interpreted as honest responses to offensive odors and not simply as frivolous exercises by neighbors who may be antagonistic towards a particular facility. The data from the community in the vicinity of the solid waste landfill site, where 282 people had responded, indicated that the age of respondents can affect their hedonic ratings of odors. Table IV correlates the hedonic values of odors for the different age groups, along with the weighted averages and the standard deviations from the average values. A comparison of these data shows that teenagers do not like the smell of cut grass (-0.7), and coffee (-2.0), whereas other age group ratings varied from +2.0 to +4.6 for cut grass, and +3.1 to +8.0 for coffee.

On the other hand teenagers do like the smells associated with fried chicken outlets (+1.7), hamburger establishments (+1.3), Chinese restaurants (+1.0), and leather jackets (+0.7).

Other age groups rated these odors to be neutral to negative with values ranging from -0.2 to -2.8 for fried chicken outlets, -1.3 to -5.0 for hamburger establishments, -0.5 to -4.2 for Chinese restaurants, and -0.3 to -5.0 for leather jackets.

An analysis of the standard deviations of hedonic ratings from the average values shows that people in the 20 to 49 age group are more representative of the community than teenagers or citizens in the over 50 age bracket.

Using standard deviations ±1.5 as an arbitrary basis, odors can be grouped into two categories. In Table V category I represents odors about whose hedonic ratings there was close agreement among all age classifications. Category II includes those odors for which there were significant differences in hedonic ratings from the different age groups.

TABLE III: Comparison of Computed and Tabulated Values of t for Various Levels of Significance

Control Group v_s	Calculated t	Tabulated Values of t			
		$\alpha^* = 0.001$	$\alpha = 0.01$	$\alpha = 0.05$	$\alpha = 0.1$
Fast Food	2.541	3.792	2.819	2.074	1.717
Landfill	2.878	3.745	2.797	2.064	1.711
Sewage	3.591	3.745	2.797	2.064	1.711
Foundry	1.562	3.745	2.797	2.064	1.711
Dravnieks	2.750	4.437	3.106	2.201	1.796
Paint Auto Upwind	1.460	3.745	2.797	2.064	1.711
Paint Auto Downwind	3.778	3.745	2.797	2.064	1.711

*Level of Significance

TABLE IV: Hedonic Ratings of Odors, Expressed by the Residents in the Neighborhood of the Landfill Site According to Their Age Groups, Along with the Weighted Average for All and the Standard Deviations from the Average Values for the Various Odors.

Odor Description	10-19	30-39	50-59	70-up	Average	10-19 ±SD	30-39 ±SD	50-59 ±SD	70-up ±SD
cut grass	- 0.7	+3.1	+4.6	+2.0	+2.5	3.2	0.6	2.1	0.5
car/truck fumes	-10.0	- 9.0	- 9.4	-10.0	- 9.3	0.7	0.3	0.1	0.7
fried chicken outlets	+1.7	- 0.2	- 1.9	- 2.0	- 0.8	2.5	0.6	1.1	1.2
sewers	- 9.3	-10.0	-10.0	-10.0	- 9.9	0.6	0.1	0.1	0.1
vinegar	- 6.3	- 6.1	- 3.7	- 2.0	- 5.3	1.0	0.8	1.6	3.3
garbage	-10.0	- 9.9	-10.0	- 9.0	- 9.9	0.1	0.0	0.1	0.9
hospital	- 6.7	- 5.6	- 5.4	- 3.0	- 5.4	1.3	0.4	0.0	2.4
fruit market	+5.0	0.0	- 1.2	+1.5	+0.5	4.5	0.1	1.7	1.0
gasoline	- 5.0	- 5.4	- 4.6	- 7.0	- 5.6	0.6	0.2	1.0	1.4
hamburger restaurants	+1.3	- 1.3	- 4.2	- 5.0	- 1.8	3.1	0.5	2.4	3.2
fresh popcorn	+5.7	+5.6	+4.4	+5.5	+5.2	0.5	0.4	0.8	0.3
granola	- 9.0	- 9.7	- 6.3	- 7.0	- 8.6	0.4	1.1	2.3	1.6
paint	- 4.3	- 5.4	- 5.6	- 4.5	- 5.1	0.8	0.3	0.5	0.4
roses	+6.0	+6.8	+7.7	+6.0	+7.3	1.3	0.5	0.4	1.3
beer	0.0	- 3.9	- 5.6	- 1.5	- 2.9	2.9	1.0	2.7	1.4
outside a chinese restaurant	+1.0	- 1.5	- 4.2	- 4.0	- 1.8	0.8	0.3	2.4	1.4
locker/dressing room	- 8.7	- 9.1	- 9.0	- 5.6	- 8.2	0.5	0.9	0.8	2.6
baking bread	+6.0	+7.4	+8.3	+5.0	+7.7	1.7	0.3	0.6	2.7
wood fire	+3.0	+4.5	+6.7	+1.5	+4.9	1.9	0.4	1.7	3.4
chocolate	+3.0	+3.8	+5.0	+5.5	+3.8	0.8	0.0	1.2	1.7
cigarettes	- 7.7	- 7.0	- 4.6	- 6.5	- 6.3	1.4	0.7	1.7	0.2
carnival	- 2.3	- 2.6	- 3.4	- 3.3	- 3.0	0.7	0.4	0.4	0.3
peanuts	0.0	+2.4	+3.8	+3.0	+2.9	2.9	0.5	0.9	0.1
barbeque	+5.7	+6.1	+3.3	+3.5	+5.7	0.0	0.4	2.4	2.2
leather jacket	+1.3	- 1.9	- 1.3	- 5.0	- 0.7	2.0	1.2	0.6	4.3
coffee	- 2.0	+4.3	+7.3	+8.0	+4.2	6.2	0.1	3.1	3.8

Table V: Two categories of odors based on agreement among the various age groups

Category I	Category II
(All Age groups Agree)	(Disagreement Among Age Groups)
car/truck fumes	cut grass
sewers	fried chicken outlets
garbage	vinegar
gasoline	hospital
fresh popcorn	fruit market
paint	ammonia
carnival	roses
	beer
	outside Chinese restaurant
	locker dressing room
	baking bread
	woodfire
	chocolate
	cigarettes
	peanuts
	barbecue
	leather jacket
	coffee

From Table V it is evident that people of all age groups agree closely on the hedonic ratings of most of the offensive odors. However, there is considerable difference of opinion about the hedonic ratings of most of the neutral and pleasant odors.

Identification of Odor Sources in a Community

The answers to Question 32 help to confirm the validity of spontaneous complaints which are generally generated by less than 15% of residential sites.

Table VI provides the responses to Question 32, "What odors in the air seriously bother you?". The descriptions and frequencies of identification of odors encountered in the various locations provide a means of defining the odorous source creating the alleged odor problems in the different communities.

From Table VI, it is apparent that people in the control group, who represent a highly heterogenous community, are bothered seriously by the smells associated with cigarette smoke, auto and truck emissions, factory pollution, gasoline, garbage, sewers, and skunks.

In addition to these odors, residents living in any neighborhood might also be bothered by the odors associated with certain industrial or commercial operations existing in their locality. For example, in the case of the community in the neighborhood of the fast food restaurant, only 2% of the

TABLE VI : Description and Frequency of Identification of Odors Encountered by Residents in the Six Communities and the Control Group

Description of Odor	Percent People Seriously Bothered						
	Control Group	Fastfood Restaurant	Sewage Treatment Plant	Paint-Auto Plant Upwind	Paint-Auto Plant Downwind	Landfill Site	Foundry
Cigarette Smoke	28	7	6	20	4	8.5	17
Auto Emissions	22	23	9	24	15	17.0	18
Pollution	22	9	4	13	4	1.8	8
Bus & Truck Fumes	21	28	26	11	12	12.6	15
Gasoline	13	8	4	4	12	5.9	.
Garbage	11	21	6	17	10	76.0	10
Sewers	9	.	11	4	17	8.0	7
Sulphur	9	9	9	6	.	1.0	.
Skunk	5	.	17	11	4	4.3	.
Strong Perfumes	4	.	.	.	.	2.0	.
Foundry	3	.	.	.	.	.	48
Amonia (bleach)	2	2	2	2	.	0.4	.
Tar Roofing (hot asphalt)	2	2	13	2	2	1.0	.
Burning Rubber	2	.	.	.	2	0.4	3
Methane	2	.	.	2	.	.	.
Sewage Treatment Plant	2	.	34	.	.	.	.
Melting Metal Fumes	3	.	.	2	.	.	.
Brewery	2	.	6	2	2	.	.
Burning Leaves	2	.	.	2	.	2	5
Factory Smoke	.	26	26	30	4	13.5	.
Burning Garbage	.	2	.	.	.	.	.
Paint Fumes	1	4	.	15	79	.	.
Fertilizer	.	4	2	4	.	2	.
Smells from Factories In Detroit	1	9	2	.	.	2	.
Smells from Garbage Plant	.	2	.	.	.	.	.
Ammonium Hydroxide	.	2	.	.	.	.	.
Smells from Zug Island	1	2	6	.	2	.	2
Dust	.	2	.	.	.	.	.
Insecticide	.	2	2	2	.	.	.
Fastfood Restaurant	1	2	.	.	.	.	.

people were bothered by the fast food odors. This community response is consistent with the total absence of complaints about fast food odors to the local air pollution control agency office. The location of the facility is in a very high traffic density region (51% annoyance by autos, buses, and trucks) which is subjected to industrial emissions from the U.S. side of the Detroit River causing annoyance among 37% of the respondents. In addition 23% of the residents were disturbed as a result of being on the direct route of garbage trucks moving to and from a private waste hauling parking area.

In the neighborhood of the sewage treatment plant, 34% of the residents surveyed were bothered by this operation. This significantly adverse responses validates the spontaneous complaints recorded by the local regulatory agency. The annoyances registered against bus and truck fumes (26%) reflects the community's proximity to a major highway. Fallout of particulate matter and odors from industrial operations across the Detroit River are responsible for the 32% adverse reactions to factory smoke and smells from Zug Island.

Residents living upwind from the automotive paint application facility were bothered more by automobile emissions (24%) and general factory smoke (30%) than by paint fumes (15%). It must be appreciated that the 15% annoyance level is sufficient to generate more than 12 complaints in six months and encourage the citizens to join a community action group. Seventy-nine percent of those surveyed in the downwind community identified the upwind paint application facility as the source of community odor problems. A well organized community action group is led by people from this area. They have enlisted the aid of municipal and provincial politicians to help solve their problem.

The 76% of the residents surveyed in the neighborhood of the landfill site who identified garbage odor as a community problem represents more individuals than the total number of complainers documented over two year period. In other communities, including the control area, garbage odors were listed as problems by less than 25% of the respondents. With 76% of those surveyed in the landfill community identifying garbage odors as a serious problem, there is no doubt that the landfill site is creating a detrimental effect on the air quality in the locality. The activities of the citizen action group that forced an environmental hearing emphasized the seriousness of the odor pollution problem.

In the neighborhood of the foundry, a citizen action group has enlisted the support of municipal and provincial politicians who encourage regular public and private meetings between the community, foundry operators and regulatory agency members. The spontaneous complaint level leading to the formation of the action group is confirmed by the 48% of those surveyed who identified foundry odors as bothersome.

Reaction to Changes in Odor Character

Question 33 was formulated specifically to assess citizen reactions to changes in the character of the odor in the community as a result of treatment of fast food restaurant odors with hypochlorite scrubbing solutions.

Table VII shows the average ratings of five specific odors by the respondents in various localities, on a scale of 1 to 5, where 1 = the best; and 5 = the least.

It is evident that the smells associated with hay, a hamburger restaurant and a swimming pool are ranked at the same level, whereas barbecues and garbage are at the extreme of the scale for the best and the least preferred odors, respectively.

The data in Table VII suggest that installation of expensive scrubbing devices using hypochlorite solution can be as objectionable or worse than the original fast food odors if excess chlorine odors are emitted to the neighborhood.

Conclusion and Recommendations

On the basis of public attitude surveys carried out in 7 different complaint boundaries, it is evident that identification of a specific source by 15% or more of the people surveyed in a locality is indicative of a community odor problem. Regulatory agencies should consider spontaneous complaints from such areas to be real indicators of poor ambient air quality.

In order to establish the validity of the complaints for legal purposes, it is recommended that regulatory agencies

- survey a representative number of residents of the community in the neighborhood of the suspected odor source

- survey an equally representative number of people in a community which is not directly under the influence of the emissions from the suspected source

- analyze the data from the two locations separately

- compare the hedonic ratings of the respondents from both locations to check if there are any differences in the reactions of people towards commonly encountered odors

- evaluate the percentage of people who are bothered by specific odors in their locality in terms of their responses to Question 32.

- investigate a source or sources if the percentage of people identifying a suspected odor source is 15% or higher.

The results of this exercise by the regulatory agency will provide the answer to the question "Is there a recognizable odor problem in the community?"

TABLE VII: Comparison of Average Ratings of Five Specific Odors Expressed by the Residents in Six Communities And the Control Group

Odor Description	Control Group	Fast Food Restaurant	Sewage Plant	Paint Auto - Plant		Landfill Site	Poundry
				Upwind	Downwind		
Barbeque	1.6	1.4	1.4	1.6	1.7	1.4	1.8
Hay	2.5	3.1	2.7	2.7	2.6	2.6	2.3
Hamburger Restaurant	2.9	2.8	2.8	3.2	3.4	2.8	3.1
Swimming Pool	3.0	3.0	2.9	2.7	2.6	2.9	2.8
Garbage	5.0	5.0	5.0	5.0	5.0	4.9	4.9

If it is established that there is a genuine odor problem in a community, it then becomes important to determine

- " how bad" is the odor?

- " how much" odor is there?

Studies are underway to quantify odors in terms of their offensiveness and subsequent evaluation of community annoyance in relation to source strength and meteorology.

Acknowledgments

The financial support provided by the Air Resources Branch of the Ontario Ministry of Environment, the Natural Sciences and Engineering Research Council (NSERC) and Imperial Oil Limited through University of Windsor Research Grants made this study possible. The help provided by Miss Dina Gnyp in the design of the survey is appreciated.

References

1. "Odors from Stationary and Mobile Sources," Board on Toxicology and Environmental Hazards, Assembly of Life Sciences, National Research Council, National Academy of Sciences, Washington, D.C., 1979.

2. G. Leonardo, "A Critical Review of Regulations for the Control of Odors," JAPCA 24:456 (1974).

3. "Environmental Hearing," Avon Township, Michigan vs. Southern Oakland County Incinerator Authority, October 19, 1983.

4. "Complaint Data Files," Air Pollution Control Division, Wayne County Department of Health, 1311 East Jefferson, Detroit, Michigan 48207 (1984-85).

5. R.L. Doty, P. Shaman, S.L. Applebaum, R. Giberson, L. Siksorski, L. Rosenberg, "Smell Identification Ability: Changes with Age", Science Magazine, 226:1441-143, December 21, 1984.

6. W. H. Prokop, "Industry View Toward Odor Regulatory Needs", National Renderers Association, Des Plaines, Illinois, presented at the 77th Annual Meeting of the Air Pollution Control Association, San Francisco, California, June 24-29, 1984.

7. "Regulatory Options for the Control of Odors", U.S. Environmental Protection Agency, Research Triangle Park, NC, EPA-450/5-80-003, U.S. Department of Commerce, National Technical Information Service, February 1980.

8. W.S. Smith, J.J. Schueneman, L.D. Zeidberg, "Public Reaction to Air Pollution in Nashville, Tennessee", JAPCA 14:418 (1964).

9. "Community Reaction to Airport Noise", Tracor, Inc., Vol. 1: Contractor Report No. 1761, NASA Contract NASW-1549, 97 pp, July 1971, Vol. 2: NASA No. CR-111-316, 245 pp, September 1970, National Aeronautics and Space Administration, Washington, D.C.

10. R. Cederlof, L. Friberg, E. Jonsson, L. Kaij, T. Lindvall, "Studies of Annoyance Connected with Offensive Smell from a Sulphate Cellulose Factory", Nord. Hyg. Tidsku, 45:39 (1964).

11. J. L. Parker, "Legal Issues Resulting from Odor Emissions", John L. Parker & Associates Ltd., Chicago, Illinois, presented at the 77th Annual Meeting of the Air Pollution Control Association, San Francisco, California, June 24-29, 1984.

12. Copley International Corporation, "A Study of the Social and Economic Impact of Odors", Phase III, Development and Evaluation of a Model Odor Control Ordinance, Report to the U.S. Environmental Protection Agency, EPA Publication No. 650/5-73-001, Contract No. 68-02-0095, Washington, D.C., February 1973.

13. S. Sudman, N.M. Bradburn, "Asking Questions - A Practical Guide to Questionnaire Design", Jossey-Bass Publishers, San Francisco, California, 1983.

14. A. Dravnieks, T. Musurat, R.A. Lamm, "Hedonics of Odors and Odor Descriptors", JAPCA 34:752-755 (1984).

15. A. Dravnieks, "Odor Character Profiling", JAPCA 33:775 (1983).

A MATHEMATICAL ANALYSIS OF ODOR THRESHOLD DETERMINATIONS

James A. Nicell, Alex W. Gnyp and Carl C. St. Pierre
Department of Civil and Environmental Engineering
University of Windsor
Windsor, Ontario, Canada

Odor threshold measurements are used by environmental regulatory agencies to place limits on odorous emissions from industrial sources. Reliable and reproducible measurements of thresholds are required to effectively enforce regulations. The popular forced–choice technique produces thresholds that are distorted significantly by the effects of chance. Also, threshold measurements are strongly influenced by panel composition. This mathematical analysis quantifies the magnitude of the error introduced by chance into a forced–choice evaluation and the influence that a single panelist has on the overall panel threshold. Results indicate that a discrimination threshold better represents a panel's ability to perceive an odor as the effects of chance and the variation in individual thresholds are reduced. Models are provided for future studies to allow optimization of test parameters including panel size and the factor relating odor concentrations at consecutive dilution levels in the odor presentation procedure.

Introduction

Tests are often conducted to quantify industrial odorous emissions in order to relate community perceptions of the odors with the offending sources, to determine the scope of odor control needed for abatement or to test compliance with local regulations.

Most procedures used today are based on threshold determinations that assume a sample of odorous gas can be described in terms of the number of dilutions required to reduce its intensity to the sensory threshold level. Such determinations depend on the olfactory responses of individuals who serve on odor evaluation panels. A threshold representing a population is designated as ED_{50}, which is an acronym for effective dosage at the 50% level. At this concentration (or dilutions) 50% of the panelists detect the odor and 50% do not[1]. This value, called the detection threshold, defines the number of odor units in a sample.

An *odor unit* (o.u.) is defined as 1 ft³ of air at the odor threshold. *Odor concentration* is the number of volumes that 1 unit volume of sample will occupy when diluted to the odor threshold. The *odor emission rate* from a source is defined as the number of odor units discharged per unit time[2]. Measurements of odor concentrations and odor emission rates provide several different approaches for the regulation of odors.

Many advances have been made in testing procedures since the implementation of thresholds for quantifying odorous emissions. Dynamic olfactometry in combination with a forced–choice technique has found the greatest popularity. While this procedure represents vast improvements over previous methods, there are still inherent problems due to the effects of panel composition and chance on the measured threshold. Since regulatory agencies require reliable and reproducible measurements of sensory thresholds as a basis for their regulations, the sources of error must be reduced to an acceptable level.

The objectives of this work are to demonstrate (1) the effects of chance upon a threshold evaluated with a forced–choice technique, and (2) the influence of an individual panelist on the threshold reported for the overall panel.

Analysis of the Forced–Choice Procedure

A forced–choice procedure requires that a panel member be presented, on each of a number of trials, a sample of air containing the odorant and one or more odor–free samples. The panelist must identify which sample contains the odorant. If the panelist is unsure, a guess is required — hence the origin of the term *forced–choice*. The concentration of the odorant is increased in each successive trial according to a geometric progression. The testing continues until the subject becomes consistently correct in the identification of the odorous sample.

If an individual begins to consistently identify the odorous sample at dilution level Z_j, then that person's detection threshold, termed BET (best estimate threshold), is assumed to be the geometric mean of Z_j and the dilutions at the previous dilution level (Z_{j-1}). The detection threshold of the panel is estimated as the geometric mean of individual panelist thresholds since for a large number of individual measurements the geometric mean closely approximates the median at which 50% of the population detects the presence of the odor. This is due to the logarithmically normal distribution of odor thresholds in the population[3]. There are a variety of other procedures for determining the overall panel threshold but this approach is the simplest and is the only method that can be treated according to the mathematical analysis that follows.

The method described depends upon forced choice decisions by panelists who often resort to guess–work for their responses. A lucky guess on the part of a panelist, who chooses correctly when deciding which sample is odorous, distorts both the evaluation of the individual's threshold and the overall panel threshold. To minimize the effects of guessing during the determination of an effective dosage for an average panel member, the concept of a *discrimination threshold* was introduced. The discrimination threshold, D_{50}, is defined as the level at which 50% of a panel can distinguish between odorous and non–odorous samples with certainty[4]. An individual's discrimination threshold is determined in the same manner as the detection threshold except that the subject must signify that his/her identification of the odorous sample was made with certainty or as a guess. Choices that are claimed to have been made with certainty are verified by checking the panelist's indication of the odorous sample.

An attempt has been made to relate the discrimination threshold, D_{50}, with the detection threshold, ED_{50}, using probability theory. The existence of such a relationship would demonstrate that the differences noted between the two thresholds are due only to the effects of chance.

Probability Model of the Forced–Choice Method

If a panelist, i, evaluates an odor using a procedure with N consecutive dilution levels arranged in an ascending series of concentration, and begins to correctly differentiate between the odorous sample and the blank(s) at dilution level j then that subject's detection threshold, expressed in dilutions, is calculated as

$$BET_i = (Z_{j-1} \cdot Z_j)^{1/2} \qquad (1)$$

where Z_j is the number of dilutions at level j. The overall panel detection threshold, ED_{50}, for n panelists is calculated from

$$ED_{50} = \left[\prod_{i=1}^{n} BET_i \right]^{1/n} \qquad (2)$$

or, alternatively, using

$$Log\ ED_{50} = \frac{1}{n} \cdot \sum_{i=1}^{n} Log\ BET_i \qquad (3)$$

If a panelist begins to correctly identify the presence of the odor at dilution level j he/she may not really be sure about the odor until the j+1, j+2, or even lower dilution levels. The consistently correct identification of the odorous port at higher dilution levels may be due entirely to chance since the forced–choice method requires a guess from the panelist when a choice is not obvious. Therefore, this panelist's discrimination threshold, D_i, will be defined by

$$D_i = (Z_{j+s-1} \cdot Z_{j+s})^{1/2} \qquad (4)$$

where s is the number of samples identified correctly due to lucky guesses before the panelist begins to discriminate the odor.

Provided that concentrations at consecutive levels are related by a constant factor, F, which is a requirement in a geometric progression, then

$$Z_j = F \cdot Z_{j+1} \qquad (5)$$

Manipulation of Equations (1), (4), and (5) leads to the following relationship between the two thresholds:

$$BET_i = D_i \cdot F^s \qquad (6)$$

The value of s cannot be predicted for a single panelist. However, for a large pool of panelists the number of individuals who have a particular value of s may be predicted using probability theory provided that s depends entirely on chance. For example, if M panelists in the pool guess s consecutive trials correctly before they begin to discriminate the odor, the value of ED_{50} can be modified to remove these guesses. This modification is accomplished by removing the M panelists' detection thresholds and replacing them with their discrimination thresholds. This adjustment to the mean detection threshold can be expressed as

$$\text{Log } X_m = \frac{1}{n} \left[\sum_{i=1}^{n} \text{Log BET}_i - \sum_{i=1}^{M} \text{Log BET}_i + \sum_{i=1}^{M} \text{Log } D_i \right] \tag{7}$$

This expression can be simplified to

$$\text{Log } X_m = \frac{1}{n} \left[\sum_{i=1}^{n} \text{Log BET}_i + \text{Log} \left[\prod_{i=1}^{M} \frac{D_i}{\text{BET}_i} \right] \right] \tag{8}$$

Substitution of Equation (6) into (8) yields

$$\text{Log } X_m = \frac{1}{n} \left[\sum_{i=1}^{n} \text{Log BET}_i + \text{Log} \left[\prod_{j=1}^{M} F^{-s} \right] \right] \tag{9}$$

$$= \frac{1}{n} \left[\sum_{i=1}^{n} \text{Log BET}_i + M \text{ Log } F^{-s} \right] \tag{10}$$

The number of panelists, M, who guess s trials correctly before they begin to discriminate can be estimated from:

$$P = \left[\frac{1}{C} \right]^s \tag{11}$$

where P is the probability of guessing s consecutive trials correctly given that there is one odorous sample and C–1 blank samples at each dilution level (trial). The number of people who guess s trials in a row correctly is then determined by

$$M = P \cdot n$$
$$= \frac{n}{C^s} \tag{12}$$

when there are n odor judges on the panel. Substitution of Equation (12) into (10) produces

$$\text{Log } X_m = \frac{1}{n} \left[\sum_{i=1}^{n} \text{Log BET}_i + \frac{n}{C^s} \cdot \text{Log } F^{-s} \right]$$

which reduces to

$$\text{Log } X_m = \frac{1}{n} \sum_{i=1}^{n} \text{Log BET}_i + \text{Log } (F^{\frac{-s}{C^s}}) \tag{13}$$

Solving for X_m and using Equation (2) to simplify gives

$$X_m = ED_{50} \cdot F^{\frac{-s}{C^s}} \tag{14}$$

Equation (14) defines a modified value of the detection threshold for panelists who have a particular value of s. However, s can vary between 0 and the number of dilution levels, N, presented to the panelist. Note that correction for s = 0 has no effect on the threshold. Therefore, the threshold must be corrected for those panelists who have values of s between 1 and N. The fully modified value of the threshold, X_f, is expressed as

$$
\begin{aligned}
X_f &= ED_{50} \times F^{\frac{-1}{C^1}} \times F^{\frac{-2}{C^2}} \times \ldots \times F^{\frac{-N}{C^N}} \\
&= ED_{50} \cdot \prod_{s=1}^{N} F^{\frac{-s}{C^s}} \\
&= ED_{50} \cdot F^{-\sum_{s=1}^{N} \frac{s}{C^s}}
\end{aligned}
\tag{15}
$$

Since the value of ED_{50} in Equation (15) has been modified to exclude all correct guesses from the threshold, the resulting value, X_f, should be an estimate of the discrimination threshold provided that the detection and discrimination thresholds differ only due to the effects of chance, as assumed earlier. Therefore, the two thresholds can be related by

$$D_{50} = K \cdot ED_{50} \tag{16}$$

with
$$K = F^{-\sum_{s=1}^{N} \frac{s}{C^s}} \tag{17}$$

This relationship suggests that the detection and discrimination thresholds are related to each other through values of F, N, and C which are parameters of the testing method. Therefore, for any given testing method with fixed F, N, and C, the two thresholds should differ only by a constant factor.

Application to a Ternary Forced–Choice Method

Equations (16) and (17) were derived on the assumption that the detection and discrimination thresholds differ only due to the random effects of chance upon individual threshold evaluations. To demonstrate the applicability of this expression to practical problems, it is necessary to verify this critical assumption by comparing values of K predicted by Equation (17) with values measured in the laboratory.

Dynamic Olfactometer. The olfactometer used in this study was purchased from the IIT Research Institute[5]. It consists of a dilution air pump, peristaltic odor pump, signal box, air rotameters, deodorizing chamber, six sets of sniffing ports, two manifolds, and Teflon sample lines. This instrument provides six dilution stations each equipped with a set of three glass sniffing ports. Two of the ports emit deodorized room air (blanks) while the third discharges the odorous gas diluted with deodorized air.

The odorous gas is delivered to the olfactometer at a rate chosen by the panel leader depending on the required range of dilutions to be used in an odor evaluation. The deodorized dilution air is supplied at a rate of approximately 10 L/min. Two manifolds divide the odor and air flows among the odor stations in specific ratios. Each port delivers between 600 and 700 mL/min of deodorized air or odorous sample. Preliminary testing showed that panelists could not sense any gas pressure on their noses and, therefore, were unable to detect variations in gas flow rates between ports. Consequently, the responses of the panelists were not influenced by the small differences in total gas flow rates that occurred from port to port.

The olfactometer was calibrated and operated according to the methods prescribed by the IIT Research Institute[5].

Test Room.   All odor tests were carried out in an odor–free environment maintained in a fully enclosed, double–walled chamber. The test room is equipped with a door, glass window, interior light, electric air cleaner and an exhaust fan for removal of odors introduced through the olfactometer. The inside walls are constructed of washable arborite. The air cleaner delivers odor–free air into the test chamber at a low flow of 100 ft³/min or a high flow of 150 ft³/min. Precautions were taken in the design of the test room to maintain an atmosphere free from distractions.

Panelist Selection and Training.   All tests during this study were performed with panels composed of 10 people who were selected at random from a total pool of 121 participants. These individuals were not selectively chosen for their abilities to detect odors. Instead, the only criterion used was that the total group must represent a mixture of both sexes and a wide range of ages.

Instructions were provided to each panelist on how to proceed with port selection and response to the odors. The seriousness of the odor testing procedure was heavily emphasized.

Theoretical Prediction of K.   The following parameters describe the design specifications for the IITRI olfactometer:

- odor concentrations increase by a factor of three between levels $(F = 3.0)$
- three ports at each dilution level $(C = 3)$
- six dilution levels $(N = 6)$

Values of C and N are fixed for this apparatus. However, the value of F can show some variation between devices. The olfactometer used in this study provided an average value of 2.88 for F. This value changes slightly when small changes in total odor flow rate are made and must be reevaluated with each calibration. Therefore, the factor relating D_{50} and ED_{50} in Equation (16) will be

$$K = 2.88^{-\sum_{s=1}^{6} \frac{s}{3^s}} = 0.455$$

Experimental Evaluation of K.   An empirical K was evaluated from thresholds determined for six pure chemical odorants in air: (1) n–butyl acetate, (2) propylene glycol monomethyl ether, (3) methyl isoamylketone, (4) isobutanol, (5) n–butanol, and (6) octane. Each chemical was evaluated by five ten–member panels who were chosen at random from the panelist pool. The overall detection and discrimination thresholds determined for each chemical are presented in Tables I and II[6].

Experimental K values were obtained from the ratio of the two measured thresholds according to

$$K = \frac{D_{50}}{ED_{50}} \tag{18}$$

when the thresholds were expressed in dilutions. The equivalent expression in terms of thresholds expressed in concentration units is

$$K = \frac{ED_{50}}{D_{50}} \tag{19}$$

due to the inverse relation between concentration and the number of dilutions. The values of K derived for each 10–member panel corresponding to each odorant are presented in Table III. The mean value of K was calculated as a geometric mean since the thresholds are logarithmic normally distributed and, therefore, their ratio must also have the same distribution form.

This analysis was extended to include information derived from industrial samples. A summary of the results for all odor samples is shown in Table IV. Note that as the olfactometer calibration changes so does the value of F.

Discussion of Results

The data in Tables III and IV show that the measured and theoretically predicted values of K are extremely close. However, some of the deviations that exist between these two values can be accounted for by the fact that in all panels participating (a total of 98) no panelists ever needed to make six guesses in a row. That is, all panelists claimed to discriminate the odors being examined before or at the lowest dilution level. Therefore, correction of K for six consecutive guesses is an over–correction. Theoretical values of K were recalculated using a correction for five dilution levels (N = 5) and are included in Table IV. The agreement between the experimental and probability–based K values improves significantly with this treatment.

In one industrial odor investigation, a number of extremely strong odor samples was examined. In this case involving 12 panels, the experimental value of K was determined to be 0.694. The mean thresholds for the panels were located somewhere between the first and second dilution levels. Therefore, an average panelist never had to make more than one guess before discriminating between the odorous and odor–free samples. This condition is confirmed by a theoretical evaluation of K for guesses of only one dilution level (N = 1) which produces a correction factor of 0.703. The difference between the theoretical and experimental values is less than two percent and are, therefore, in extremely good agreement.

The excellent correlation between the predicted and measured values of K indicates that the detection threshold is the same as the discrimination threshold when it is corrected for the effects of chance. Therefore, a detection threshold determined using a forced–choice technique represents a biased measurement of the discrimination threshold.

An additional numerical evaluation was performed to confirm this conclusion. If a detection threshold is affected by chance then its reproducibility should be lower than that of the discrimination threshold. Such a difference was confirmed for all six pure chemical odorants when a comparison was made between the variance in measurements of detection and discrimination thresholds for all panelists. Table V shows that the variation among panelists is consistently lower for the discrimination thresholds than for the detection thresholds. However, the reproducibility of the thresholds between panels did not differ significantly as shown by the standard deviations between panel evaluations given in Tables I and II. Perhaps since the effects of chance are also reproducible they do not introduce significant variations into the overall panel thresholds.

These results indicate that the discrimination threshold is a more reliable measurement of a panel's ability to perceive an odor. Consequently, the concentration of an odor in odor units (o.u.) should be based on the dilutions required to reduce the odor concentration to the discrimination threshold value.

Two alternatives are available for the determination of discrimination thresholds. The threshold measurement technique can be modified to provide direct evaluation or a forced–choice technique can be used in conjunction with a statistical method for removing the effects of chance from threshold estimates.

The first alternative has been used in this study and has been proven to be effective. The procedure involves six basic steps:

1.   Use an odor sample presentation technique involving dilution levels arranged according to an ascending geometric progression of odor concentration.

2.   Start at the highest dilution level and ask the panelist to decide whether he/she can discriminate between the odorous sample which is mixed among a number of odor–free samples.

3.   If the panelist cannot perceive the odor then he/she can proceed to the next dilution level without making a choice. If the subject claims to discriminate then an indication of which of the samples is odorous must be provided.

4.   Continue the above steps until all panelists have examined the full range of dilutions.

5.   When reducing the data, a panelist is considered to have perceived the odor at a particular dilution level provided he/she has claimed to discriminate and has correctly identified the odorous sample at that level and at all higher concentrations (lower dilutions).

6.   Determine the overall panel threshold using an accepted technique.

The second alternative for establishing the discrimination threshold by using a statistical method for eliminating the effects of chance is not as simple. There are a number of models available to accomplish this elimination but they involve testing each individual more than once to get an estimate of the proportion of people who are guessing[4][7]. This procedure, while effective, increases the time required for completing an odor test. The extended time requirement represents a severe drawback over the alternative of measuring the discrimination threshold directly.

Equation (17) demonstrates that the parameters describing the testing procedure (C, F, and N) have an influence on the threshold values produced from a forced–choice method.

The number of samples presented to panelists at each dilution level, C, has a significant effect on the correction factor, K, as shown in Figure 1. However, practical constraints dictate that increasing C above a value of three is a questionable means of reducing the uncertainty in a forced–choice technique. For example, values of C above three may significantly increase olfactometer costs and the duration of odor tests. There would also be a reduction in equipment portability.

The value of C is also critical when measuring the discrimination threshold directly since panelists must still demonstrate that they can perceive the odor by identifying the odor sample among one or more blanks. As more blanks are used, the error in verification will be reduced.

The measurement error is also influenced by the factor relating the number of dilutions between consecutive levels, F. Figure 1 demonstrates that by decreasing the value of F the effects of chance on the final threshold value are decreased. However, practicalities dictate that F must be chosen to ensure that an adequate range of dilutions is represented by the N dilution levels of the testing procedure. Therefore, an optimum value of F must be chosen. This choice will be discussed in more detail in a later section.

Variations of N values have a negligible influence on K when the N values are greater than 4 as shown in Figure 2 for C values of two, three and four. This figure also demonstrates that as N decreases, the correction factor approaches unity. This trend does not mean, however, that by reducing N the error inherent in the forced–choice method will be reduced. Instead, the results must be interpreted to mean that the effect of panelists making consecutive correct guesses becomes insignificant after guessing at four dilution levels. For example, the values of K (with F = 3.0 and C = 3) calculated using N = 5 and N = 100 differ by less than 2 percent. Therefore, any test that does not measure a discrimination threshold directly but, instead, relies on the statistical elimination of the effects of chance using Equations (16) and (17) must produce a corrected threshold which is located somewhere after the fourth dilution level in order to be valid. If the corrected threshold is located at some higher dilution level then correction of the original threshold for lucky guesses will produce questionable results. It is also necessary to have one or more dilution steps after the fourth dilution level so that confirmation of panelists' claims of discrimination can be made. Thus, a system with six dilution levels should be adequate to meet these requirements.

Note that for tests designed to use a forced choice between one blank and one odorous sample with a factor of 10.0 relating consecutive dilution levels, the resulting value of K would be 0.01. The correction factor, K, represents a measure of the mean error introduced by chance and, therefore, the value for an individual panel could be either larger or smaller than this mean. Thus, with a correction factor of 0.01, it would not be surprising if some reported threshold measurements differed by several orders of magnitude. In such a severe case, it would be justifiable to conclude that the effects of chance contribute more to the measurement of the threshold than the panelists' abilities to perceive the odor. This condition cannot be tolerated if sensory thresholds are to remain as a basis for regulating odorous emissions from industrial sources.

Effects of Individual Panelists

One of the major problems with any odor evaluation technique is the high variability in the sensitivities of the individual panelists. For example, when a detection threshold is established by a panel of 10 people it is possible that if one of those panelists was replaced by another person the same detection threshold would not be obtained, despite the fact that the panels were nearly identical. The same variations would be expected for determinations of the discrimination threshold.

The effect of exchanging one panelist for another can be quantified using expressions similar to those developed previously. In the analysis that follows, the detection threshold was used as a basis. An analogous derivation can be performed using the discrimination threshold in place of the detection threshold.

A panel of n members would produce a threshold evaluated from

$$\text{Log ED}_{50} = \frac{1}{n} \left[\sum_{i=1}^{n} \text{Log BET}_i \right] \tag{2}$$

If the same test was performed using all but one member of the original panel who was replaced by another panelist with a threshold that occurs d dilution levels earlier, the resulting threshold would differ from the original by

$$\text{Log ED}_{50n} = \frac{1}{n} \left[\sum_{i=1}^{n} \text{Log BET}_i - \text{Log BET}_j + \text{Log BET}_{j-d} \right] \tag{20}$$

175

Equation (20) can be simplified to

$$\text{Log ED}_{50n} = \frac{1}{n} \left[\sum_{i=1}^{n} \text{Log BET}_i + \text{Log} \left[\frac{\text{BET}_{i-d}}{\text{BET}_j} \right] \right] \tag{21}$$

Substitution of Equation (5) into (21) produces

$$\text{Log ED}_{50n} = \frac{1}{n} \left[\sum_{i=1}^{n} \text{Log BET}_i + \text{Log F}^d \right] \tag{22}$$

Solving Equation (22) for ED_{50n} and using Equation (2) to simplify gives

$$\text{ED}_{50n} = \text{ED}_{50} \cdot \text{F}^{d/n} \tag{23}$$

A change, Δ_{50}, can be defined as a measure of the effect that the new panelist has on the old panel threshold by correctly identifying the odor d dilution levels earlier than the original panelist, according to

$$\Delta_{50} = \frac{\left[\text{ED}_{50n} - \text{ED}_{50} \right]}{\text{ED}_{50}} \times 100\%$$

This equation can be simplified using Equation (23) to produce

$$\Delta_{50} = \left[\text{F}^{d/n} - 1 \right] \times 100\% \tag{24}$$

Accordingly, if the replacement panelist on a panel of 10 members detects the odor 2 ports earlier ($d = 2$) than the replaced panelist (using the IITRI olfactometer with $F = 3.0$), the change in the threshold, Δ_{50}, is

$$\Delta_{50} = \left[3.0^{2/10} - 1 \right] \times 100\% = 24.6\%$$

Therefore, the replacement panelist would cause an increase in the original detection threshold by 24.6%.

Minimum Number of Panelists for a Given Tolerance

The analysis conducted in the previous section can be used to determine the number of panelists required to reduce the effect of a replaced panelist to an acceptable level. Equation (24) can be solved to provide n in terms of Δ_{50}, F, and d according to

$$n = \frac{d \cdot \text{Log F}}{\text{Log} \left[1 + \frac{\Delta_{50}}{100} \right]} \tag{25}$$

If a tolerance, T, is defined as the allowable degree of control that a panelist is allowed to exercise by detecting or discriminating the odor at one higher dilution level

$(d = 1)$ then it follows that

$$n_t = \frac{\text{Log } F}{\text{Log } \left[1 + \frac{T}{100} \right]} \tag{26}$$

where n_t is the minimum number of panelists required to meet the tolerance value, T.

For example, if the effect of the single panelist is to be reduced to a 10% tolerance level when using the IITRI olfactometer ($F = 3.0$), the minimum number of panelists required would be

$$n_{10} = \frac{\text{Log } 3.0}{\text{Log } \left[1 + \frac{10}{100} \right]} \cong 12$$

It must be appreciated that 12 is the number of panelists required to reduce the effect of only one dilution level difference between the thresholds of replaced and replacement panelists. If there were a greater difference between the two subjects ($d > 1$), the number of panelists required to accomplish the same tolerance would be

$$n_{td} = d \cdot n_t \tag{27}$$

when using a replacement panelist whose threshold differs from the replaced panelist by d dilution levels.

Although there is, as yet, no truly objective basis for defining an acceptable value of T, it is possible to choose a tolerance such that one person on a panel of 10 has only a 10% effect on the threshold or one person on a panel of 20 has only 5% effect. Such a tolerance definition can be stated in terms of

$$T = \frac{100}{n} \tag{28}$$

Substitution of Equation (28) into Equation (26) and solving for F produces

$$F = \left[\frac{n + 1}{n} \right]^n \tag{29}$$

Equation (29) defines the factor, F, relating consecutive dilution levels as a function of the number of panelists. Substitution of a range of values for n shows that this equation is a very insensitive function of n and provides values of F that are always very close to 2.5. This trend suggests that the optimum value of F which reduces the degree of control of an individual panelist to a reasonable level, defined by Equation (28), is 2.5. It should be noted that this factor was calculated using only a preliminary estimate of an acceptable tolerance value. The determination of the ideal tolerance level is beyond the scope of this study but is recommended for further consideration.

Conclusions

The basic objective of odor testing procedures is to establish the level at which an individual truly perceives an odor. This information can then be used to determine a threshold which represents an odor concentration at which 50% of odor panel members will perceive the odor and 50% will not.

A detection threshold established with a forced–choice technique is distorted by the random effects of chance. It has been shown that the elimination of these chance effects produces a discrimination threshold which is a better representation of a panel's ability to perceive an odor. The discrimination threshold can be quantified either by statistically resolving the effects of chance when using a forced–choice technique, or by measuring the discrimination threshold directly using the procedure developed during this study.

Equations have been derived to provide a measure of the effect of a single panelist on the evaluation of panel thresholds. These equations can be used in future studies to optimize important test parameters such as panel size and the factor relating consecutive dilution levels in the odor presentation procedure. The determination of an acceptable tolerance for individual panelist effects on overall thresholds is recommended for further study.

Acknowledgments

Financial support from the Ontario Ministry of the Environment and the Natural Sciences and Engineering Research Council of Canada (NSERC) is greatly appreciated.

References

1. A. Dravnieks, F. Jarke, "Odor threshold measurement by dynamic olfactometry: significant operational variables," J. Air Pollut. Control Ass. 30, pp. 1284–1289. (1980).

2. P.N. Cheremisinoff, R.A. Young, Pollution Engineering Handbook, Ann Arbor Science, Michigan. 1976, pp. 275–279.

3. A. Dravnieks, M. Meilgaard, W. Schmidtsdorff W., "Odor thresholds by forced–choice dynamic triangle olfactometry: reproducibility and methods of calculation," J. Air Pollut. Control Ass. 36, pp. 900–905. (1986).

4. E. Poostchi E., A.W. Gnyp, C.C. St. Pierre, "Comparison of models used for the determination of odor thresholds," Atmospheric Environment 20, pp. 2459–2464. (1986).

5. IIT Research Institute, Instructions for Dynamic Triangle Olfactometer 1977 Model, Chicago, Illinois. (1977).

6. J.A. Nicell, Preliminary Assessment of the Odor Impact Model as a Regulatory Strategy, M.A.Sc. Thesis, University of Windsor, Windsor, Ontario. (1986).

7. S. Viswanathan, G.P. Mathur, A.W. Gnyp, C.C. St. Pierre, "Application of probability models to odor threshold determinations," Atmospheric Environment 17, pp. 139–143. (1983).

Notation

BET_i	best estimate threshold (detection threshold) of panelist i
C	number of samples presented at each dilution level of which 1 is an odorous sample and C-1 are blanks
d	number of dilution levels difference between thresholds of replaced and replacement panelists
D_{50}	discrimination threshold of a panel of odor judges
D_i	discrimination threshold of panelist i
ED_{50}	detection threshold of a panel of odor judges
ED_{50n}	new panel detection threshold after a single panelist is replaced
F	factor relating the odor concentration at consecutive dilution levels
K	factor relating discrimination and detection thresholds
K_e	experimentally determined value of K
K_t	theoretically predicted value of K
M	number of panelists with a particular value of s
n	number of individuals on an odor panel
n_t	number of panelists needed to meet a given tolerance, T
n_{td}	number of panelists needed to meet a tolerance, T, when a replaced and replacement panelist have thresholds that differ by d dilution levels
N	maximum number of consecutive dilution levels at which guesses can occur
P	probability of guessing s consecutive trials correctly given that there is one odorous sample and C-1 blank samples at each dilution level
s	number of odorous samples correctly identified by chance before a panelist discriminates
T	influence that a panelist can exercise by detecting or discriminating the odor at one higher dilution level
X_f	fully modified threshold for all values of s
X_m	threshold modified for a particular value of s
Z_j	dilutions of a sample at level j
Δ_{50}	change in a panel threshold resulting from the replacement of a panelist

Table I. Detection thresholds for six pure chemical odorants
determined using five ten–member panels.

Odorant	ED_{50} (mg/M^3)	$Log(ED_{50})$	SD of $Log(ED_{50})$
1	1.17	0.068	0.08
2	107.	2.028	0.07
3	0.76	−0.119	0.15
4	3.38	0.529	0.16
5	3.54	0.549	0.18
6	54.8	1.739	0.12
Concentrations reported at 10^{0}C and 101.3 kPa			

Table II. Discrimination thresholds for six pure chemical
odorants determined using five ten–member panels.

Odorant	D_{50} (mg/M^3)	$Log(D_{50})$	SD of $Log(D_{50})$
1	3.12	0.494	0.14
2	183.	2.262	0.06
3	1.56	0.191	0.18
4	7.24	0.859	0.14
5	8.15	0.911	0.21
6	115.	2.061	0.13
Concentrations reported at 10^{0}C and 101.3 kPa			

Table III. Experimentally determined values of K using six pure chemical odorants.

Odorant	Values of K from Panel No.					Mean
	1	2	3	4	5	
1	0.42	0.33	0.27	0.43	0.46	0.37
2	0.54	0.59	0.61	0.74	0.47	0.58
3	0.46	0.45	0.51	0.57	0.41	0.48
4	0.52	0.46	0.47	0.41	0.48	0.47
5	0.37	0.46	0.30	0.58	0.52	0.44
6	0.72	0.28	0.58	0.43	0.48	0.47
Overall K value calculated as a geometric mean						0.46

Table IV. Comparison of experimental and theoretical values of K using three calibrations of the IITRI olfactometer.

Calibration	F	K_e	K_t (N=6)	K_t (N=5)
No. 1	2.88	0.460	0.455	0.459
No. 2	2.77	0.473	0.469	0.473
No. 3	2.88	0.457	0.455	0.458

Table V. Comparison of variations in panelist detection and discrimination thresholds.

Odorant	Mean of Log (BET_i)	Variance	Mean of Log (D_i)	Variance
1	0.068	0.258	0.497	0.178
2	2.027	0.164	2.263	0.107
3	−0.118	0.252	0.203	0.200
4	0.528	0.519	0.858	0.289
5	0.549	0.257	0.910	0.197
6	1.739	0.172	2.063	0.087

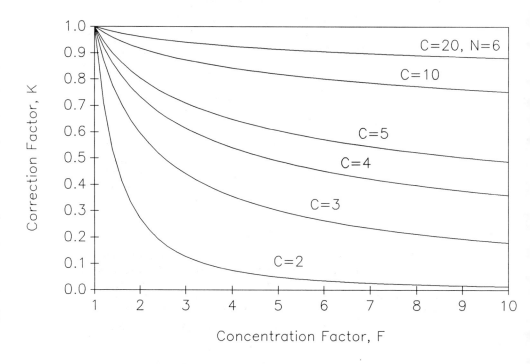

Figure 1.   Variation of K as a function of F and C with N=6.

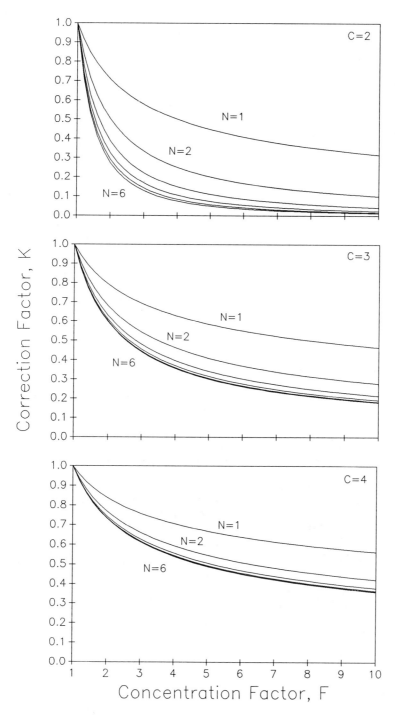

Figure 2. Variation of K as a function of F and N for fixed C values.

ODOR THRESHOLDS FOR CHEMICALS WITH
ESTABLISHED OCCUPATIONAL HEALTH STANDARDS

Samuel S. Cha
TRC Environmental Consultants, Inc.
East Hartford, Connecticut

A report has been prepared by TRC Environmental Consultants, Inc. (TRC) for American Industrial Hygiene Association (AIHA) In 1988. The report was intended to serve as a chemical odor threshold reference for use by industrial hygienists and other health or safety professionals.

Odor threshold values are presented in virtually every health and safety text or reference manual. Attempts to put these values to practical use, however, can often result in confusion over which values from which text to employ. This occurs partially because the thresholds reproduced in the literature are often from historical rather than updated sources. Additionally, threshold values from different current sources can vary for the same substance over several orders of magnitude. Regardless, with time, threshold values come to be considered physical constants instead of the statistical best estimate value they represent. The confusion created by this variation limits the utility of odor threshold values for the industrial hygienist. AIHA believes that the critique of odor threshold literature reported here is a needed first step in addressing this situation. In addition, the mean estimates of odor thresholds developed from acceptable data sources and presented herein should be useful to health and safety professionals.

This presentation describes the report and presents samples of odor threshold tables.

1.0   INTRODUCTION

This report is intended to serve as a chemical odor threshold reference for use by industrial hygienists and other health or safety professionals. It has been prepared for the American Industrial Hygiene Association (AIHA) by TRC Environmental Consultants, Inc. (TRC).

Odor threshold values are presented in virtually every health and safety text or reference manual. Attempts to put these values to practical use, however, can often result in confusion over which values from which text to employ. This occurs partially because the thresholds reproduced in the literature are often from historical rather than updated sources. Additionally, threshold values from different current sources can vary for the same substance over several orders of magnitude. Regardless, with time, threshold values come to be considered physical constants instead of the statistical best estimate value they represent. The confusion created by this variation limits the utility of odor threshold values for the industrial hygienist. AIHA believes that the critique of odor threshold literature reported here is a needed first step in addressing this situation. In addition, the mean estimates of odor thresholds developed from acceptable data sources and presented herein should be useful to health and safety professionals.

There are a number of threshold value compilations available, specifically, Van Gemert (1977 and 1982), Fazzalari (ASTM D548A, 1978), Billings and Jones (1981), Amoore and Hautala (1983) and, most recently, Ruth (1986). Some, such as Amoore and Hautala (1983), transform the original data for use while others, such as Ruth (1986), record the range of threshold values. None of these compendia, however, critique the threshold values on the basis of their experimental design. Such a critique is necessary to produce values useful from a health and safety viewpoint. Threshold values reported from the first half of this century were probably not obtained under the same conditions of methodological precision taken for granted today. Additionally, some values are reported from many interdisciplinary sources in which the intent is not threshold measurement per se. Therefore, sensory experimental criteria are not considered a necessary factor in design. Critiquing the experimental odor threshold determinations reported in the literature provides a basis for developing a best estimate of odor threshold and represents a necessary refinement of the data available in odor threshold compilations.

The report presents a single value representing the best estimate of the odor detection or recognition threshold for odorant chemicals for which a Threshold Limit Value (TLV) has been published by the American Conference of Governmental Industrial Hygienists (ACGIH). These TLV's are published in a booklet entitled, "Threshold Limit Values and Biological Exposure Indices for 1986-1987." The reader should be aware, however, that the listing of odorous chemical thresholds contained in this report is by no means a compendium of all odorous chemicals or odor threshold data. There are 913 chemicals with odor thresholds listed in the Compilation of Odor Threshold Values in Air and Water, published jointly by the National Institute for Water Supply and the Central Institute for Nutrition and Food Research, Netherlands in 1977 and updated in 1982. This document is the major source of odor threshold data reviewed for this report. There are 680 compounds with TLVs listed in the ACGIH booklet. However, there are only 183 chemicals which have both a published odor threshold and a TLV. This is primarily because most of the odorous chemicals (odorants) are not included in the TLV booklet.

185

Odor threshold research is an evolving science. As the technology of olfactometry and analytical measurement advances, threshold experiments are conducted and reported in the literature. The result is that any threshold compilation can be slightly out of date at the time of its publication. Additionally, even as comprehensive a volume as prepared by Van Gemert (1977 and 1982) does not contain all threshold values. Threshold determinations from other disciplines or periodicals often come to light at later dates. Therefore, threshold value estimates should be reassessed periodically to assure their continuing usefulness to health and safety professionals.

Table 5-1 of the report includes the range of experimentally determined odor thresholds for each odorant, the published TLV and the best estimate for the detection or recognition odor threshold. The best estimate of the odor threshold was determined by applying criteria (discussed in Section 4.0) to the experimentally determined values listed in the original publication. Of the 183 odorants with a TLV, only 102 had thresholds determined by techniques that met the evaluation criteria.

The critique of each published experiment is given in Table 5-2 of the report. A listing of published thresholds organized by chemical in alphabetical order is given in Table 5-3 of the report.

Section 2.0 of this report presents background material on odor perception and odor properties. Section 3.0 describes the literature search and review procedure. In Section 4.0 a brief review of odor threshold methodology is given. In this review, the relative importance of factors associated with (1) odor panelist selection and training techniques, (2) the apparatus used in preparing, presenting and quantifying the odorant concentration, and (3) the presentation method is discussed. The results of the evaluation of the literature and the summaries of the threshold material are given in Section 5.0. In this presentation, only examples of each table are shown. The entire report is currently in the process of being printed by the American Industrial Hygiene Association.

2.0 ODOR PERCEPTION

A brief review of the sensory properties of odor and some of the attributes of human olfactory response is presented to facilitate understanding of odor threshold values.

2.1 The Dimensions of Odor

The sensory perception of odorants has four major dimensions: detectability, intensity, character, and hedonic tone. Odorant detectability (or threshold) refers to the theoretical minimum concentration of odorant stimulus necessary for detection in some specified percentage of the population, usually the mean. Threshold values are not fixed physiological facts or physical constants but are a statistical point representing the best estimate value from a group of individual responses. As such it may be an interpolated concentration value and not necessarily one that was actually presented. Two types of thresholds are evaluated: the detection threshold and the recognition threshold. The detection threshold is the lowest concentration of odorant that will elicit a sensory response in the olfactory receptors of a specified percentage of a given population. In test procedures it is the minimum concentration of stimulus detected by X percent of the panel members. This is usually defined as the mean, 50% of the population;

186

however, it is sometimes defined as 100% (including the most insensitive) or 10% (the most sensitive). Additionally, Russian literature defines detection thresholds as absolute thresholds, i.e., the lowest concentration that will produce a measurable physiological change, e.g., as an electroencephalogram (EEG) response, in the most sensitive human subject.

The detection threshold is identified by an awareness of the presence of an added substance, but not necessarily recognized as an odor sensation. The recognition threshold is defined as the minimum concentration that is recognized as having a characteristic odor quality by X percent (usually, 50%) of the population. It differs from a detection threshold in the subject's awareness of an odor sensation, and is often defined as the point at which a specific odor character description may be attributed.

Odor _intensity_ refers to the perceived strength of the odor sensation. Intensity increases as a function of concentration. The relationship of perceived strength (intensity) and concentration is expressed as a psychophysical power function as follows:

$$S = K\ I^n$$

where

    S = perceived intensity of sensation
    I = physical intensity of stimulus (odorant concentration)
    n = slope of psychophysical function
    K = y-intercept

The slope and intercept of a function vary with type of odorant. This has important implications for the perception and control of odors. Odors with high slope values dissipate more quickly with dilution and are consequently easier to control. Odors with low slope values are more difficult to eliminate as they are perceivable at lower levels of concentration.

The third dimension of odor is the odorant _character_, in other words, what the substance smells like. A recent American Society for Testing and Materials publication (ASTM data series DS 61) presents character profiles for 180 chemicals using a 146-descriptor scale. The scale includes such terms as fishy, hay, nutty, creosote, turpentine, rancid, sewer, ammonia, etc.

The fourth dimension of odor is _hedonic tone_. Hedonic tone is a category judgement of the relative pleasantness or unpleasantness of the odor. Perception of hedonic tone outside the laboratory is influenced by such factors as subjective experience, frequency of occurrence, odor character, odor intensity, and duration.

2.2 Properties of Olfactory Functioning

Human response to odorant perception follows a number of characteristic patterns associated with sensory functioning. In determining safety margins it is important to remember that a threshold value is a statistical point known as a mean value. Olfactory acuity in the population follows a normal distribution of sensitivity. Most people, assumed to be about 96% of the population, have a "normal" sense of smell. Two percent of the population are predictably hypersensitive and two percent insensitive. The insensitive range includes people who are anosmic (unable to smell) and hyposmic (partial smell loss). The sensitive range includes people who are hyperosmic (very sensitive) and people who are sensitized to a particular odor through repeated exposure. Individual threshold scores can be distributed around the mean value to several orders of magnitude.

Although sensitivity is normally distributed, it is not constant across odorants or individuals. Different odorants each have individual normal curve distributions falling around their mean. Similarly, individuals vary in their sensitivity to different odorants. A person may be hyposmic to one odorant and hyperosmic to another. This variation occurs in specific anosmia, and is often caused by repeated exposure to a particular odor. It is not uncommon among chemists or other workers who have had daily exposure to an odorant over a period of years.

Another sensory property of odor that can cause confusion in organoleptic (i.e., sensory as opposed to analytical) odor identification is that odor character may change with concentration. For example, butyl acetate has a sweet odor at low concentrations, taking on its characteristic banana oil odor at higher intensities. Carbonyl sulfide has a "fireworks" or "burnt" character at concentrations below 1 part per million and "rotten egg" character at higher levels. This, along with individual variability, accounts for discrepancies in odor character reports. The odor character descriptors in this paper are based on a combination of reports in the literature and experience in odor investigation.

3.0 THE LITERATURE SEARCH AND REVIEW

In order to set a manageable scope to this project, AIHA specified two publications as a basis for review and critique of compiled odor threshold values:

> Van Gemert, L.J. and A.H. Nettenbreijer. Compilation of Odor Threshold Values in Air and Water. CIVO-TNO, Netherlands, 1977, and Van Gemert, L.J. Compilation of Odor Threshold Values in Air. Supplement IV. CIVO-TNO, Zeist, Netherlands, 1982.

> Oregon Lung Association. Warning Properties of Industrial Chemicals. Oregon Lung Association, Portland, Oregon.

These sources will henceforth be referred to as "Van Gemert" and "WPIC", respectively.

3.1 Odorant Chemical Selection

The initial task in the literature review process was to identify those chemicals whose odor thresholds are especially important in the field of industrial hygiene. Accordingly, all chemicals listed in TLVs Threshold Limit Values and Biological Exposure Indices for 1986-1987 were compared with Van Gemert and WPIC for odor threshold values. All chemicals with reported odor thresholds and TLVs were included in this project.

The task involved a preliminary review of reference citations from the two threshold compilations. Van Gemert contains 366 references (inclusive of supplement), the majority of which are primary references. The WPIC document contains a reference list noted as "not intended to be a complete source listing". Of the 25 references listed only two are primary sources. These two sources (Little, 1963 and May, 1966) are included in the Van Gemert compilation. The remaining sources are reference volumes such as training manuals, guides, handbooks, and encyclopedias. The set of unpublished University of California-Berkeley (UCB) data referenced in WPIC are theoretical data transformations, and do not represent primary experimental work. Based on the review of the literature references, the WPIC document threshold values and references are not included in this study.

A literature search was conducted to obtain the 366 references listed in the Van Gemert compilations for threshold values in air. Publications not already available were obtained through local university libraries or national and government sources.

3.2 Preliminary Literature Review (Phase I)

As previously described, the work scope for this project established the use of the Van Gemert compendium and its updates as the major reference source to identify literature for review. As a preface to the literature review the reader should keep in mind two considerations. First, the compilation of odor threshold values is truly a formidable task encompassing both an interdisciplinary and world-wide search. Second, although Van Gemert does not attain perfection as a source, it is by far the best compendium of threshold values published to date. Van Gemert collected data from a wide variety of countries (i.e., Australia, Japan, Russia, Czechoslovakia, The Netherlands, Germany, France, and the United States) which had not previously appeared in American compendia, extracted thresholds from a wide variety of disciplines (i.e., Industrial Hygiene, Psychology, Sensory Evaluation, Food Technology, Clinical Medicine, Air Pollution Control, Engineering, Chemistry) and encompassed a century of research. The intent here is not to criticize Van Gemert but to bring some understanding and order to the large number of threshold values cited, so that they may have a practical application.

A two phase review was conducted on the acquired sources. The review consisted of determining if the source was a primary odor experimental paper and then critiquing the experimental odor papers according to the criteria described in Section 4.0.

During the first phase of the review, four categories of sources were identified as being other than primary threshold measurement experiments and, therefore, were not critiqued. Each of these categories was given a code identifier ranging from C1 to C4. Some sources were either omitted or not reviewed. These were designated as D and E codes respectively. The categories and their description follow.

3.2.1 Rejected Value Based on Review (Code C)
3.2.1.1 Secondary Source (Code C1)
Secondary sources are identified as papers in which an odor threshold value, "noticeable odor", or "detectable odor" is mentioned, but either not experimentally determined in the paper or referenced. For example, Davis (1934) notes in passing "the odor of carbon tetrachloride can be detected with as little as 0.5 mg to a liter", but there is no indication if the statement is based on experience or another source.

3.2.1.2 Incidental Reference (Code C2)
The incidental reference is slightly different from the secondary source in that experimental work was conducted, but not with odor thresholds. Usually this involved noting the presence of an odor at a given concentration during analytical work or worker exposure sampling.

For example, Berck (1968) states that "the characteristic odor of phosphine (PH_3) was readily discerned by sniffing 1 or 2 cc of discharge from a syringe containing 0.002 mg of PH_3 per liter."

3.2.1.3 Passive Exposure - Workplace (Code C3)
A number of studies were conducted in the work environment to determine worker exposure levels to a variety of substances at differing concentration levels. The presence of an odor (along with eye irritation and other physiological manifestations) was noted during the air sampling procedure for multiple concentration levels. For example, Hollingsworth et al. (1956) reports "a faint odor was noted at 15 to 30 ppm, and the odor became strong at 30 to 60 ppm".

3.2.1.4 Passive Exposure - Experiment (Code C4)
Certain odor threshold procedures resemble test chamber experiments designed to determine the permissible limits of worker exposure to various substances. However, in many of the latter a prime factor in odor threshold testing (but not exposure testing) is violated. Subjects report noticing an odor at the lowest concentration level, as in a study by Carpenter (1937) where "As a preliminary to the first exposure the concentration in the chamber was built up to 50 ppm and three of the subjects entered the chamber ... all three were able to detect the odor." Concentration levels below threshold are required to determine either a detection or recognition threshold. If experiments did involve concentration levels below odor detection, they are included in the threshold critique.

3.2.2 Omitted Source (Code D)
A number of sources are not included in the review and critique. These are:

- unpublished data (D1),
- personal communication (D2),
- anonymous references (D3),
- omitted reference per Van Gemert 1982 (D4),
- pre-1900 references (D5), and
- references with compounds that do not have TLV's (D6).

3.2.3 Sources Not Reviewed (Code E)
A major difficulty in this study was the acquisition and review of the old and foreign language sources. The majority of sources not acquired (Code E2) are either old, foreign periodicals, or theses. Similarly, 100 percent of the documents not reviewed (Code E1) are foreign language articles. Some of the languages are German, Czechoslovakian, Swedish, French, Japanese, and Russian. The final status of the literature review is presented in Table 3-1.

3.3 The Literature Critique (Phase II)
Primary experimental papers identified by review in Phase I were critiqued in Phase II. One hundred eighty-eight sources of odor thresholds were evaluated. Of these, 19 percent (35) survived Phase II critique and yielded acceptable odor threshold values. These selected sources (A codes) provide the odor threshold data used in developing the odor threshold best estimates discussed in Section 5.0 A description and summary of the codes used is presented in Table 3-1.

Of the 183 odorant chemicals which have TLV's, only 56 percent (102) were found in odor threshold data sources which survived Phase I (A and B codes). Another 27 percent (50) of these had odor threshold values from

TABLE 3-1

FINAL STATUS OF LITERATURE REVIEW AND CRITIQUE

Code and Status	Number of Sources	Percent
CRITIQUED PAPERS (PHASE II)		
A. Accepted Value Based on Critique	35	10
B. Rejected Value Based on Criteria or Other Factors		
0. Criteria	48	
1. Water Threshold	1	
2. Minimum Perceptible Value	74	
3. Water Threshold – Air Conversion	2	
4. Intensity	1	
5. Insufficient Methodology	27	
	153	42
REVIEWED PAPERS (PHASE I)		
C. Rejected Value Based on Review		
1. Secondary Source	39	
2. Incidental Reference	10	
3. Passive Exposure – Workplace	7	
4. Passive Exposure – Experiment	16	
	72	20
D. Omitted Source		
1. Unpublished data	6	
2. Personal communication	9	
3. Anonymous references	2	
4. Omitted in Van Gemert 1982	3	
5. Pre-1900 references	9	
6. References with compounds that do not have TLVs	16	
	45	12
E. Sources Not Reviewed		
1. Located but not reviewed	18	5
2. Not located	43	11
Total	366	100

sources which were not primary experimental work (C and D codes) and so could not be critiqued.

The remaining 18 percent (32) were from 28 E code sources. Consideration of these citations suggests that no further search/review is warranted. Eight of the 28 sources are in Russian and would be expected to be B2 codes. Fourteen of the 28 sources are from pre-1950 sources. Of the critiqued A-D code papers only 7.5 percent of the pre-1950 sources had acceptable A code methodology. Therefore, only one more acceptable source would be expected from continued search for the 13 remaining pre-1950 sources. The remaining six of the 28 sources were published after 1950. Of the other post-1950 sources only 12 percent were categorized as A codes. Therefore, no further A codes would be expected from the six remaining sources. Based on these findings and a consideration of the cost/benefit ratio in searching for these sources, literature search/review was ended.

4.0 ODOR THRESHOLD METHODOLOGY

4.1 The Variability of Threshold Values

Odor threshold measurement has interested researchers for a century. Over this period hundreds of threshold measurements along with nearly as many measurement techniques have been reported in the literature. Threshold compilations such as Van Gemert, Verschueren (1977) and Fazzalari (1978) contain threshold values from sources published in the early 1900's and before. Reported threshold values may vary by a factor of a million or more for one compound. For example, the reported values for n-butyl alcohol range from 0.000188 to 0.000000145 g/l (Amoore and Hautala, 1983).

The fact that threshold values may vary widely as do their methods has often been recognized. Factors affecting threshold measurement (Punter, 1983) include stimuli flow rate, olfactometric systems, age and type of panelist, instruction and threshold procedure, and panelists' experimental experience.

Other important factors contributing to threshold value variability are the purity of chemical compound, the type of threshold (detection or recognition) determined, and the stimulus itself (water vapor or gas vapor). These last two factors make the practice of pooling thresholds questionable at best. Considering the sources of variability it is understandable that published threshold values differ.

4.2 Criteria for Acceptability of Odor Threshold Measurement Techniques

A set of criteria elements considered essential to any modern threshold determination procedure was developed (see list below). The papers with published odor thresholds (listed in Table 5-2) were evaluated in terms of their conformity to these criteria. The criteria are summarized as follows:

a. Panel size of at least six per group
b. Panelist selection based on odor sensitivity
c. Panel calibration
d. Consideration of vapor modality (air or water)
e. Diluent in accord with compound
f. Presentation mode that minimizes additional dilution (ambient) air intake
g. Analytic measurement of odorant concentration

h. Calibration of flow rate and face velocity (for olfactometers)
i. Consideration of threshold type (detection or recognition)
j. Presentation series that reduces olfactory fatigue
k. Repeated trials
l. Forced-choice procedure
m. Concentration step increasing by a factor of two or three.

A detailed description of the criteria follows. They are divided into three sections: the panel, presentation apparatus, and presentation method.

4.2.1 The Panel
a. Panel Size
In order to approximate the distribution of olfactory sensitivity in the population it is preferable to use a large number of subjects or, since this is often impossible, a smaller group selected to represent the general population. Accordingly, to replicate the distribution curve shown in Section 2.2 it is preferable to use a larger panel with fewer trials rather than a small panel (e.g., 2 or 4 experts) with many trials. Additionally, panels of fewer than six judges reduce precision for a reliable mean value. Repeatability for individuals' threshold results are poor ($\pm$ 18 percent); therefore, results should not be based on the repeated observations of less than six panelists. However, there is a point beyond which more panelists become superfluous. One study found that a pooled group of ten with one trial produced the same thresholds as a group of thirty-six with five trial presentations (Punter, 1983).

b. Panel Selection
Prospective panelists should be evaluated for olfactory sensitivity to the chemical compounds in question. This will insure that the panel will not include judges with general or specific anosmia. An early version of an ASTM threshold procedure (ASTM 1391-57 Syringe Dilution Method) recommended testing with only two all-purpose odorants, vanillin and methyl salicylate. Subsequent studies showed that these compounds did not rate panelists properly. Panelists should be evaluated with a compound selected to represent the particular chemicals associated with the industrial process under investigation, rather than with two standard compounds.

Physiological and/or personal factors to be considered when selecting a panel include smoking, drug dependency, pregnancy, sex, and age. Smokers should be excluded from the panel even though the effect of smoking on olfactory acuity is unclear. Studies have reported results ranging from definite to no effect from smoking (see Commetto-Muniz and Cain, 1982 for discussion).

Drug dependency and pregnancy are known to reduce and elevate odor perception, respectively (Amerine, Pangborn, and Roessler, 1965). Anosmia due to drug dependency would be discovered during screening. Similarly, prospective panelists being treated with high levels of medication would be screened and omitted from the panel. Pregnant women should be excluded as a precautionary measure.

As with smoking, results of investigations of changes in olfactory acuity due to age and sex are in disagreement. The common conception has been that women are more sensitive than men and that sensory acuity decreases with age. However, this may be too simplistic an explanation. Recently, the approach has been to separate odor sensitivity from odor identification ability (e.g., see Doty, Shaman, Applebaum, Giberson, Sikorski, and Rosenburg, 1984 for changes with age; Cain, 1982 for differences between sexes).

With unlimited resources a panel with a sensitivity range resembling a normal curve could be selected. However, this is seldom possible in research. Panel screening and calibration helps in effectively controlling for reduced olfactory perception, whether due to age or sex.

c. Panel Calibration
Panel odor sensitivity should be measured over time to monitor gross individual discrepancies and maintain panel consistency. Individual variability is $\pm$ 18 percent while person to person variability can differ by four orders of magnitude. A daily rating of an n-butanol wheel olfactometer would provide a quick and accurate measure of individual and group variability.

4.2.2 Presentation Apparatus
d. Vapor Modality
Vapor modality, i.e., whether the odor measured is in the form of a gas-air mixture or vapor over an aqueous or other solution, is determined by the test purpose and in turn determines the presentation method. The majority of reported thresholds are gas-air measurements. Therefore, some criteria for the apparatus will pertain directly to gas-air instead of vapor over an aqueous solution.

e. Diluent
The diluent, whether liquid or gaseous, should be consistent with the chemical compounds tested and not influence odor perception. For example, diluent air may be filtered through activated carbon, be unfiltered room air, or dry nitrogen. Liquids used as diluents include water, diethyl phthalate, benzyl benzoate and mineral oil. The selected diluent is determined by the test purpose and practical considerations of the compound. Additionally, the relative humidity of the diluent should be controlled at approximately 50 percent.

f. Presentation Mode
Vapors are inhaled from openings of varying size. Some of these allow ambient air to be inspired along with the sample, thereby increasing the dilution factor by an unknown amount. Common delivery systems are 1) nose ports held under the nostrils, 2) vents into which the whole head is inserted, 3) flasks into which the nose is inserted, 4) syringes that impinge vapor into the nose, and 5) whole rooms into which the odorant is injected. In general, an opening that allows insertion of the nose or the whole head is desirable as it reduces the intake of ambient air.

g. Analytic Measurement
The concentration of odorant as it reaches the panelist should be measured accurately. The capability to measure odorant concentration has only occurred recently. Therefore, a major problem with early threshold studies and a drawback of some modern studies is the absence of such analytic devices.

h. System Calibration
Important system calibrations include flow rate and face velocity. The flow rate of odorant through the system should be of sufficient volume to fully stimulate the olfactory receptors. Flow rates on individual olfactometers vary from 0.5 lpm to 9 lpm or more. This

disparity in the flow rate has been found to cause a 4-fold difference in threshold values. Odorant flow rate should be at approximately three liters per minute although researchers differ in their opinion of a "best" flow rate. Flow rate then becomes an important consideration in the critique. The face velocity refers to the rate at which the odor is flowed at the panelist and should be maintained at a flow barely perceptible by the panelist.

4.2.3 Presentation Method
i. Threshold Type
Thresholds may be either of two types, detection or recognition. The detection threshold is defined as the lowest concentration at which a specified percentage of the panel (usually 50%) detects a stimulus as being different from odor-free blanks. The recognition threshold is the lowest odorant concentration at which a specified percentage of the panel (again, usually 50 percent or the median) can ascribe a definite character to the odor. In general, recognition thresholds are approximately three to five times higher than detection thresholds (Hellman and Small, 1974). The type of threshold measured is dependent on the test purpose. For example, detection thresholds are of greater interest in basic research while recognition thresholds are of greater value to the food industry. Recognition and detection thresholds are differentiated in this report.

j. Concentration Presentation
Concentration presentation order is an important factor in the presentation method, as olfactory adaptation occurs rapidly. After three minutes of exposure to an odorant, perceived intensity is reduced about 75 percent (Bartoshuk and Cain, 1977). A common method to control for this is to present concentrations in ascending order (from weaker to stronger concentrations, or greater to less dilution) or to allow for long periods between presentations. Descending and random presentation series do not control for adaptation unless specific steps are taken to eliminate it. Recognizing the need to control for adaptation effects in random or descending patterns of presentation, researchers apply various methods such as presenting one concentration per day (Dixon and Ikels, 1977) or using different subjects at each concentration step (Gundlach and Kenway, 1939).

k. Number of Trials
Individual test-retest reliability for threshold values is low (Punter, 1983). Determinations should be repeated for reliability. Additionally, computing the mean across panelists' scores will reduce individual variability.

l. Forced-Choice
A forced-choice procedure minimizes anticipation effects for thresholds by eliminating false positive responses. Panelists choose between the stimuli and one or two blanks. This procedure, from Signal Detection Theory (Green and Swets, 1966), reduces the probability of a subject criterion bias, false positives and negatives.

Forced-choice is a comparatively new procedure. Therefore, it was not stringently applied as a criteria. An earlier method, presenting a stimuli and blank as a paired comparison, was also included in this category. Both methods reduce anticipation effects.

m. Concentration Steps

In determining odor threshold values, the odorant should be presented successively at concentration intervals no more than three times the preceding one. It has generally been found that concentrations higher or lower by 25 to 33 percent are perceived as different. Indeed, olfactory sensitivity may be more acute. In a carefully controlled study by Cain (1977) the average perceptible difference between concentrations was 11 percent, ranging from 5 percent to 16 percent for different compounds. Intervals larger than 30 percent (a 3-fold step size) fail to give proper indication of the discriminatory capacity of the nose.

4.3 Critique of Odor Threshold Measurement Techniques

Table 5-2 presents an annotated list of the primary experimental odor threshold references which survived Phase I review. If the thirteen element criteria, described in Section 4.2 above, were applied to the published data, none of the odor threshold values could be accepted. Therefore, these papers were reviewed for their overall adherence to experimental procedures which address the response characteristics of the human olfactory system.

Three criteria considered vital to olfactory experimentation were particularly weighted to determine the procedure's acceptability:

- a concentration presentation procedure designed to eliminate olfactory fatigue;

- a maximum of a 3-fold concentration interval separating stimuli in the series to reflect human discrimination ability; and

- panel size of at least five judges to represent the range of olfactory sensitivity.

Papers that did not account for these criteria in their experimental design were not accepted. Individual papers presented such a variety of experimental design solutions that each is given a footnote to clarify the critique. The notes should be consulted to resolve apparent discrepancies in the review. For example, a random presentation series was accepted when concentration levels were evaluated by different subjects (Gudlach and Kenway, 1939) but not when presented sequentially (Stone, 1962).

Acceptability of the thresholds is indicated in the last column of Table 5-2. The value for odor thresholds presented in Table 5-1 is the geometric mean of the values accepted in accordance with the criteria.

5.0 ODOR THRESHOLD VALUES

During this study, 188 primary sources of odor threshold measurement data were evaluated. Of these, 153 publications were excluded as unacceptable for the reasons described in Sections 3.0 and 4.0. As mentioned previously, the matching of the chemicals with published TLV's and the odor thresholds listed in Van Gemert yielded 183 compounds. Only 102 of these compounds had odor threshold values that met the evaluation criteria. These acceptable values are from 35 sources.

5.1 Critiqued Odor Threshold Values

The geometric mean value or recommended best estimate for the odor threshold for each of the compounds is given in Table 5-1, pp. 30-58.

The table has the following format:

- Compound name and synonyms
- Formula
- Molecular weight
- Geometric mean odor threshold value (recommended best estimate/based on Van Gemert air odor threshold references)
- Type of threshold (detection or recognition)
- Odor character
- Range of acceptable values
- Range of all published values.

Geometric means were computed for the mean odor threshold values. This is a common practice in sensory evaluation, as it accounts for the wide range of response over several orders of magnitude. The means were rounded off to two significant digits. Where values were given as a range, the geometric mean of the two points was taken for the threshold.

In some cases, the mean value for detection is higher than the mean value for recognition. This is a result of pooling of several data sets for the geometric mean.

Odor character descriptors in Table 5-1 are based on reports in the literature and experience in odor investigation. The intensity level at which the character is determined is seldom given in the sources reviewed. Since odor character can change with intensity, it should be remembered that the character reported may differ from source to source. The purpose here is to include an observation on the odorant character to accompany the threshold value.

5.2 Critique of Published Odor Threshold Measurements

An annotated list of the critiqued references for the threshold values is presented in Table 5-2. Threshold methodologies are evaluated according to each of the thirteen criteria discussed in Section 4.0.

Acceptability of the thresholds for inclusion in the recommended values is presented in the last column of the table and is based on the critique requirements discussed in Section 4.3.

The table is organized in the following format:

1. Name of first author and date
2. Panel size
3. Panel selection criteria, i.e., trained, screened, etc.
4. Panel calibration
5. Vapor modality, usually air; however, in a few cases water vapor or water
6. Diluent; unless specified otherwise in the paper, it was assumed to be air
7. Presentation mode, type of instrument at interface
8. Analytic measure; gc, gc/ms or other analytic techniques
9. Flowrate
10. Threshold type, R = Recognition, D = Detection, MP = Minimum Perceptible, I = Intensity
11. Concentration series, A = Ascending, D = Descending, R = Random, V = Variable, U-D = Up-Down Series
12. Trials, greater than one trial
13. Forced-choice
14. Concentration Interval, less than or equal to a 3-fold step size

15. Notes – Further explanatory information on critique acceptance
     or rejection follows the table
16. Code – Source Code
17. Critique Acceptance

Each study was reviewed for compliance with the criteria. If sources
that do not seem to fit the critique are accepted, the reasons for the
acceptance are discussed briefly in notes at the end of the table. For
example, Dixon and Ikels (1977) used an Up–Down presentation series, but
avoided adaptation effects by presenting only one concentration per day.

The Russian papers were reviewed for threshold methodology. Only 2%
of those reviewed present a value comparable to a type of threshold
calculated in the West. Instead, the minimum perceptible concentration
value, i.e, the lowest concentration detected by the most sensitive
panelist, was presented. This value is not a group mean nor a
"sensitive" population mean. Those papers referring to the "standard"
method (Ryazanov, Bushtueva, and Novikov, 1957) are listed as Code B2,
and incomplete methodologies are supplemented from the Ryazanov method.
Since one or two other methods (e.g., Khachaturyan, 1968) are employed,
each article was critiqued separately.

Conversely, the well known Arthur D. Little chamber study (Leonardos,
Kendall, and Barnard, 1969) was not accepted because, among other
considerations, it determined 100 percent recognition thresholds. This
is the point at which all subjects recognized the odor, even the most
insensitive, rather than the mean population level of 50 percent
recognition as defined in the other studies.

As is clear in Table 5-2, many papers were not included. It must be
emphasized that critique rejection is not a reflection on the overall
paper or researcher. In many cases, the presented threshold values were
not the paper's true focus and are included only as supportive evidence.
For example, in a review of five papers by the late Andrew Dravnieks, a
pioneer in odor measurement techniques, only two of them were accepted.
The rejected papers were written for gas chromatography research or
theoretical chemistry. Although they presented threshold values, the
methodology was not given and the values in the paper were presented only
as examples. Some investigators listed in Table 5-2 do not appear in
Table 5-3. Their studies did not present chemicals from the TLV list.

5.3 Odor Threshold Values
All published odor threshold values for the 183 chemicals with TLV's
are presented in Table 5-3. The table is in the following format:

1. Chemical name and synonyms

2. All threshold values from Van Gemert in both mg/m^3 and ppm

3. Odor threshold values reported as either "detection" (d) or
   "recognition" (r)

4. A code letter denoting the reason for acceptance or rejection of
   the published value. The code explanation is:

   A. Accepted value based on critique
   B. Rejected value based on critique
   C. Rejected source based on review

D. Omitted source

E. Source not reviewed

For the table, the ppm values of the compounds were converted from the mg/m^3 concentrations as given in Van Gemert. Conversion of units from mg/m^3 to ppm was based on the molecular weight of the compound and the known volume of a perfect gas or vapor at standard temperature and pressure (STP).

The sequence within the table is as follows. For each compound first are presented the C-E code sources, i.e., those papers reviewed in the first phase that were determined not to be primary sources. Second are the A and B code sources, those that were determined to be primary references and were critiqued according to the criteria. Finally, the A code sources only, i.e. those threshold values which met the modified criteria, are underlined. The progressive fine tuning of threshold value acceptance is dramatically apparent.

6.0 ACKNOWLEDGEMENT

This project was supported by funds from American Industrial Hygiene Association. Janet Hooper, now with Richardson-Vicks, was the principal investigator of the project. Richard Duffee of Odor Science & Engineering, Harry Guy of Syntex, and Carol Dupraz provided valuable assistance during the project.

SAMPLE
TABLE 5-1

Critiqued Odor Threshold Values

Compound Name Synonyms	Formula	M.W.	TLV (ppm)	Geometric[a] Mean Air Odor Threshold (ppm)	Type of Threshold[f]	Odor Character	Range of Acceptable Values[e] (ppm)	Range of All Referenced Values (ppm)
Acetaldehyde Ethanal	C_3H_4O	44.05	TWA=100 STEL=150	0.067	d	pungent/ fruity	*	0.0028-1,000
Acetic Acid Ethanoic Acid	$C_2H_4O_2$	60.05	TWA=10 STEL=15	0.15	d	pungent	*	0.010-31
Acetic Anhydride	$C_4H_6O_3$	102.09	C=5[c]	<0.14 0.36	d r	sour acid	* *	0.12-036
Acetone 2-Propanone	C_3H_6O	58.08	TWA=750 STEL=1000	160[d] 130	d r	sweet/ fruity	20-653 33-699	0.40-800
Acetonitrile Methyl Cyanide; Ethanenitrile	C_2H_3N	41.05	TWA=40 STEL=60	1,160	d	etherish	*	<40-1,161
Acetylene Ethyne	C_2H_2	26.02	Appendix E[b]	none		gassy	none	226-2,584
Acetylene Tetrachloride See: 1,1,2,2-Tetrachloroethane								
Acroleaic Acid See: Acrylic Acid								
Acrolein 2-Propenal	C_3H_4O	56.06	TWA=0.1 STEL=0.3	1.8	d	pungent	*	0.022-1.8
Acrylic Acid Glacial Acrylic Acid; 2-Propenoic Acid; Propene Acid; Vinyl Formic Acid;	$C_3H_4O_2$	72.06	TWA=10	0.092 1.0	d r	rancid/plastic/ sweet	* *	0.092-1.0

a Based on critiqued Gemert references.
b Appendix E = simple asphyxiant per TLVs
c C = ceiling limit
d D:R result of values from different papers.

e * = one value
f d = detection threshold
  r = recognition threshold

SAMPLE
TABLE 5-2

Critique of Published Odor Threshold Measurements

Source	Year	Panel Size	Panel Select. Criteria	Panel Cali-bration	Vapor Mod-ality	Presentation Apparatus Diluant	Presen-tation Mode	Analytic Measure	Flowrate	Threshold Type	Presentation Method Conc. Series	Trials	Forced Choice	Conc. Inter-nal	Note	Code	Critique Accept.
Adams	1968	114-789	No	No	Air	Pure Air	Odor Hood	Yes	2-5 lpm	D	A/D/R	Yes	No	Yes	1	A	Yes
Akhemedov	1968	4	Yes	No	Air	Carbon Filtered	Cylinder	Yes	15 lpm	MP	ng	Yes	Yes	Yes	2	B2	No
Alibaev	1970	25	Yes	No	Air	Carbon Filtered	Cylinder	Yes	15 lpm	MP	ng	Yes	Yes	Yes	2	B2	No
Allison	1919	ng	ng	ng	Air	Pure Air	Glass Funnel	No	28 lpm	D	ng	ng	No	ng	32	B5	No
Andur	1953	14	ng	ng	Air	Air	Face Mask	Yes	ng	R	ng	ng	No	Yes	32	B5	No
Amoore	1977	>10	ng	ng	Water	Water or Buffered Water	Flask	No	Static	D	D	ng	Yes	Yes	3,35	B	No
Amoore	1978	>10	ng	ng	Water	Water or Buffered Water	Flask	No	Static	D	D	ng	Yes	Yes	3	B	No
Andreescheva	1964	29	Yes	Yes	Air	Carbon Filtered	Cylinder	Yes	15 lpm	MP	ng	Yes	Yes	Yes	2	B2	No
Andreescheva	1968	26	Yes	No	Air	Carbon Filtered	Cylinder	Yes	15 lpm	MP	ng	Yes	Yes	Yes	2	B2	No
Appell	1969	ng	ng	ng	Water	Water	Bottle	ng	Static	MP	ng	ng	ng	Yes	4	B	No
Babin	1965	ng	ng	ng	Air	ng	ng	ng	ng	ng	ng	ng	ng	ng	5	B5	No
Baikov	1963	-	-	-	Air	ng	-	-	-	MP	-	-	-	-	29	B2	No
Baikov	1973	28	-	-	Air	-	-	-	-	MP	-	-	-	-	29	B2	No
Basmadzhieva	1968	13	ng	ng	Air	ng	ng	ng	0.2-0.6 lpm	MP	ng	Yes	Yes	Yes	2	B2	No
Belkov	1969	ng	ng	ng	Air	ng	ng	ng	ng	ng	ng	ng	ng·	ng	5	B5	No
Berzins	1967	18	ng	ng	Air	ng	ng	ng	ng	ng	ng	ng	ng	ng	5	B5	No
Bezpalkova	1967	23	-	-	Air	-	-	-	-	MP	-	-	-	Yes	25	B2	No
Blinova	1965	9-10	ng	ng	Air	ng	Gas Mask	ng	ng	MP	ng	Yes	ng	ng	31	B2	No

ng = not given   na = not applicable

201

SAMPLE
TABLE 5-3

Reported Odor Thresholds From All Code Sources

Compound Name and Synonyms	Source	Code	Type of Threshold	Odor Thresholds mg/m³	ppm
Acetaldehyde Ethanal	**Phase I Rejected/Unreviewed Sources**				
	Zwaardemaker 1914	E2	d	0.7	0.39
	Backman 1917	E1	r	0.062-0.075	0.034-0.042
	Takhirov 1974	E1		0.49	0.27
	Anon. 1980	D3	d	0.0027	0.0015
	Anon. 1980	D3	r	0.027	0.015
	Phase II Critiqued Sources				
	Pliska 1965	B		1.800	1.000
	Katz 1930	A	d	0.12	0.062
	Golmekler 1967	B2	d	0.012	0.0067
	Leonardos 1969	B	r	0.38	0.21
	Teranishi 1974	B3		0.041	0.023
	Hartung 1971	B5		0.005	0.0028
Acetic Acid Ethanoic Acid	**Phase I Rejected/Unreviewed Sources**				
	Passy 1893b, 1893c	D5	d	5-10	2.0-4.1
	Grijns 1906	E2		49-76	20-31
	Backman 1917	E1	r	4.8-5.0	2.0-20
	Grijns 1919	E2		2	0.82
	Mitsumoto 1926	C1	r	0.074-0.57	0.030-0.23
	Henning 1927	E1	d	3.6	1.5
	Morimura 1934	E2	r	1.82-1.91	0.74-0.78
	Jung 1936	E2	d	0.025	0.010
	Jung 1936	E2	r	0.05	0.020
	Stone 1963c	E2	d	3.9	1.6
	Endo 1967	E2	r	6.5	2.6
	Punter 1980	D1	d	0.09	0.037
	Phase II Critiqued Sources				
	Hesse 1926	B5	r	0.6	0.24
	Stone 1965	B	d	4.2	1.7
	Takhirov 1969	B2		0.60	0.24
	Leonardos 1969	B	r	2.5	1.0
	Homans 1928	A	d	0.37	0.15
Acetic Anhydride	**Phase I Rejected/Unreviewed Sources** No C-E Codes			—	—
	Phase II Critiqued Sources				
	Takhirov 1969	B2		0.49	0.12
	Hellman 1973a, 1974	A	d	≤0.6	≤0.14
	Hellman 1973a, 1974	A	r	1.5	0.36

ODOR PANEL SELECTION, TRAINING AND UTILIZATION PROCEDURES –
A KEY COMPONENT IN ODOR CONTROL RESEARCH

Martha A. O'Brien
Odor Science & Engineering, Inc.
57 Fishfry Street
Hartford, Connecticut 06120

Significant advancements have been made in the field of olfactometry in
recent years. In an effort to standardize odor measurement techniques
much attention has been devoted the variables associated with sampling
procedures and instrumentation specifications. We have found less
attention paid to the important, and often overlooked, variables that must
be considered in odor panel selection. The evaluation of odors by an odor
panel is a key component to any odor investigation. The objectives of
odor investigations vary and therefore panel selection and training
methods will vary. In odor pollution control work, in addition to being
able to accurately quantify the odor level (D/T) of emission samples, it
is necessary to obtain quality data regarding panelists judgement of odor
intensity levels (based on the n-butanol scale) at various
concentrations. Together these data supply the basis for determining the
degree of control required to obtain a desired level of odor reduction.
We have found that proper panel selection and training procedures can have
the greatest effect on the resulting odor data upon which important odor
control decisions are based. This paper presents and discusses data which
has lead to the development of a proposed methodology for the proper
selection, training and utilization of odor panelists in odor control
research.

INTRODUCTION

Resolving an odor problem in which community complaints arise against an alleged source(s) involves a systematic approach involving both problem definition and control. Once the relationship between source emissions and ambient levels downwind are established, the question then becomes one of determining what constitutes a nuisance and how much control is required to eliminate the nuisance.

A key element of all odor investigations involves the quantification of source emissions. Samples are collected and brought to a panel of human observers for the evaluation of various sensory parameters. Studies have shown that aging (storage) of samples prior to presentation of the sample to an odor panel can result in odor unit values significantly lower than the true value.[1] The experience of Odor Science & Engineering, Inc. (OS&E) has shown that evaluation of samples should take place within four hours to prevent sample deterioration and thus erroneously low measurements. Avoiding excessive delays between sampling and analysis necessitates performing the odor evaluations in close proximity to the sources being sampled. Thus, an odor panel must be recruited locally.

ODOR PANEL RECRUITMENT

Perhaps the first consideration is what population to recruit the odor panelists from. In many cases, the panel consists of plant personnel such as administrative staff who are not directly exposed to plant process odors. If enough members of this group are dedicated and available for the duration of the on-site testing periods, the logistics in panel arrangements are simplified. This population is especially desirable if the odor evaluations are taking place at a location on plant grounds. As desired sampling times are often dependent upon unpredictable process operations and/or meteorological conditions, the plant-personnel panel can be assembled daily on an as needed basis rather than pre-schedulded evaluation periods which may result in inefficient use of panelists' time while they wait for samples to arrive or an excessive delay between sampling and analysis.

Odor panelists can also be recruited from local temporary employment agencies and/or market research firms. These sources can usually provide large numbers of people to serve as potential odor panelists from their pool of temporary staff upon relatively short notice. Panel evaluations using this group are most often held at a location away from plant property. Without knowing which particular local facility the samples were obtained from (at least at the beginning of the study), the members of the odor panel represent an unbiased population of local residents.

Panels can also be recruited from members of the affected community. As this is ultimately the population that must be satisfied, it is often desirable to use them as panelists. Utilizing the people who are actually complaining about the odor situation can also have a great public relations value. The people generally react positively to the fact that someone is paying attention to their complaints and that they are somehow involved in a study whose effort it is to better their situation. By using the actual community complainants as panelists the sensitivity of the affected community is then indicated and the objectionable level of odors in that community can be determined.

A second consideration is how many panelists to use. The number of panelists used in evaluations can have significant effects on the resulting measurement values especially if they are not selected on the basis of their olfactory acuity for the odorous material to be measured. For most field studies a panel of 8-10 people is desirable so that some simple statistical tests can be applied to the results. In general, a panel of eight to ten panelists selected in accordance with their ability to detect the odorants of interest, but not for specific homogeneity, will yield results within 20% of the mean of repeated measurements of the same sample by the same panel.[2,3] Smaller panels are sometimes used by necessity and are often adequate for source screening evaluations. Although the results may lack statistical validity, the use of small panels of less than 8 people, the members of which are properly trained, screened, and consistent in their evaluations, is useful in accomplishing many of a study's objectives.

PANEL SCREENING/TRAINING PROCEDURES

Once the pool of potential odor panelists is obtained and the desired panel size determined, the selection criteria of the individual panelists are vital to the quality of data obtained. All candidate panelists must first be screened for general olfactory acuity, for their ability to detect the odor qualities of interest and for their ability to match odor intensities. Based on research done by OS&E's staff for the U.S. Environmental Protection Agency we have developed a standard procedure for odor panel screening and training. The highlights of this procedure are outlined below:

1. Candidates are first tested for their ability to detect n-butanol vapors above an aqueous solution of n-butanol in odorless bottles. Initially, a sub-threshold concentration of n-butanol in water is presented to the potential panelists along with two bottles containing only water. The candidate is required to sniff the headspace of each bottle and pick the bottle containing the butanol solution. A series of similar triangular presentations are made in an ascending series with the aqueous butanol concentration doubling at each presentation. The n-butanol vapor concentrations in the headspace and their relative ratings on the butanol intensity scale are shown on Table 1.

TABLE 1
CONCENTRATIONS OF N-BUTANOL USED IN
DETERMINATION OF ODOR INTENSITY

N-BUTANOL INTENSITY SCALE	N-BUTANOL CONCENTRATION	
	IN WATER PPM (WT)	IN AIR PPM (VOL)
1	150	15
2	300	30
3	600	60
4	1,200	120
5	2,500	250
6	5,000	500
7	10,000	1,000
8	20,000	2,000

A forced choice procedure based on Signal Detection Theory is used, i.e., the candidate must choose one of the bottles even if unable to detect an odor in any of the bottles. The concentration at which the candidate first correctly selects the odor-containing bottle and correctly selects the odorous stimulus at each higher concentration is that individual's odor threshold for butanol. This procedure is essentially that described in ASTM Standard Procedure E544 (Standard Recommended Practices for Referencing Suprathreshold Odor Intensity).

Each candidate must also compare the odor intensity of an "unknown" aqueous solution of butanol, actually a level 4 odor sample, to the eight (8) bottles of aqueous butanol solutions which constitute the butanol scale. To be selected as a panelist the candidate must match the intensity of the sample within one scale unit, i.e. either bottle 3, 4, or 5 on the butanol scale.

2. The second screening procedure used by OS&E involves familiarization of the potential panelists with the dynamic olfactometer that will be used for the evaluation of the actual field samples. (Research has shown that panels unfamiliar with the olfactometers can produce results low by as much as 50%.[1]). A Tedlar bag is inflated with air containing 200 ppm of n-butanol vapor. This concentration is approximately 100 times the odor threshold of n-butanol. This sample is then analyzed by the panel using forced-choice triangular dynamic dilution olfactometry. Each individual's detection and odor recognition thresholds are determined and compared to the panel mean.

During the actual evaluation of field samples the panel is calibrated each day with a similar butanol sample of 200 ppm in the same manner. In this way the performance of the entire panel and of individual panelists can be checked on a daily basis.

3. The final screening test of the potential panelists involves having them evaluate samples of odorous emissions collected in Tedlar bags from the sources to be quantified during the study. This is to test their ability to detect the odor character(s) of interest to the study. We have found that there are no universal odorants that can be used to classify people in terms of their olfactory acuity. Each of us has different odor sensitivities to different odor characters. Thus, it is mandatory to test potential panelists for their ability to smell the odors of interest in any given application. Several plant sources are used for screening samples to document individuals sensitivities to these odor characters.

SELECTION CRITERIA

The selection criteria for the individual panelists again depend on the objectives of the particular study. In many cases, panels are chosen to represent the responses of a larger "normal" population of people. In these situations, both the most sensitive (hypersensitive) and least sensitive (hyposensitive) individuals would be screened from the group. In other situations, however, the most sensitive individuals are chosen to represent the responses of perhaps a less tolerant population. In either case, all selected candidates must be able to detect butanol, at least at an intensity equivalent to 3 on the butanol scale, can match butanol intensities within one scale unit, and are able to detect the odor characters to be evaluated during the study.

USE OF ODOR PANEL DATA

Using an olfactometer, a sample is diluted with various volumes of odor-free air and presented at known dilutions to each odor panelist for sensory evaluation. This procedure allows for quantification of the odor level of the sample in terms of dilution-to-threshold ratio (D/T). Additional data supplied by each panelist include their perception of the intensity (based on the n-butanol intensity scale) of the odors when presented at various dilution levels. These data are crucial in order to establish the unique relationship between D/T ratio and intensity for each emission source.

The relation between odor intensity and D/T ratio is known as Steven's Psychophysical Law. It states that the intensity of a sensory response is proportional to the magnitude of the stimulus raised to some power. For odor, this relation is expressed as:

$$I = aC^b$$

Where   I = odor intensity on butanol scale
               a = a constant
               C = the odor concentration (D/T)
               b = an exponent whose value usually varies between
                    0.2 and 0.8

The Steven's Law relation, also known as a dose-response function, should
be determined from the panel data for every odor measurement related to
ambient odor impact so that D/T ratios can be converted to intensity
values, which in turn can be converted to probable objectionability
levels. It is also critical in odor control work as it determines the
rate at which odor intensity decreases as the odor concentration is
decreased either by atmospheric dispersion or an odor control device. The
function is therefore used in predicting the reduction in odor
concentration which is required to bring odor intensity to a desired
level, judged not objectionable.

HOW RESULTS DIFFER BASED ON THE SELECTION OF ODOR PANELISTS

The data derived from the odor panel are what is used to determine what
constitutes an objectionable odor level. This fact stresses the
importance of proper panel selection.

Previous studies have shown that odor panels comprised of members who have
not undergone any training or screening exercises can produce results low
by as much as a factor of 2.[1] Table 2 presents some recent OS&E data
from large odor panel screening sessions which also illustrate the
importance of proper training and screening of odor panelists. The D/T
values of the screening samples are listed comparing those calculated with
all (pre-screened) panelists responses to those who were actually selected
as final panelists based on all screening criteria. The standard butanol
screening samples were 24-31% lower with the unscreened panel. The
variability between panel groups with actual source samples is much
greater ranging from very little difference (4%) up to 73% lower values
with an unscreened panel. In most cases the screening process included
elimination of those members found to be very insensitive to the specific
plant odors which ultimately lead to a more sensitive panel.

TABLE 2
VARIABILITY OF RESULTS BASED ON ODOR PANEL SELECTION
UNSCREENED PANEL MEMBERS VS. SCREENED PANEL MEMBERS

| ODOR SOURCE | ODOR LEVEL (D/T) | | % DIFFERENCE BETWEEN PANEL RESULTS |
	AS MEASURED BY ALL POTENTIAL PANELISTS	AS MEASURED BY SELECTED ODOR PANELISTS	
BUTANOL SCREENING SAMPLE	95	125	24
BUTANOL SCREENING SAMPLE	91	132	31
WWTP PRIMARY EFFLUENT CHANNEL	21	78	73
SLUDGE COMPOSTING PILE	103	125	18
AUTOMOTIVE PAINT OVEN STACK	191	184	4

A recent OS&E project for a sludge composting operation in Pennsylvania
afforded us the opportunity to have a large odor panel of trained and screened
individuals comprised of two sub-groups. Several of the members were
residents from the neighborhood adjacent to the plant. These individuals were
very familiar with the plant odors and frequently complained of objectionable
levels of plant odors at their homes. The other group (non-residents) was
comprised of City staff, members of the regulatory agency, and plant
personnel.

All members of the panel were qualified in terms of acceptable sensitivities
to both the butanol and selected plant emission screening samples. The
comparison of the two groups' D/T values is shown in Table 3. As illustrated,
the variability among the sub-groups in the screened panel is quite low.
Additionally, to be selected all panelists were adept in matching butanol
intensities.

TABLE 3
VARIABILITY AMONG SUB-GROUPS WITHIN SELECTED PANEL
RESIDENT VS. NON-RESIDENT

ODOR SOURCE	ODOR LEVEL (D/T)		% DIFFERENCE BETWEEN GROUPS
	RESIDENT	NON-RESIDENT	
BUTANOL SCREENING SAMPLE	80	77	4
COMPOST EMISSIONS	256	233	9

The difference that developed between the sub-groups was their intensity ratings during the evaluation of plant emission samples. Although both groups matched the intensity of butanol samples comparatively, a general trend was noted that on most plant emission samples, the resident members of the panel tended to rate the intensity of the perceived odors at least one scale unit higher on the butanol intensity scale in comparison to the non-resident panel members. Examples of this are illustrated in Table 4.

TABLE 4
COMPARISON OF INTENSITY RATINGS
RESIDENT VS. NON-RESIDENT

ODOR SOURCE		AVERAGE INTENSITY RATING (1-8)	
		RESIDENT PANEL MEMBERS	NON-RESIDENT PANEL MEMBERS
BUTANOL SAMPLE	10 X Dilution	2.5	2.6
COMPOST SAMPLE A	50 X Dilution	3.5	2.8
	25 X Dilution	4.6	3.4
COMPOST SAMPLE B	25 X Dilution	4.3	3.5
	10 X Dilution	5.7	4.6
COMPOST SAMPLE C	25 X Dilution	3.9	2.9
	10 X Dilution	5.5	4.0

This finding was of great importance in determining the degree of control required at the plant sources to preclude objectionable odor levels in the community downwind. In this case we found the residents were not sensitized physiologically in that they were able to detect the odor at lower levels in comparison to any other population, yet they were sensitized psychologically in the sense that once they detected the particular odor character that was the problem at their homes, it quickly reached what they considered to be an objectionable level. Their data showed that they were basically intolerant to any levels significantly above their odor recognition thresholds.

The importance of proper panel selection is again critical when dealing with a truly sensitized population. The term "sensitized" refers to the heightened ability to pick out a certain odor from the normal background, especially one that the individual finds objectionable. If the population in the community has become sensitized to a particular odor because of frequent exposure to relatively intense odors the job of obtaining unobjectionalbe (non-complaint) odor levels is exceedingly difficult. If this situation has developed, the sensitized community members must be used to establish the "acceptable" odor level. The dose response functions should be established for every emission source by using this odor panel. The level of odor determined to be at an objectionable level (intensity of about 3 on the butanol scale) by these people then becomes the maximum level that can occur in the community resulting from the emission sources sampled.

Figure 1 illustrates the same odor source sample evaluated by both a "sensitized" population of community residents as panel members vs. "non-sensitized" panel members from the general population. If the odor level which corresponds to a butanol intensity level of 3 is chosen as the maximum tolerable ambient odor level, we find the targets are quite different for these two populations. What may be deemed acceptable for the "normal" population may be far from an acceptable level for a sensitized population. Control targets and thus alternatives for control are different in many cases.

212

FIGURE 1.

DOSE RESPONSE FUNCTIONS

SENSITIZED VS. NON-SENSITIZED POPULATIONS

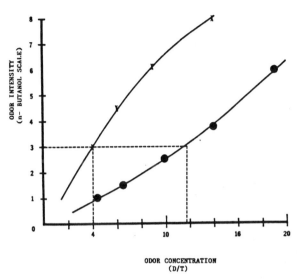

ODOR CONCENTRATION
(D/T)

X = DATA DERIVED FROM "SENSITIZED" POPULATION

● = DATA DERIVED FROM "NON-SENSITIZED" POPULATION

SUMMARY

In conclusion, proper panel selection and correct interpretation of the data
can have a profound effect on the resulting odor data upon which important
odor control decisions are based.   Facilities can spend large sums of money in
an effort to eliminate a community odor problem, yet if the control targets
chosen were not applicable to the population who define the community odor
problem, then the odor problem continues.   Once consideration is paid to which
population you desire as panelists, proper screening and training of these
individuals is critical to obtain quality data.   The successful alleviation of
an odor problem then relies upon proper interpretation of the data to ensure
that the correct degree of control required is determined and appropriate
control alternatives are recommended.

REFERENCES

1.) J.P. Wahl, R.A. Duffee, W.A. Marrone; "Evaluation of Odor Measurement Techniques", Summary Report to the U.S. EPA, EPA Contract 68-02-0662, Program Element 1A1010, 1974, Report No. 1.

2.) R.A. Duffee, J.P. Wahl, W. Marrone, and J.S. Nader; "Defining and measuring objectionable odors". In Pollution Engineering Techniques - An Overview Introduction the the Problem. Ann Arbor Science Publishers, 1974.

3.) A. Dravnieks, D.M. Benforado, W.S. Cain, R.A. Duffee, and R.D. Flesh. "Measurement methods". In Odors From Stationary and Mobile Sources. National Academy of Sciences, Washington, D.C. 1979.

A MATCHING-STANDARDS METHOD FOR
EVALUATING ODOR ANNOYANCE

John E. Amoore and Robert S. O'Neill
Olfacto-Labs
3065 Atlas Road, Suite 109A
Richmond, California 94806

At the 1988 Annual Meeting we proposed a "Decismel Scale" for comparative measure of odor levels and sensitivities. This could serve as a basis for evaluation and coordination of different systems for measuring odor intensity, in much the same way that the decibel scale is used for evaluating sound levels. Using the nose for measuring an odor level, creates the necessity to correlate perception by the nose with an appropriate standard.

Experiments have been conducted with squeeze-bottles of odor samples, which are used on-site for comparison with ambient odors. A selection of a standard odor series, with several calibrated odor intensity levels, serves as an aid for evaluation of odors involved in complaints. If one (or more) of the standards is considered representative of the ambient odor, then an intensity scale is provided, consisting of squeeze-bottles representing 7 odor levels. The evaluator sniffs them, in ascending order of strength, and selects the closest match with the overall intensity of the ambient odor.

Evaluations are done in real time; or may be used to evoke a remembered odor type and level from previous experience. Professionals are assisted with their ability to identify the major odor types and intensity levels. Various lines of reasoning are presented that support a recommendation that an unpleasant odor level of 5 times average detection threshold as a level of odor annoyance for regulatory purposes.

Introduction

Effective ambient odor control regulations include some specified procedure for odor evaluation. Since odorous compounds are diverse and numerous, it is impractical to define odor limits in terms of ambient concentrations for named pollutants (except in rare instances such as geothermal H_2S). An odor characterization including both an hedonic and trigeminal calculation will greatly assist in establishing meaningful odor quality regulations. Most States and air pollution control Districts having quantitative odor standards use a dilution-to-threshold procedure. Often this is cumbersome, and requires many special procedures to cope with the acknowledged wide variability of odor sensitivity among the population.

Employing the nose as an odor intensity comparator creates the opportunity to evaluate the ambient odor with a given standard odor, presented in a reproducible manner at a calibrated concentration. Persons with widely differing sensitivities (when able to smell the odors) should reach concordant results when matching an unknown to a given standard. In other words, at the point of match, the standard and the unknown should be perceived with equal intensity; even though that intensity may be low for some observers, intermediate for others, and high for a few. Calibrating a specific set of odor standards avoids the problem of having to constantly calibrate a great many different observers!

The Weber's Law ratio for the just noticeable difference between two supra-threshold odorant concentrations has been reported[1] to be about 30% or even less. This difference in intensity should be adequate for guiding the application of odor regulations. In practice, it is preferable to provide a reproducible series of standard odor concentrations, and ask the subject to say which one best matches the ambient odor in respect of intensity. If the observers find it easier to match odors that are similar in quality, then a standard odor series might be constructed with an appropriate odorous compound, or mixture, so as to obtain a closer qualitative match between the standard and the ambient odor.

A question arises as to what kind of numerical scale may be used for these odor intensity standards. Our Firm has been experimenting with the "decismel scale" of odor thresholds and odor level comparisons, proposed at the 1988 Annual Meeting[2]. This scale is closely modeled on, and follows a similar mathematical form to, the well-known decibel scale of hearing sensitivity and sound levels:

$$\text{Odor level in decismels} = 20 \times \log_{10} \left(\frac{\text{vapor concentration}}{\text{reference concentration}} \right)$$

The reference concentration (0 decismels, or 0 dS) is set at the mean odor detection threshold for healthy 20-year-olds (by analogy with the clinical definition of the "zero hearing threshold level"). Increasing odorant concentrations, and hence odor intensities, are represented by increasing decismel values (Table I). Correlations between the decismel scale and several other conventional scales of odorant concentration have been discussed earlier[2].

Experimental

The above definition of the decismel scale presents a more focused reference to the experimenter, because it operates fairly independently of the method used to present the standard odors. Air-dilution olfactometers, sniff-bottles, squeeze-bottles or scratch-sniff tapes, could all serve the purpose, so long as they provide a controlled and reproducible scale of odorant concentrations to the nose. We have been using 2-oz. polypropylene squeeze-bottles for odor presentations, because we have considerable experience in their manufacture for clinical olfactory testing, and also because they are portable, unbreakable, and quite practical.

The bottles are carried in sectional cardboard or aluminum boxes. Each bottle contains 15 ml of an odorless diluent, in which an appropriate concentration of odorant is dissolved. The liquid is absorbed in an inert felted material, having a large surface area for odorant equilibration with the air in the bottle. Opening a flip-top exposes the orifice. Holding the orifice close to the nostrils and squeezing the bottle while sniffing, delivers the odorized air to the nose. The subject notes the character and/or intensity of the odor as directed. After use, the top is closed. Having the odor compounds absorbed in a felt matrix helps avoid spills. Depending upon the chemical nature of the odorant, up to 1000 squeezes of air may be taken from the bottle before there is a significant loss of odorant. Again depending upon the odorant, room-temperature shelf life can be 6 months or more.

For establishing the decismel scale for a given odorant, we prepare a series of 5 dS odor increments Each concentration step is 1.778 times (the fourth root of 10) the concentration of the one before. Each odorant-containing bottle has a paired blank, containing diluent without any added odorant. The individual odor detection threshold for each subject is determined in a conventional manner (descending concentration series, with three double-blind correct identifications at each concentration)[3]. For accuracy, 25 or more subjects should be tested. Applying a regression analysis of threshold measurement against the subject's age, the mean threshold for healthy 20-year-olds is interpolated. From these data the required chemical concentration of odorant in the diluent is calculated for arriving at the 0 dS standard, and hence the concentrations for all other dS odor levels. The whole process can be repeated with each odorant, since thresholds vary over an extremely wide range for different substances.

For field work with ambient odors and naive subjects, we assemble eight or more odor samples that may be relevant, ambiguous, or deliberately irrelevant, to a given ambient odor. If an odorant is selected as similar to the ambient odor, we set up a scaling series of eight bottles, presenting the given odor at 10-decismel intervals. That is, a blank, followed by -5 dS, then +5 dS, 15 dS ... etc. up to 55 dS.

Results

The use of this system of ambient odor characterization may be illustrated by its application in the vicinity of an industrial plant which emitted a characteristic odor. About four years earlier the plant had discontinued the odorous process, but our objective was to reconstruct an evaluation of the odor character and intensity when it had been in operation.

We selected eight odorants (each at 25 dS) for the initial screening test. Table II shows a simplified version of the questionnaire. We wanted to find out if the neighborhood residents could detect each odorant, whether they could identify the odors, and whether any of the odors smelled like the Plant used to when it was operating. Among the eight odors, two were selected because they closely resemble known odorous constituents of the process. Two more were clearly unrelated to the Plant operation. The remaining four odors might or might not have any bearing on the situation

The subjects were asked to sniff each odor in turn and note its degree of resemblance to any odors they associated with the operation of the Plant, according to a scale of similarity extending from "3" for just like the Plant, to "0" for unlike the Plant. The results obtained from the questionnaire were as expected for some of the odors. Irrelevant odors gave average scores less than 0.05, while odorants resembling the Plant scored to 1.1 on this scale of similarity. There were also additional resemblances noted by the subjects.

The second phase of the questioning made use of two scaling kits, each containing 8 squeeze-bottles. Each covered the same range of odor levels, in 10 dS intervals from -5 dS through +55 dS. As indicated in the lower section of Table II, the subjects were asked if they noticed an odor from the Plant, when they were in or near their home. If so, they were asked to sniff the bottles in the first odor kit, in ascending concentration sequence, to demonstrate how strong the odor had seemed to be. They were asked to match on the basis of intensity of odor, and to ignore the fact that the quality of the odor in the bottles (odor type) might not resemble the odor of the Plant, in their opinion. Then they were asked to repeat this retrospective odor intensity matching with the second odor scale.

During the years when the Plant was in peak operation, the neighborhood homes had individual wells for their water supply (a public water supply was installed later). Parallel questions were directed to the well-water at the subject's home. Had he or she noticed any odor in the water, and if so, did it smell like either (or both) of these sample odors from the scaling kits? Once again, a matching of intensity between the recollected odor, and one of the scale of squeeze-bottles, was requested. Many of the responses to these inquiries were that none of the scale bottles delivered an odor as strong as that associated with the Plant, or with the well-water.

Clearly it would be desirable to find some way of checking on the accuracy of the neighborhood's collective memory of the odor intensities. The odorous industrial process had been discontinued four years earlier. Nevertheless, we were able to obtain two current well water samples that were quite odorous. Samples were collected from Well A that was used for garden irrigation, Well B was in disuse and required priming to collect samples. These were iced and carried to a distant city for olfactory evaluation.

The 2-oz polypropylene squeeze bottles are treated to render them virtually odorless. In one of these was placed 15 ml of the well water (without any felt). The new subjects were instructed to sniff the water in the test squeeze-bottle, and compare it with the first and second odor scaling kits, so as to pick out the closest intensity matches (also ignoring any differences of odor quality). The averaged results are given at the top of Table III, together with data for a sample of fresh public-service tap water from the second city. The well water samples in real time scored from 31 to 42 dS, which agrees rather well with the 29 to 35 dS obtained from the 4-year old collective memory of the Plant neighbors (foot of Table II).

For comparison, we also conducted with the same subjects in the second city a conventional water-dilution threshold test (foot of Table III). The water samples were presented in clean 125 ml Pyrex-glass conical (Erlenmeyer) flasks, each capped with an inverted polypropylene 1-oz medicine cup. The volume of water in each flask was 50 ml. All the "blank" flasks contained essentially odorless distilled water purchased from a grocery store. For the threshold measurements, the well water samples were diluted with distilled water in a binary series (1, 1/2, 1/4, 1/8 ... etc). Six dilutions of each water sample were presented, paired with "blank" flasks, in a descending concentration series. The flasks were numbered for identification and randomized for blank position. Randomization was reversed when half the subjects had been tested.

Thresholds of individual panelists were not determined;   just the panel
average was obtained. By convention, the flask dilution yielding closest to
75% correct choices (50% is the chance level) is taken to represent the
threshold dilution. The threshold dilutions (TON) are shown in the last line
of Table III. It may be noted that the threshold dilutions for the two well
water samples are considerably greater than would be expected by reference to
Table I for decismel equivalents.

Conclusions

The odor intensity assessment on a contemporary well-water sample agreed
with the remembered experience of the neighborhood residents from four years
ago. Olfactory memory persists from childhood into advanced years for odors
with fundamental associations.   This applies to the quality of odor such as
Grandma´s cookies or disinfectant soap. What is quite gratifying, is the
conclusion that the memory is quite good also for intensity. Nevertheless,
one needs a memory assist, and here is where the odor demonstration samples,
both qualitative and quantitative, were very helpful.

We have used this procedure for other ambient odor investigations, in
which the professional evaluates the area in real time.   The squeeze-bottles
are used for odor comparisons in the field.   With practice, the intensity
scale of a particular odorant can be memorized quite confidently at 10 dS
intervals.   For ambient odors assessment, where odor fatigue develops
insidiously and irresistibly, it is essential for the professional to take
precautions to avoid continuous exposure of his own olfactory sense. Just a
few inhalations are enough to develop significant odor fatigue.   One should
refrain from using the squeeze-bottles too frequently in contaminated air,
because the bottles themselves suck back the surrounding air, and will slowly
pick up the odor.

It is well known that experts such as perfumers and wine-tasters have
learned to recognize hundreds or thousands of odor types.   Chemists and
probably other professionals could undoubtedly learn to recognize and
remember dozens of odors; but with the use of appropriate sets of odor
standards one does not have to strain the memory. This capability may well
prove valuable for the investigative work of identifying the likely source of
an ambient odor. It may not be necessary for the simpler task of evaluating
the intensity of an odor. The human mind seems to have a fairly strong grasp
of the concept of intensity, and is capable of making reproducible intensity
comparisons not only between different qualities of odor[4], but even cross-
modally between different senses (smell, taste, sound, light)[5].

Addendum: Quantitative Assessment of Odor Annoyance

There is a growing need for selecting objective standards to define a
threshold or criterion of annoyance by ambient odors. Ultimately this has to
be done by a committee or a legislative body. Nevertheless, it would be more
satisfactory for all concerned if some scientific consensus could be reached.
A few years ago one of us[6] was asked by the California Air Resources Board to
develop a justifiable position. In the interests of contributing to the
discussion which is expected to develop at this Conference, an excerpt from
this Report is reproduced (loc. cit., pp. 30-33):

"Scattered through the literature there have been some observations that
certainly provide some indication of the likely magnitude of odor annoyance.
Considering that the odor detection threshold concentration, both for pure
substances and mixtures, is an accessible and relatively well understood
measurement, this may be a useful point of reference for assessing any

incremental odor stimulus that is capable of causing annoyance, particularly for unpleasant odors.

"Five References address this problem in various ways (Table IV). Winkler[7] found that the odor recognition threshold of hydrogen sulfide is approximately 10 times its detection threshold. He argued that unless an odor was recognizable, it could not very well be categorized as objectionable. Hence the recognition threshold might serve as a lower limit for an emission standard. Among the petrochemicals examined by Hellman[8] there were 34 that were described as unpleasant in hedonic tone. The average ratio between recognition and detection threshold for these compounds is 3.1.

"Another approach is that of Adams[9] who showed that un-trained observers could, on average, characterize the odor of hydrogen sulfide as unpleasant, at only 1.6 times its odor detection threshold concentration. In later work (NCASI[10]) on the more complex effluent gases from Kraft pulp mills, it was found that workers employed in, or persons economically associated with, the mill plant or offices did not consider the Kraft odor unpleasant until it reached an average of 6 times the mean detection threshold. This level of tolerance, by workers accustomed to the odor, might be interpreted as an upper limit for an acceptable ambient level.

"The most thorough work on this topic, with a multi-faceted questionnaire and analysis of responses, has been conducted by Winneke and Kastka[11] among people living near factories which are sources of odor pollution. Using their criteria of occasional odor-induced headache or nausea in about 50% of the exposed population, it appears that from 5 to 15 odor units of pollution are capable of causing annoyance. There may of course be a moderate level of tolerance, or odor fatigue, among residents in an industrial city. Winneke and Kastka also developed more comprehensive averaged values for degrees of annoyance based on larger sets of sensory, emotional and physiological responses, that seem to run in parallel with the headache and nausea evaluations.

"As a first attempt at quantifying the annoyance/detection ratio for unpleasant odors, it seems appropriate to give equal weight to all six of these estimates. The geometric mean for the ratio, at the foot of Table IV, is 5.3 odor units. This implies that when an unpleasant odor reaches an average concentration only about 5 times its detection threshold for the individual observer, it will very likely be recognized (so long as his attention is not distracted), it will be judged unpleasant, and it will cause undesirable emotional, social and physiological reactions among approximately 50% of people thus inadvertently exposed."

The above quotation is relying on averages. There is in fact a wide range of olfactory sensitivities in the population. This factor is illustrated in Figure 1, in which the lower part is modified from an earlier publication[12]. The standard deviation is a measure of the variability of odor sensitivity in the population. From prior measurements of the individual detection thresholds of hundreds of persons to dozens of chemicals, the standard deviation for persons aged 20 to 70 years averages close to ± 2.0 binary dilution steps. The slope of the detection line in Figure 1 is based on this standard deviation. It indicates what percentage of the population is able to detect, under good laboratory test conditions, various multiples (or sub-multiples) of the mean (50%) threshold concentration, here entered as 1.0. (As indicated by the broken lines, this generalization becomes uncertain for some small percentage of the population that has a general or specific anosmia).

Newly added to this chart is a provisional "annoyance line", which has been drawn parallel to the detection line, and displaced uniformly to five times higher concentration. (If some other annoyance threshold multiplier is under consideration, it can be entered instead at the appropriate position). Evidence as to the slope of the annoyance line is under current study. It has been drawn temporarily as parallel to the detection line. For specific catagories of odors the position of this annoyance slope will change. The value of this illustration is that is permits Committees, Legislators, etc. to read off the expected percentages of the population that will experience annoyance from any particular detection threshold multiplier, that they might be considering for a criterion standard of odor annoyance.

A large proportion of olfactometric research is devoted to the laboratory odor detection threshold. That is, for fully attentive, well instructed test subjects doing their best to succeed in a test of their olfactory sensitivity. Nevertheless, people in their everyday lives, work and recreation are not normally particularly attentive to odor. Furthermore, if their attention is distracted by concentrating on some other occupation or task, they are still less responsive to odor. These points were investigated by Whisman et al.[13]. The "warning line" on Figure 1 is based on their "mis-directed" test to delineate the necessary properties of warning odorants for bottled gas. From further work by Whisman's group on "semi-directed" and "un-directed" tests, it appears that the "noticeability threshold" for odor detection by an inattentive (but not distracted) person is at about 4 times their "directed" test threshold, which is the standard laboratory detection threshold. An interesting conjecture from this line of reasoning is that the noticeability threshold and the annoyance threshold (for an unpleasant odor) may in fact be the same.

Those who are interested in standardization face a little more deliberation on an additional issue. Referring once more to the lower part of Figure 1, it is constructed on the basis of the mean threshold for the population. In most scientific work on odor this effectively means the working age population, say from ages 20 through 65 or 70, with an average usually between 40 and 45. This is quite appropriate for developing odor regulations for the benefit of the general population. Nevertheless, there is a well documented decrease in olfactory sensitivity with age[14]. The same applies to the sense of hearing, and the zero hearing threshold level for the decibel scale is based on the average auditory sensitivity of healthy 20-year-olds. In setting up a parallel for the decismel scale[2], we adopted the same standard, with zero decismels representing the average olfactory threshold for age 20. The difference of mean sensitivity between age 20 and age 42 is approximately one binary step (doubling of the threshold concentration; increase of 6 dS in threshold).

Illustrating this discrepancy, we have added to the top of Figure 1 a histogram representing the individual olfactory thresholds in decismels for 100 persons (ages 5 to 83) to the odor of phenylethyl methyl ethyl carbinol[15] (used for clinical smell testing). The "0 dS" axis is offset 6 dS (one binary step) to the left of the "1 multiple of threshold" axis. This point is by definition also one Odor Unit, which is itself the basis for a great deal of applied research in flavor science, water quality and air pollution. A similar discrepancy (of 9 decibels) has caused difficulties for practitioners of audiometry (based on the age 20 zero hearing threshold level) and acoustics (based on a sound pressure level of 0.0002 microbars).

Taking the proposed unpleasant odor annoyance ratio as 5 times mean odor detection threshold (5 odor units), we may calculate as follows: $20 \times \log_{10}(5.0) = 13.98$, which is very close to 14 decismels. Add on 6 dS for the 22 year difference in baseline age between the decismel and odor unit scales, and we get $14 + 6 = 20$ dS. Hence an odor level of 20 decismels of an unpleasant odor is the equivalent reading for the average threshold of odor annoyance.

There is yet another matter that needs to be considered, and that is the relationships between odor concentration and odor intensity. Both the odor unit scale and the decismel scale are in effect simply chemical concentration scales, with the refinement of being "anchored" at the mean odor detection threshold for each chemical compound or mixture. Realistically, however, different odorants at equal multiples of their detection threshold do not necessarily yield equal odor intensities (see Reference[16] for a review of experimental values for the odorous intensity slopes).

As a standard substance for providing a reference scale of odor intensity, Moskowitz et al.[17] recommended 1-butanol in 1974. Since then a fair number of assessments of ambient air samples have been made with reference to the "butanol scale". Should this compound be adopted, by agreement or by default, to provide a standard scale of odor intensity? Whatever substance is chosen, there will be a further need to specify the method of presentation, because a given threshold multiple may not deliver the same odor intensity impression from a squeeze-bottle as from a full flow air dilution olfactometer.

References

1.   H. Stone, J. J. Bosley, "Olfactory discrimination and Weber's Law," Percept. Motor Skills 20: 657-665. (1965).

2. J. E. Amoore, R. S. O'Neill, "Proposal for a unifying scale to express olfactory thresholds and odor levels: the "decismel scale," Proc. Air Poll. Control Assoc. Ann. Mtg. Paper No. 78.5, 21 pp. (1988).

3. J. E. Amoore, B. G. Ollman, "Practical test kits for quantitatively evaluating the sense of smell," Rhinology 21: 49-54. (1983).

4.   J. E. Amoore, D. Venstrom, "Sensory analysis of odor qualities in terms of the stereochemical theory," J. Food Sci. 31: 118-128. (1966).

5. S. S. Stevens, "The psychophysics of sensory function," Amer. Sci. 48: 226-253. (1960).

6.   J. E. Amoore, "The perception of hydrogen sulfide odor in relation to setting an ambient standard," Final Report to California Air Resources Board, ARB Contract A4-046-33, 46 pp. (1985).

7.   K. Winkler, "Zur Diskussion Gestellt: Imissionsgrenzwerte zur Verhinderung von Geruchsbelastigungen," Wasser Luft Betrieb 19, 411. (1975).

8. T. M. Hellman, F. H. Small, "Characterization of the odor properties of 101 petrochemicals using sensory methods," JAPCA 24: 979-982. (1974).

9. D. F. Adams, F. A. Young, R. A. Luhr, "Evaluation of an odor perception threshold test facility," Tappi 51 (3): 62A-67A. (1968).

10.   NCASI, "Evaluation of the use of humans in measuring the effectiveness of odor control technology at the source," Atmospheric Quality Improvement, Technical Bulletin No. 56. National Council of the Paper Industry for Air and Stream Improvement, New York, 1971.

11. G. Winneke, J. Kastka, "Odor pollution and odor annoyance reactions in industrial areas of the Rhine-Ruhr region." In Olfaction and Taste VI, pp. 471-479. Ed. by J. Le Magnen and P. MacLeod. Information Retrieval, London, 1977.

12. J. E. Amoore, E. Hautala, "Odor as an aid to chemical safety: odor thresholds compared with threshold limit values and volatilities for 214 industrial chemicals in air and water dilution," J. Appl. Toxicol. 3: 272-290. (1983).

13. M. L. Whisman, J. W. Goetzinger, F. O. Cotton, D. W. Brinkman, C. J. Thompson, A New Look at Odorization Levels for Propane Gas. Bartlesville Energy Research Center, Bartlesville, OK, 1977.

14. D. Venstrom, J. E. Amoore, "Olfactory threshold in relation to age, sex or smoking," J. Food Sci. 33: 264-265. (1968).

15. J. E. Amoore, S. Steinle, "A graphical history of specific anosmia." In Genetics of Chemical Sensing and Communication Ed. by C. J. Wysocki and M. R. Kare. Marcel Dekker, Inc. (in press).

16. F. Patte, M. Etcheto, P. Laffort, "Selected and standardized values of suprathreshold odor intensities for 110 substances," Chem. Senses Flavour 1: 283-305. (1975).

17. H. R. Moskowitz, A. Dravnieks, W. S. Cain, A. Turk, "Standardized procedure for expressing odor intensity," Chem. Senses Flavour 1: 235-237. (1974)

Table I. Basis of the "decismel scale" of odor levels. For formula, see Introduction section of text.

Multiple of threshold[a]	Decismels (dS)
1000	60
316	50
100	40
31.6	30
10	20
3.16	10
1	0
0.316	-10
0.1	-20

[a] Defined as the average odor detection threshold for healthy 20-year olds.

Table II. Retrospective olfactory evaluation of an industrial plant.

1. SCREENING:- Did the Plant, when it was operating, smell like any of these samples? (each squeeze-bottle is 25 dS)

Just like[3]   Quite like[2]   A bit like[1]   Unlike[0]

Odor class   Score

Musky
Scallops
Smoky
Solvent
Camphor
Flowery
Mothballs
Gassy

2. ODOR INTENSITY:-

Did you notice an odor from the Plant, when you were in or near your home? Use these graded samples to show how strong the odor seemed to be.

Air, avge. dS

First odorant scaling   >55
Second odorant scaling   ~55

Did you notice an odor in the well-water at your home? Did it smell like either (or both) of these samples? Show me how strong the odor seemed to be.

Water, Avge. dS

First odorant scaling   ~35
Second odorant scaling   ~29

Table III. Olfactory evaluation of water samples (by distant subjects).

1. ODOR INTENSITY SCALE:-

SQUEEZE-BOTTLES: A, B and C are one-quarter filled with water! Please keep them up-right and don't shake them. The two sets of eight bottles in this test provide scales of odor strengths or intensities. Please compare the water with the scales and note which bottle (in each series) closely matches the intensity of the water odor.

	Well A	Well B	City water
First odorant scaling, dS	40	31	8
Second odorant scaling, dS	42	32	3
Dilution to threshold, TON	1500	250	8

2. ODOR THRESHOLD DILUTION:-

GLASS FLASKS: There are three sets composed of six pairs. Each set represents one of three water samples. In each set the odor strength decreases as you work from left to right. In each pair of flasks one contains pure water, and the other contains a certain amount of odor. Raise the caps, smell the contents, and decide which flask has the odor (or strongest odor) in the pair. You may swirl the flasks but don't shake them. Circle below which one of each pair has the stronger odor (a Table was provided). Please replace the caps on the same flasks from which they came.

Table IV. Tentative basis for estimating the ratio of annoyance threshold to detection threshold for unpleasant odors.

Odorous substance	People tested	Judgment criterion	Ratio	Ref.
Hydrogen sulfide	Laboratory panel	Odor recognition	10	7
34 Unpleasant petrochemicals	Laboratory panel	Odor recognition	3.1	8
Hydrogen sulfide	County fair visitors	Judged unpleasant	1.6	9
Kraft stack gases	Paper mill workers	Judged unpleasant	6	10
Phenolics	Factory neighbors	Occasional headache, nausea	5	11
Hydrocarbons & sulfur compounds	Factory neighbors	Occasional headache, nausea	15	11

Geometric mean ratio = 5.3

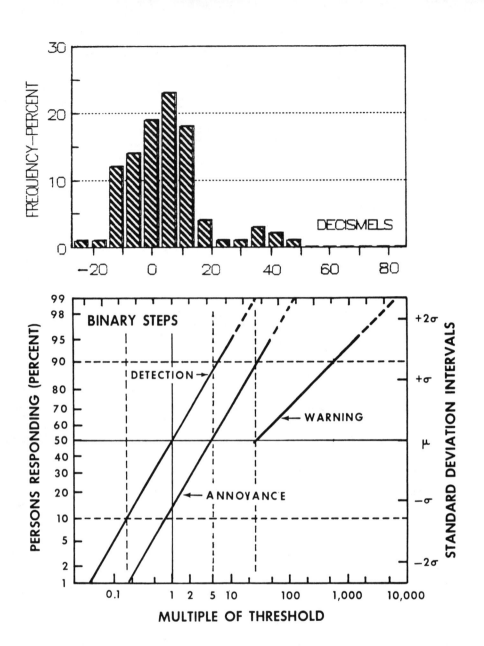

Figure 1. (Lower chart). A practical guide to the quantitative interpretation of odor detection and odor annoyance. Modified from Ref.[12] (Fig. 2, with addition of annoyance line). The coordinates are log/probability, so care is required in interpolating between marked intervals. The detection line represents the performance of fully attentive persons under good laboratory conditions. The annoyance line is drawn on the assumption that for any given person, his or her threshold of odor annoyance (for an unpleasant odor) is at 5 times his threshold of odor detection.

(Upper chart). Individual odor detection thresholds, of 100 persons ages 5 to 83 yr, for the odor of phenylethyl methyl ethyl carbinol. Modified from Ref.[15] (Fig. 4, with change of magnification to make the x-axis concentration scales coincide).

MEASUREMENT OF ODOR ANNOYANCE BY POPULATION PANELS LIVING IN AN INDUSTRIAL AREA

M.L. PERRIN, M. JEZEQUEL
Commissariat à l'Energie Atomique
IPSN/DPT/SPIN
Laboratoire d'Olfactométrie
BP 6, 92265 FONTENAY AUX ROSES CEDEX, France

Following numerous complaints, the DRIR (Direction Régionale de l'Industrie et de la Recherche) Provence-Alpes-Côte d'Azur decided to undertake an olfactometric study on the east bank of the Etang de Berre, in the south-east of France in order to evaluate the level of on-site odorous pollution and the degree of odor annoyance experienced by the local population.

The study involved :
- measurements of environmental odor intensities in order to draw odor maps of the area corresponding to different meteorological situations,
- measurements of odorous sources in order to evaluate their environmental impact,
- a survey of members of the local population who over a one-year period completed a questionnaire regarding the annoyance occasioned by the odors. The survey results were used to calculate odor annoyance index for each sector and for various meteorological conditions, and to plot odor annoyance compass cards.

INTRODUCTION

Following numerous complaints from locals near the Etang de Berre in south-east France, notably those on the north-east bank of the pool, the Direction Régionale de l'Industrie et de la Recherche decided to undertake an olfactometric study of this highly industrialised area, where a decrease in pollution levels had not reduced odor annoyance.

The study involved cooperation between local volunteers from Berre, Rognac, Velaux and Vitrolles (figure 1) and olfactometry specialists. The former assessed the degree of odor annoyance and the latter measured the level of odorous pollution at the site. In the event of real olfactory annoyance, the aim was to define potential sources of odor so as to reduce or even eliminate them.

METHODOLOGY

1. Survey

To evaluate the level of annoyance provoked by odors (1), we used local volunteers who assessed their environment at a precise time of day over a one-year period. They were asked to leave their homes and note their responses to a very simple question, such as "if you smell something, is the smell which you perceive now not annoying, a little annoying, ... extremely annoying ?", on a reply card. Volunteers were free to make additional comments on the back of the card.

For each data set for a given day at a set time, the individual scores were used to calculate an odor annoyance index which, by definition, was the sample mean of the individual responses. In practice, the individual responses were rescaled using an arbitrary scale ranging from 0 to 100, with discrete intervals equal to 25. Individuals who perceived no odor were scored 0 on this scale, those who were "slightly annoyed" were scored 25, and so on. The odor annoyance index was zero when nobody was annoyed by the odors and 100 when all participants were extremely annoyed.

2. Odor measurements in the environment

Environmental measurements are intended to provide as quantitative an evaluation as possible of the odor level in a given zone. The local odor level was measured using the odor intensity evaluations of a panel of qualified experts assumed to be a representative sample of the population (2).

Odor intensity was measured using reference values ; the test atmosphere value was compared in-situ by the experts with the intensities of a standard range of concentrations of pyridine chosen as pure reference compound. The n determinations of intensity given by the experts were each expressed as a concentration of pyridine in water.

For each field measurement campaign, the data were collected from all the experts, the mean intensity values were determined and an odor map was drawn for the area.

3. Odor measurements at emission sources

The olfactometric measurements of sources of odor are intended to characterise the concentrations of emissions and evaluate their environmental impact (5).

Samples for olfactometric analysis are collected differently depending on whether the sources are channelled or not (3). When the gas effluent is or can be channelled, the odorous gas is sampled in the duct and is introduced into a bag made of unreactive material. If emission is from a surface, a storage tank for example, the odorous liquid is sampled and introduced into a caisson swept by air of known flow rate. The flow rate of odorous products per unit surface area is then determined. When multiplied by the surface area of the tank, this flow rate provides the basis for calculations.

Laboratory batch measurements of odors involve dilution of the sampled gas with clean, odourless air in order to determine the dilution factor at the threshold of perception of a panel of experts (4). An olfactometer is used following a well-defined operational procedure.

The dilution factor at the threshold of perception of an odor is, by definition, the degree of dilution for which 50 % of positive replies are received from a panel of experts.

Given the dilution at the perception threshold, and the emission rate of the source, it is possible to calculate the odor emission rate and, using odor dispersion tables, the sector of continuous odor perception.

RESULTS

1. Evaluation of odor annoyance experienced by the local population

The response cards from 150 volunteers received between July 1st, 1987 and July 1st, 1988 were sorted into seven geographical sectors. The calculated values of the annoyance index are indicated in table I.

Figure 2 shows the 18-direction odor annoyance compass cards which were plotted for each sector, using mean annual values for the annoyance index expressed as a function of wind direction. It can be seen that these index depend greatly on the geographical location of the sectors.

In the sectors most distant from potential sources (5 and 7), low annual values were recorded, in contrast to sector 4, where annoyance index were high regardless of wind direction, or sector 1, where high values were only recorded for a well-defined wind direction.

Table I shows that sectors 3 and 6 were similar in terms of annoyance frequency and intensity, the differences in compass cards being due to distinct configurations between the industrial zone and the sector.

2. Environmental odor measurements

Environmental olfactometries were performed by six experts who moved over the site in 11 measurement campaigns, thus covering the most frequent regional meteorological situations. Natural odors (flowers, grass, ...), urban odors (essentially car exhaust) and industrial odors were measured and the results are given in table II.

Appart from Velaux, where industrial odors were virtually always absent whatever the weather conditions, there was virtually always one area where an industrial odor was recorded.

The least penalising meteorological situation for the test site was a south-east wind, for which no industrial odors were noted anywhere. Outside this wind sector, an industrial-type odor was always evident at Berre, whereas at Rognac and Vitrolles odors were perceived with south-west, west and nord-west winds.

3. Odor measurements at emission sources

Comparison of the on-site measurements of odor intensity and the annoyance compass cards with a map of industrial installations indicated potentially odorous sources. The odor emission rates were measured and the distances of continuous perception in the environment were estimated (table III).

The results given in table III confirm that several sources of industrial-type odors may by perceived on the site at levels at least equal to that of the threshold of perception, and may explain the odor intensities measured in the environment and justify the odor annoyance levels reported by the local inhabitants.

CONCLUSIONS

A one-year olfactometric study has been carried out on the east bank of the Etang de Berre in south-east France using 150 local volunteers panel. Collaboration with local authorities, local factories, where samples were taken, and volunteers was excellent. The results reveal close correspondence between annoyance reported by local inhabitants and the data provided by olfactometric analysis performed by experts in the environment and at emission sources. This provides the justification required to local industries for action to reduce odorous emission by certain sources and thus contribute to improvement of the environment.

REFERENCES

1. E.P. KOSTER et al., "Direct scaling of odor annoyance by population panels ", VDI Berichte 561 - Odorants. (1985)
2. M.F. THAL, P. ZETTWOOG, "Dix ans d'expérience de mesures olfactives dans l'industrie et dans l'environnement", VIème Congrès mondial pour la qualité de l'air, Paris. (1983)
3. Norme AFNOR PR X43-104. "Qualité de l'air. Atmosphères odorantes. Méthodes de prélèvement". En cours de publication
4. Norme AFNOR NF X43-101. "Qualité de l'air. Méthode de mesurage de l'odeur d'un effluent gazeux. Détermination du facteur de dilution au seuil de perception". (1986)
5. M.L. PERRIN, M.F. THAL, M. JEZEQUEL, "Standardization of olfactometry in France", Proceedings of the 81st Annual Meeting of APCA : 88 - 78.2. (1988).

Sector	0	<5	<10	<20
1 Rognac : north	47	58	68	82
2 Rognac : Centre-east	30	55	80	93
3 Rognac : Centre-west	21	40	65	87
4 Berre	1	7	18	38
5 Vitrolles south	68	78	86	96
6 Vitrolles north	30	45	66	90
7 Velaux	55	80	90	98

Table I - Percentages of measured odor annoyance index

Wind direction	BERRE	ROGNAC CENTRE	ROGNAC NORD	VITROLLES
Variable weak wind	1 to a	X	1	X
60°-80°	a to H	X	X	X
120°-160°	X	X	X	X
180°-220°	1	1	1 to a	X
240°-280°	1	a to H	1 to a	1
280°-320°	1	1 to a		1

Table II - Intensities of industrial-type odors for meteorological
conditions encountered during 11 on-site measurement campaigns

X = no industrial-type odor
1 = low intensity
a = average intensity
H = high intensity.

Odour source	Threshold dilution R_S	Odour emission m^3/s at $20°C$	Lower and upper distances defining the sector of continuous perception of the odor (= d in m)	
			Good conditions of diffusion Pasquill. class C Wind speed : 4 m/s	Poor conditions of diffusion Pasquill. class E Wind speed : 2 m/s
Carbon black	775	$8,2x10^3$	< 100	< 100
Industrial waste incinerator	2 630	$3,6x10^4$	< 100	< 100
Industrial waste incinerator	2 690	$3,7x10^4$	< 100	< 100
Refined bitumen tank	$2,7x10^5$	$9x10^3$	100 < d < 250	100 < d < 1 000
Expanded polystyrene reactor outlet	$2,4x10^7$	$2,8x10^6$	> 3 000	> 10 000
Industrial waste reception disposal	3 760	$9,2x10^5$	100 < d < 3 600	> 10 000
API pool	2 750	$8,1x10^5$	100 < d < 3 500	> 10 000
Thickening basin for industrial biological purification	3 390	$3,7x10^5$	100 < d < 2 300	> 10 000
Solvant neutra- lisation pit	1 120	$3,1x10^4$	100 < d < 600	100 < d < 2 300

Table III - Values of the dilutions at the threshold of perception of odor and odor emission rates. Environmental impact forecasts

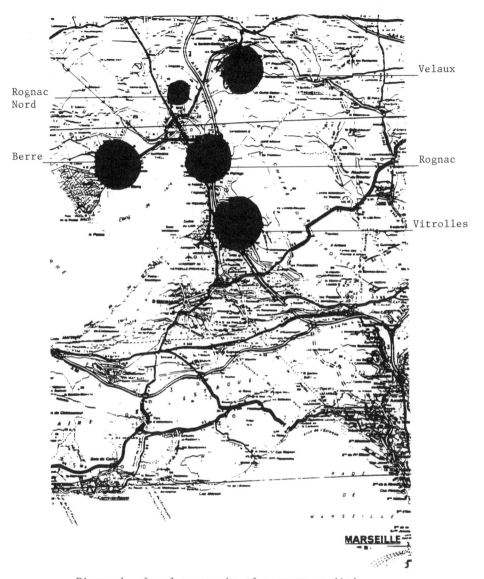

Velaux

Rognac
Nord

Berre

Rognac

Vitrolles

MARSEILLE

Figure 1 – Local geography of sectors studied

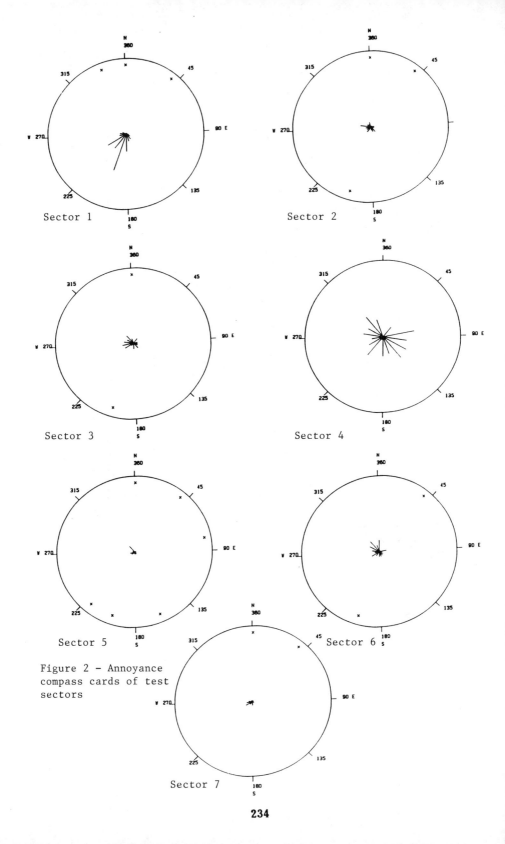

Figure 2 - Annoyance compass cards of test sectors

Sector 1

Sector 2

Sector 3

Sector 4

Sector 5

Sector 6

Sector 7

234

USE OF ODOR THRESHOLDS FOR PREDICTING
OFF-PROPERTY ODOR IMPACTS

Marcia T. Willhite
S. Thomas Dydek
Research Division
Texas Air Control Board
Austin, Texas

Air pollution odor causes many complaints and the discomfort it can cause implies that it be evaluated as an adverse health effect of air pollution. As a regulatory agency with the responsibility of protecting the public from adverse effects of air pollution on health and welfare, the Texas Air Control Board (TACB) must evaluate the odor nuisance potential of emissions from facilities not yet constructed. Computer dispersion modeling of proposed emissions can predict a worst-case off-property concentration on a compound-by-compound basis. Some decision must be made as to whether that worst-case concentration would be odorous enough to be a nuisance. Published odor thresholds are typically used in making this evaluation.

The weaknesses of this approach (widely variable thresholds depending on test protocol, consideration of a single compound at a time, relationship between a lab-derived detection threshold and real-world odor perception, etc.) are presently outweighed by the regulator's need to make decisions based on a somewhat inadequate database. A guideline for using published odor thresholds for predicting off-property odor impacts has been developed using the relationship between detection threshold and "complaint" levels from a recent odor impact model study. Compounds with unpleasant odors which exceed three times their odor threshold have the potential of causing odor nuisances. For chemicals with pleasant or neutral hedonic tones, a nuisance may not occur until concentrations are five times the odor threshold. Further odor science investigation is needed to fill in gaps in the database useful to regulators; meanwhile, the approach to odor nuisance prediction used by TACB is useful until more accurate tools are available.

Introduction

Air pollution odor is difficult to evaluate because of the many facets (thresholds, intensity, character, hedonic tone) that contribute to its perception and the fact that only a very subjective instrument can truly measure it, namely the human olfactory system. Yet odors cause many complaints and the discomfort odors can cause implies that they be evaluated as health effects of air pollution. As a regulatory agency with the responsibility of protecting the public from conditions of air pollution which may adversely affect health and welfare, the Texas Air Control Board (TACB) typically must evaluate the odor nuisance potential of air emissions. Published odor thresholds are usually relied upon in such evaluations. Several weaknesses exist in this type of approach: 1) the values of published odor thresholds can vary depending on the method used for their derivation, 2) the consideration of each compound separately doesn't allow for additive or synergistic effects of odors, and 3) a laboratory-derived threshold may not be an accurate predictor of what concentration would be a nuisance in the real world. However, for agencies constrained by limited personnel and resources, this approach is useful until more accurate tools are available.

The TACB Approach to Odor Potential Evaluation

The Effects Evaluation section of the TACB reviews predicted impacts from facilities seeking pre-construction permits. Part of this process is to assess whether odorous (as well as toxic) conditions may result off company property. Computer dispersion modeling of emissions can predict a worst-case off-property concentration on a compound-by-compound basis for use in making the evaluation. This predicted concentration is compared to published odor detection thresholds. If the predicted concentration is below that odor threshold, we feel it is unlikely that nuisance odor conditions would occur. If the predicted concentration is higher than the odor threshold we must decide whether it would actually be odorous enough to be considered a nuisance condition.

To assist us in this effort, we have compiled a database of published odor thresholds and other odor information for about 600 compounds. For many compounds there is wide variability in published odor thresholds. Typically we use the lowest odor threshold unless it seems unreasonable in comparison to similar compounds or if the testing protocol seems questionable.

Strengths/Weaknesses of Published Odor Thresholds For Regulatory Use

Most published lists of odor data describe various thresholds derived under laboratory conditions. Generally, an odorant is diluted with air and presented to an individual for decision. For a detection threshold, the decision is whether odor is present or not, the aim being to determine at what concentration odor is noticeable. Other thresholds often determined are 50% and 100% recognition thresholds, i.e. that concentration of odorant which 50% or 100% of subjects can recognize (describe). There are various methods of physical presentation of the sample and format of presentation with different sources of error, such as: loss of odorant in delivery method, odor fatigue of the test subject, bias towards or against a perception of odor, olfactory sensitivity of the subject, temperature, humidity, background odor, etc. In recent years some standardization of methods has occurred. A published odor threshold for a compound is usually the average of many individuals' thresholds for that compound.

Is a detectable level of odor the same as a nuisance level? What is the relationship between lab-derived values, "real-world" values and nuisance values? Logically, concentrations that are detectable to a trained odor judge, focused on perceiving an odor, are likely different (and lower) than a level detectable by an average citizen under ambient conditions. Unfortunately, threshold determinations under ambient conditions are seldom done. One study found that detection thresholds for ethyl mercaptan and thiophane were 26 times higher if the attention of the subjects was directed away from odor perception (1). The author considered this a good model for odor detection under occupational conditions.

Odor perception under ambient conditions is confounded by many factors. First, the chemical itself may change under atmospheric interaction between the source and the receptor, which might affect its odor properties. Second, other odors received concurrently may be additive, synergistic or counteractive to the odorant of concern (2). Third, adaptation to odors occurs which tends to raise the detection threshold at different rates for different odors (3). Further, detection of an odor can be distracted by other sensory input received simultaneously. So, published detection thresholds are probably only a crude estimate of the odor potential of a compound under ambient conditions.

Properties of Compounds Related to Nuisance Prediction

Characterization of the many facets of odor (odor intensity, acceptability, hedonic tone, and recognizability) can be found and utilized in a general way to assist in assessing the potential off-property odor impacts. For regulatory work, these properties must be related back to a concentration to be useful. For example, field studies to develop different "olfactometers" for

verifying existence of nuisance conditions are plentiful. These
are mainly designed to measure odor intensity: a trained odor
judge who has learned to scale intensities of some reference
chemical uses a device to rate intensity (under ambient conditions) of odorous emissions. Published lists of what ambient
concentrations of compound or what intensity levels could be described as a nuisance are apparently unavailable.

Odor acceptability is a property closely related to an odorant's
nuisance potential. Individual response to odor can be affected
by attitudes towards and experience with air pollution. For this
reason, a quantifiable relationship between detection and nuisance based on concentration is difficult. A relationship between detection, recognition and annoyance thresholds has been
theorized, with the annoyance value measured by a survey of residents expressing annoyance caused by odors (4). No published
test of this relationship correlating ambient concentrations with
expressed annoyance was found.

Odor recognition is the ability to describe the odor; a recognition threshold is that concentration at which the subject can
apply one of many descriptors (e.g. sweet, pungent, musty, burnt/
smoky, etc.) Recognizability seems to be more related to intensity, an intrinsic property of the odor which is related to concentration. Steven's Law, $I=kC^n$, is one description of this
relationship, where k and n (approximate range = 0.15 to 0.8) are
characteristic of each odorant (5). However, quantification of
these factors for most chemicals of concern to us has apparently
not been done.

At concentrations above detection threshold levels, intensity vs.
concentration varies with the compound. A given increment in
concentration may cause the intensity (or perceived magnitude) of
one substance to increase markedly and that of another to increase only slightly. For example, a 200-fold change in concentration caused a 15-fold change in the intensity of 1-propanol
but only a two-fold change in the intensity of n-amyl butyrate
(6). So the odor of sub-stances such as n-amyl butyrate which
are intense over a broad range of concentrations will likely be
less easily controlled simply by controlling the levels emitted.
Again, identification of which substances fall into this category
has apparently not been done for any broad range of chemicals.

Some general statements correlating structure with recognizability have been made (7). Straight chain aliphatics have the
highest recognition thresholds, and that level decreases with increasing molecular weight. This trend holds true for most classes of straight chain molecules. The functional groups of small
molecules have dominating influence on their threshold, and can
intensify or reduce each other's effects depending upon their
position in the molecule. A double bond usually reduces the
threshold odor concentration, except with aldehydes. Chemicals
with branched-chain structures are usually more odorous than
their straight-chain isomers.

Relationship Between Odor Threshold and Annoyance Levels:
A Basis for Prediction

To regulate for the prevention of air pollution odor nuisance
conditions, it is important to have some notion of the relationship between the levels at which odor can just be detected (odor
threshold) and the levels which could be nuisance levels. Unfortunately this type of data is not available for many air pollutants. An exception to this is a recent study (8) in which
panels of odor judges evaluated "detection", "complaint", and
"annoyance" levels of 40 odorous compounds. The "complaint"
levels were taken to be those which could represent a condition
of odor nuisance. Table 1 shows the 40 chemicals tested and
their 50% detection and 50% complaint concentrations (50% of the
panel either detected the odorant or complained about it at the
levels given).

It would have been convenient if there were a constant ratio between detection level and complaint level, but this is not the
case. The 50% complaint to 50% detection ratios were calculated
and they ranged from 1.8 to 60.9. The mean ratio for all chemicals was about seven and the median ratio was four. Because
several ratios were much larger than most, the median value is
probably the preferable measure. This implies that people will
complain about odors when they reach about 4 times the odor
threshold.

The type of odor a chemical has is a consideration in determining the potential for odor nuisance. In odor research parlance,
the type of odor is determined by its "quality" and "hedonic
tone". Table 2 gives the odor qualities and hedonic tones of the
chemicals under study. Common sense and experience dictate that
human reactions to chemicals having pleasant or neutral odors
will be different from their reactions to chemicals with unpleasant odors. This idea was tested using these data. Median ratios
of complaint to detection levels were calculated for these two
types of odorants. Three chemicals had no assigned hedonic tones
and these were not evaluated. The results for the other chemicals are given in Table 3 (neutral to pleasant odor) and Table 4
(unpleasant odor). For the former group the median ratio was
about five whereas for the latter, it was about three. This
confirms the common sense impression that people are more likely
to complain about levels of chemicals with unpleasant odors at
lower levels than they will about chemicals with pleasant or
neutral odors. The difference is not great, however.

These findings have implications for air pollution regulation.
Assuming that the median ratios can be applied to all odorous
chemicals, it can be said that levels of chemicals with unpleasant odors which exceed three times their odor threshold have the
potential of causing odor nuisances. For chemicals with pleasant
or neutral hedonic tones, a nuisance may not occur until odor
levels exceed five times the odor threshold. There are several
weaknesses in this approach. Many compounds may not be well-
represented by using the median values derived here. Until we do

have information about the complaint and detection levels for other chemicals, this would seem to be the best available information. This is one example of the fact that as regulators we do not have the luxury of having all of the data we would like to have. However, when reviewing potential emissions from a facility not yet in operation a decision about its odor nuisance potential must be made. When no other information exists, these guidelines to evaluating odor nuisance potential are the best approach to use.

Conclusion

The TACB currently uses an approach to predicting odor nuisance potential of emissions based on comparison of the predicted off-property concentrations of individual compounds to published odor thresholds. This approach may under- or over-estimate real-world odor potential because it does not consider the combined effect of mixtures nor any transformation of the odorant under ambient conditions. There are many sources of variability in published odor information which can also lead to inaccurate odor potential estimations.

Given that regulators must depend upon what information is available, a guideline to the use of published odor thresholds has been developed. For compounds with pleasant or neutral hedonic tones, an odor nuisance would be unlikely to occur until the ambient concentration is five times the respective odor detection threshold. For compounds considered unpleasant, a three-fold exceedance is likely to be a nuisance.

With few exceptions, limited resources require that regulatory agencies depend upon the research community to fill in the gaps in the odor information database. Some of the gaps that could be filled to increase the accuracy of odor nuisance potential evaluation by regulators are: 1) the relationship between odor detection and complaint thresholds for mixtures, 2) characterization of additive, synergistic and antagonistic interactions of odorous air pollutants under ambient conditions, 3) quantification of the relationship between intensity and concentration for common air pollutants and 4) information about odors from specific industrial processes.

References

1. J.F. Amoore, E. Hautala, "Odor as an aid to chemical safety: odor thresholds compared with Threshold Limit Values and volatilities for 214 industrial chemicals in air and water dilution," _J. Applied Toxicology, 3 (6)_: 272-290. (1983).

2. J. Ruth, "Odor thresholds and irritation levels of several chemical substances: a review," _American Industrial Hygiene Assoc. J. (47)_: A142-A151. (1986).

3. P.N. Cheremisinoff, R.A. Young, _Industrial Odor Technology Assessment,_ Science Publishers, Ann Arbor. 1975, pp. 52-53.

4. reference #3, pp. 70-73.

5. A. Dravnieks, H.J. O'Neill, "Annoyance potentials of air pollution odors," _Am. Industrial Hygiene Assoc. J. 40 (2)_: 85-95 (1979).

6. "Odors from Stationary and Mobile Sources," National Academy of Sciences, Washington, DC, 1979.

7. K. Vershueren, _Handbook of Environmental Data on Organic Chemicals,_ 2nd ed. Van Nostrand-Reinhold, New York. 1983, pp. 57-58.

8. "Procedure for the Determination of Odour Impact Models by the Binary Port Odour Panel Method," Ontario Ministry of the Environment, Toronto, 1988.

Table I. 50% detection and 50% complaint levels for various air contaminants (8).

AIR CONTAMINANT	50% DETECTION (ug/m3)	50% COMPLAINT (ug/m3)
biphenyl	2.3	9.5
hydrogen sulfide	7	19.5
methyl acrylate	58	160
isopropyl benzene	500	5100
ethoxyethanol acetate	500	7900
isobutyl acetate	630	3200
PGMMEA*	660	2800
ethyl hexanol	900	3000
methyl amyl ketone	1300	4850
styrene	1500	5800
Solvesso 100	1600	6000
butoxyethanol, 2-	1900	10500
heptanol	1900	5900
ethyl benzene	2000	16000
methylmercaptoaniline	2200	4800
methyl styrene	2200	9800
Solvesso 150	2200	4900
furfural	3000	10800
methacrylic acid	3100	7800
methyl methacrylate	3200	7100
EGMHE**	3300	20000
propionic anhydride	3500	11700
xylenes, mixed	3700	14000
methyl nonyl ketone	3800	8700
methyl ethyl ketone	3900	49000
chlorobenzene	4600	22400
kerosene	5800	25100
methyl isobutyl ketone	7100	28000
diisobutyl ketone	9100	43700
ethoxyethanol	12600	575400
toluene	14000	47000
ethylene glycol	15000	35000
hexylene glycol	20000	36000
ethanol	23000	$1 \times 10^{+6}$
acetone	32000	200000
butyl alcohol, t-	41000	140000
dioxane	68000	180000
diacetone alcohol	83000	224000
nitromethane	95000	490000
heptane	100000	350000
decane, n-	170000	$1 \times 10^{+6}$
naphtha	370000	690000

* PGMMEA is propylene glycol monomethyl ether acetate
** EGMHE is ethylene glycol monohexyl ether

Table II. Odor qualities and hedonic tones for various air contaminants (8).

AIR CONTAMINANT	QUALITY OF ODOR	HEDONIC TONE
biphenyl	peculiar	pleasant
hydrogen sulfide	rotten eggs	unpleasant
methyl acrylate	sharp, fruity	unpleasant
isopropyl benzene	sharp	unpleasant
ethoxyethanol acetate	sweet, musty	pleasant
isobutyl acetate	sweet, ester	pleasant
PGMMEA*	none found	neutral to unpleasant
ethyl hexanol	musty	pleasant to unpleasant
methyl amyl ketone	none found	sl. unpleasant
styrene	sharp, sweet	unpleasant
Solvesso 100	none found	sl. unpleasant
butoxyethanol, 2-	sweet, ester	pleasant
heptanol	sweet	pleasant
ethyl benzene	aromatic	unpleasant
methylmercaptoaniline	none found	none found
methyl styrene	sweet, aromatic	pleasant
Solvesso 150	none found	sl. unpleasant
furfural	almonds	sl. unpleasant
methacrylic acid	none found	unpleasant
methyl methacrylate	sweet, sharp	unpleasant
EGMHE**	none found	none found
propionic anhydride	none found	sl. unpleasant
xylenes, mixed	sweet	neutral to pleasant
methyl nonyl ketone	none found	sl. unpleasant
methyl ethyl ketone	sweet, sharp	neutral to unpleasant
chlorobenzene	sweet, almond	unpleasant
kerosene	none found	unpleasant
methyl isobutyl ketone	sweet, sharp	sl. unpleasant
diisobutyl ketone	sweet, ester	pleasant
ethoxyethanol	sweet, musty	neutral
toluene	sour, burnt	unpleasant
ethylene glycol	sweet	sl. unpleasant
hexylene glycol	none found	sl. unpleasant
ethanol	sweet	neutral
acetone	sweet, fruity	pleasant to neutral
butyl alcohol, t-	camphor-like	unpleasant
dioxane	sweet, alcohol	neutral
diacetone alcohol	sweet	pleasant to unpleasant
nitromethane	mild, fruity	neutral
heptane	gasoline-like	neutral
decane, n-	none found	neutral
naphtha	none found	none found

* PGMMEA is propylene glycol monomethyl ether acetate
** EGMHE is ethylene glycol monohexyl ether

Table III. Complaint to detection ratios for compounds
with pleasant to neutral hedonic tones.

CHEMICAL	50% COMPLAINT LEVEL 50% DETECTION LEVEL
dioxane	2.6
heptanol	3.1
heptane	3.5
xylenes, mixed	3.8
biphenyl	4.1
diisobutyl ketone	4.8
methyl styrene	4.4
isobutyl acetate	5.0
nitromethane	5.1
butyoxyethanol, 2-	5.5
acetone	6.3
decane, n-	7.9
ethoxyethanol acetate	15.8
ethyoxyethanol	45.7
ethanol	60.9
MEAN	11.7
MEDIAN	5.0

Table IV.   Complaint to detection ratios for
compounds with unpleasant hedonic tones.

CHEMICAL	50% COMPLAINT LEVEL / 50% DETECTION LEVEL
hexylene glycol	1.8
methyl methacrylate	2.2
Solvesso 150	2.2
methyl nonyl ketone	2.3
ethylene glycol	2.3
methacrylic acid	2.5
hydrogen sulfide	2.8
methyl acrylate	2.8
propionic anhydride	3.3
toluene	3.3
butyl alcohol, t-	3.4
furfural	3.6
methyl amyl ketone	3.7
Solvesso 100	3.7
methyl isobutyl ketone	3.9
styrene	3.9
PGMMEA	4.2
kerosene	4.3
chlorobenzene	4.9
ethyl benzene	8.0
isopropyl benzene	10.2
methyl ethyl ketone	12.6
MEAN	4.2
MEDIAN	3.4

IV. Odor Measurements, Modeling and Technology

STANDARD TEST METHOD FOR EVALUATING THE ODOUR POTENTIAL OF AUTOMOTIVE COATINGS

M.R.E. Rix
Air Emissions and Control Section
ORTECH International
2395 Speakman Drive
Mississauga, Ontario
L5K 1B3

In this test method the potential for odour emissions from automotive coatings is determined using a set of procedures designed to evaluate both spraying and baking operations. Emissions from a laboratory spray booth are sampled for odours and hydrocarbons during spraying of test panels to prescribed specifications. Then emissions from a laboratory oven are sampled as the same test panels are baked.

Odour samples are evaluated for detection and recognition thresholds and for intensity. Standard tests for viscosity, VOC content and density are performed on the paint. Paint application data are recorded to enable the calculation of transfer efficiencies, booth/oven solvent split ratios, and to allow the odour emission rate to be related to paint weight, total paint sprayed or paint film thickness.

Replicate tests on one paint, used as a control in the method, have indicated that results are reproducible. Tests on over 30 other paints show a wide range in odour emission potential.

INTRODUCTION

Odour emissions are one of the main environmental concerns of automotive plants. At some locations continued operation has been threatened due to pressure from community groups and regulatory agencies. Emission control devices are expensive to operate on the high volume, low concentration emissions from paint spray booths. Control strategies tend, therefore, to reduce the amount of odour generated by increasing paint transfer efficiencies or modifying paint formulations.

The modification of paints may appear to be an attractive solution to the paint odour problem, since no change in operations and no capital investment is usually required. However, implementation of this approach has many difficulties. Formulations have often been changed by removing solvents from a paint using a "hit list" of known odorous solvents. This reduces the number of solvents available to the paint formulator, and does not always achieve the desired goal, since some published odour threshold values used to generate the "hit list" are unreliable. This approach also overlooks potential interactions of solvents in a formulation, and the effects of minor impurities in solvents on the odour emissions.

A test procedure is required to determine whether a paint formulated to reduce odour actually achieves that goal. Various procedures have been used in the past, but many lacked adequate attention to some details. Various paint application parameters, in particulate, may have a significant effect on the reliability of the test procedures.

In 1988, the Ford Motor Company designed a detailed test procedure to evaluate the odour emission potential of paint formulations.[1] This procedure required expertise both in paint technology and emission monitoring, and ORTECH International, with capabilities in these areas, was selected by Ford to implement the method.

This paper describes the procedure, as implemented at ORTECH, which has been used now to evaluate over 30 paints of various types from several suppliers.

METHODOLOGY

The test procedure involves spraying steel test panels to prescribed specifications in a laboratory spray booth while sampling emissions from the spray booth exhaust for odour and total hydrocarbon concentrations. Emissions are sampled in a similar way during baking of the same panels in a laboratory oven. For each test, three runs are performed in one day.

Paint Application and Baking

Table I lists the paint parameters recorded for each test. The paint characterization tests are performed at the beginning of the test day. The paint pot is filled with paint and weighed to within 0.1 gram. Steel test panels, usually three per run, are each weighed to within 0.1 mg. The panels are sprayed with the number of coats required at an appropriate application rate to achieve the required coating thickness. The test panels and paint pot are then weighed again. An industry standard DeVilbiss spray gun is used. The spray booth is a standard water curtain type, but is operated without water in order to eliminate any effects of the water on the emissions. After the required flash-off period for the paint type, the panels are placed in the electrically-heated forced draft oven for baking.

Measurement of any of the paint parameters requires experience with standard paint test procedures and equipment. The application of paint to the required thickness is particularly difficult. Earlier tests involved manual spraying of the panels, however a Spraymation automatic test panel spray unit is currently being used.

Emission Monitoring

Table II lists the emission parameters measured at the spray booth and bake oven. Odour samples are collected in 30 L Tedlar gas sample bags using a rigid sampling lung and vacuum pump. A steady, high sampling rate is required at the spray booth to obtain a sample over the 60 to 150 second spray period. A lower sampling rate is required for the 17 to 20 minute make period.

Total hydrocarbon concentrations are measured continuously during spraying and baking using Ratfisch RS5 or RS55 heated flame ionization analysers. An electronic data acquisition system is used to simplify integration of the data over the test period. The total hydrocarbon concentration in each Tedlar bag sample is also recorded, as a comparison to the integrated continuous analysis.

The volumetric flowrates of the spray booth and oven are measured periodically and following any changes to damper settings.

Odour Evaluation

The odour samples are evaluated within four hours of collection in a specialized odour evaluation facility at ORTECH, using a nine member odour panel and dynamic olfactometry. The odour evaluation facility is designed with an air exchange rate of 20 air changes per hour. Makeup air to the room is maintained at 24°C and 35% relative humidity and is filtered through activated charcoal. Furnishings and finishes are designed to minimize odour adsorption and emission.

The olfactometer is a binary system and is not operated in a forced choice mode. The volumetric flowrate to each nose piece is approximately 7 litres per minute. Samples are evaluated over a series of dilutions, starting at a high dilution, and proceeding in step decreases in dilutions by 1.41 times per step. At each dilution panelist responses are recorded for detection, recognition of the odour as characteristic of paints or solvents, and intensity, using a 0 to 9 point category estimation scale.

The dilution to detection threshold and dilution to recognition threshold are calculated from the odour panel responses.

Data Evaluation

Table III lists some of the values that may be calculated from the test results. Some of these are useful in evaluating the execution of the test (transfer efficiencies, etc.), and others are useful in quantifying the odour potential of the paint. it has been difficult to decide which of the various calculated values is most appropriate to describe the odour emission potential. The value currently used is one that eliminates the effect of slightly varying transfer efficiencies, and normalizes data to the amount of paint solids. The spray booth odour emission potential is given as the number of cubic meters of air at the odour detection threshold produced by spraying one gram of paint solids, and is calculated as follows:

$$\frac{\text{Dilution to odour detection threshold x volumetric flowrate x spray time x 100}}{\text{wt of paint sprayed x \% solids}}$$

The corresponding value for the bake oven is the number of cubic meters of air at the odour detection threshold produced by baking one gram of paint solids, calculated as follows:

$$\frac{\text{Dilution to odour detection threshold x volumetric flow x bake time}}{\text{wt of dry paint on panels}}$$

Because of the nature of the olfactory response, geometric means rather than arithmetic means are used to obtain a final value from the three replicate runs per test.

RESULTS

Over 30 different paints have been evaluated to date, including eleven triplicate runs of a black single shade paint used as a control. Most of the odour, typically 90 to 95%, is generated from spray booth emissions, and since these emissions are more difficult to control, attention in this discussion will be limited to the spray booth data.

Figure 1 shows the variation in spray booth odour emission potential in replicate tests with the control paint between February 1988 and May 1989. Initially, replicate tests were very consistent, but some higher numbers were obtained in the summer and fall of 1988. Some of this variation appeared to be due to variability in the odour panel evaluations, although some may also have been due to deterioration of the control paint over that period. While this variation in results obtained using the control paint is discouraging, the range of values obtained in tests of other paints suggests that the procedure is still very useful.

Figure 2 shows the values obtained for a variety of types of paints. Each data point represents the geometric means of three runs of a given paint sample. Values range overall from less than 300 to over 15,000. Within one category, basecoats, the range is also very large.

CONCLUSIONS

This work has demonstrated that it is possible, with the right equipment and expertise, to obtain a good estimate of the odour emission potential of a paint formulation. At present it appears that slight differences in odour potential cannot be reliably assessed. The range of odour emission potentials obtained in tests to date show that paints can be produced which have much less odour than others currently in used. Actual formulations are proprietary, and no correlations have been made between solvent composition and odour potential. However, it is clear that paint formulation is a means of reducing automotive paint odours.

REFERENCES

1. R.E. Devlin, "Paint Odour Evaluation Procedures", Ford Motor Company Stationary Source Environmental Control Office, Dearborn, Michigan. January 7, 1988.

Table I   Paint parameters recorded in tests.

Paint Character
    Paint type (basecoat, clearcoat etc.)
    VOC content
    Density
    Solids content
    Viscosity
Paint Application
    No. of panels sprayed
    No. of coats
    Initial panel weights
    Wet panel weights
    Initial and final paint pot weights
    Spray gun fluid flow
    Atomization air pressure
    Flash–off time
Paint Baking
    Oven temperature
    Bake time
    Paint thickness
    Dry panel weights

Table II Emission parameters recorded in tests.

Spray Booth and Bake Oven
    Continuous total hydrocarbon response
    Dilution to odour detection threshold
    Dilution to odour recognition threshold (recognition as a paint
      or solvent)
    Perceived intensity (category estimation)
    Volumetric flowrate
    Total hydrocarbon concentration in bag sample
    Sample time

Table III Calculated values from paint and emission data.

Transfer efficiency
Solvent loss from spray booth
Solvent loss from oven
Solvent split ratio, booth to oven
Total hydrocarbon concentration at odour detection threshold
Dilution to odour detection threshold per gram solvent emitted
Dilution to odour detection threshold per gram dry paint sprayed
    (spray booth)
Dilution to odour detection threshold per gram dry paint baked
    (bake oven)
Cubic metres of odour at the detection threshold per mil paint
    thickness
Cubic metres of odour at the detection threshold per gram paint
    solids applied (spray booth)
Cubic metres of odour at the detection threshold per gram paint
    solids sprayed (spray booth)
Cubic metres of odour at the detection threshold per gram paint
    solids baked (bake oven)

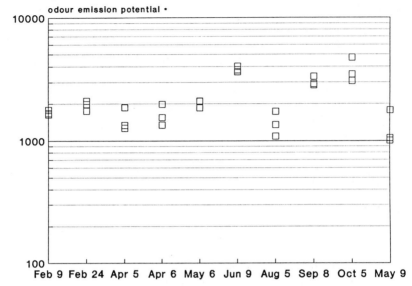

Figure 1   Replicate tests of control paint at spray booth.

• cubic metres of air at the odour detection threshold per gram of paint solids sprayed

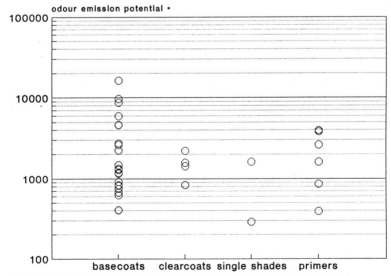

Figure 2   Range of odour emission potential by paint type.

• cubic metres of odour at the odour detection threshold per gram of paint solids sprayed

ODOR EXPERIENCE/RESEARCH AT METROPOLITAN
WASTE CONTROL COMMISSION

Robert Polta
Thomas Wahlberg
Mark Yonan
Air Quality Division
Quality Control Department
Metropolitan Waste Control Commission
St. Paul, Minnesota

The Minnesota Pollution Control Agency (MPCA) regulates point source odor emissions based on odor concentration units (OCU) and the elevation of the discharge above ground level. The MPCA regulations require that ASTM Method D-1391-57 be used to determine OCU values. Because ASTM indicated withdrawal of the D-1391 standard in 1986, the MWCC initiated an evaluation of dynamic, forced choice, triangle olfactometry.

Several experiments were conducted to characterize both the within and between laboratory variability using the olfactometer procedure. A large experiment was conducted to characterize the variability of odor sensitivity in the population and to determine the attributes of an individual which may impact that sensitivity. Finally, comparative data were collected on over 250 stack gas samples to determine the relationship between the OCU values generated by the old syringe dilution procedure and the newer olfactometer method.

Introduction

The Metropolitan Waste Control Commission (MWCC) owns and operates 11 wastewater treatment plants in the seven county area around, and including, the cities of Minneapolis and St. Paul. The MWCC also owns approximately 500 miles of interceptor sewer along with the associated pumping and metering stations, 63 and 172 respectively. Most all of these facilities were initially constructed in semi-isolated areas; however, because of recent development of adjacent properties, many of the facilities are now surrounded by residential and/or commercial properties. Because of this, the MWCC has become more interested in odor characterization and odor control technology during the past several years.

The Minnesota Pollution Control Agency (MPCA) regulates point source odor emissions based on odor concentration units (OCU). The OCU is defined as the number of dilutions to threshold using the old syringe dilution technique (ASTM D-1391-57). The MPCA rules limit discharges from stacks $\geq$50 ft to 150 OCU; whereas, discharges nearer the ground are limited to 25 OCU. In no case can the odor emission rate (OCU x gas flow rate in scfm) exceed 1,000,000.

If the MWCC violates the above limits, the agency may be subject to regulatory action and will certainly suffer some public ridicule. In addition, real odor problems may be observed on adjacent properties if the dispersion characteristics are not favorable. Because of the potential consequences, the MWCC does not want to be in the position of failing the MPCA limits because of the variability associated with the testing procedures. An investigation was thus initiated to determine the reliability of the procedures used to characterize the OCU of point source emissions.

Methods

At about the time this investigation was initiated (1985) ASTM indicated that the D-1391 standard would be withdrawn in 1986. The MWCC's experiences with the method to that time had not been well documented; however, there was evidence that significant variability could be introduced while preparing dilutions of high strength samples as well as during the presentation of the dilutions to panel members. The literature available at that time, as well as discussion with several experienced odor researchers, convinced MWCC staff that the method of choice for determining OCU was dynamic, forced choice, triangle olfactometry. As a result, the first of two IITRI olfactometers was purchased and put into service.

Odor Strength

The odor strength, or dilution to threshold, was determined by both the syringe dilution technique as per ASTM D-1391-57 as well as the IITRI olfactometer. In the latter case, the analyses were generally conducted according to the second draft of the standard method prepared by the ASTM Task Group E 18.04.23 on Odorous Air Pollution and dated April 24, 1986; and, the resulting threshold data were reported in terms of best estimate threshold (BET) as per ASTM E679-79.

Round Robin Experiments

Two round robin tests were conducted in 1986, using the olfactometer procedure, to characterize both within lab and between lab variability. The first experiment consisted of two tests in June 1986 (19th and 26th) at each of two laboratories (MWCC and consultant). The samples consisted of 50/50 mixtures of a known concentration of butanol in hydrocarbon free air and additional hydrocarbon free air, both supplied by Scott Specialty Gases.

The ratio was controlled by precision rotometers on each cylinder; and, the resulting mixture was stored in new tedlar bags purged with the mixture. The gas samples were made up the day prior to the test. The nominal butanol concentrations were 47.4 ppm and 54.4 ppm (v/v) for the two tests.

Each laboratory received two samples on each test day and replicate analyses were conducted for each sample. The order of presentation of samples and replicates ($\Sigma n=4$) was random. The number of panel members was between 7 and 10 for all tests.

The second experiment was conducted on August 22 and involved the same consultant, a local industry, as well as the MWCC. Each of the three laboratories received two butanol samples, prepared as above, with nominal concentrations of 54.4 ppm butanol. In addition, each laboratory received two tedlar bags containing a sample of gas from the discharge stack for a sewage sludge incinerator. These six sample bags were filled sequentially at the incinerator site and distributed to the three labs on a random basis. As above, all samples were prepared/collected the day prior to the test.

Population Variability

On November 28 and 29, 1987 a large experiment was conducted at a regional shopping mall to determine the relationship between odor sensitivity and the characteristics of potential panel members. Butanol at a nominal concentration of 50 ppm was administered to a total of 245 individuals on each of two IITRI olfactometers. The butanol was made up as previously described the day before testing and stored in large tedlar bags. Each of the individuals tested completed a questionnaire addressing: age, sex, smoking status, use of perfume/cologne, etc.

Results

Round Robins

The results of the first experiment are summarized in Table I. The MWCC data demonstrate much more variability than the consultant's and the BET values are also much higher. If the replicate analyses for each day are combined via the geometric mean of the BET values, the hypothesis that the mean MWCC BET value equals the mean consultant value can be tested. For a type I error (α) of 0.01 (probability of rejecting the hypothesis when, in fact, it is true = 0.01) the hypothesis is rejected using the students t-distribution.[1] This significant difference between the two labs may have been due to subtle differences in procedures and/or equipment, although they were nominally identical. The panel makeup could also have been responsible for all or part of the between lab variability.

The results of the second experiment are presented in Table II. Here again there are large differences between the consultant's and the MWCC's values for the BET of the butanol sample, with the industry's values falling in between. For the stack sample, however, the consultant's and the MWCC's values are very close (670 and 705 respectively on geometric mean basis) and the industry's value (312) was substantially lower.

When all of the BET values in Table I are considered together the log mean BET is 55 and the standard deviation is 0.42 on a log basis. The butanol data presented in Table II yield a log mean BET of 90 and a standard deviation of 0.39 on a log basis; whereas, the stack gas samples yield a log mean BET of 542 and a standard deviation of 0.17 on a log basis. Dravnieks, et al.,[2] reported a standard deviation of 0.37 (log basis) for four panels of nine different individuals evaluating butanol. It thus appears that the results of this experiment are consistent with the published data.

For both experiments the standard deviation of the individual panelist's BET values, on a logarithmic basis, varied between 0.19 and 1.02 and averaged approximately 0.5 and 0.6 for the butanol and stack samples respectively.

Population Variability

The raw data and detailed analyses are available in an MWCC report.[3] When all of the tests are considered en masse the resulting log mean BET value is 1.6306 and the log standard deviation is 0.554. The average difference between the BET values for the two olfactometers was -0.0636 with a standard deviation of 0.6559, both on a log basis. For $\alpha=0.05$, the hypothesis of equality between the two instruments cannot be rejected. The data summarized in Figure 1 demonstrate that the distribution of sensitivities in the population is log normal, at least for butanol.

Previous work by Koe and Brady[4] with H_2S and a panel size of 21 also demonstrated a log normal distribution with a standard deviation of 0.5 on a log scale. Brown, et al.,[5] collected odor threshold data for eight compounds using a 60 member odor panel. Although the authors did not calculate the standard deviation of the individual values they did provide sufficient raw data to do so. The compounds and standard deviations (log basis) are as follows: 2 mercapto ethanol - 0.4, 1 methol - 0.4, ethylene dichloride - 0.44, α propylbenzylalcohol - 0.4, d camphor -0.87, formic acid - 0.4, potassium cyanide - >>1, and ammonia - 0.55. Plots of hundreds of odor panel tests for n in the range of 5 - 12 individuals, including those described in the previous section, demonstrate that the log normal distribution is appropriate for categorizing the data.

The data collected during the experiment were uniformly scattered both within and between days. This indicates that the olfactometers and butanol standard did not change significantly as a function of time during the experiment.

Each of the 245 individuals was characterized by age ($\leq$19, 20-39, $\geq$40), sex (m,f), smoking status (yes, no), and day tested (1,2). Simple hypotheses were tested using the t statistic as summarized in Table III. Based on these tests, individuals that are 40 years of age or older are significantly less sensitive to butanol odors. In addition the individuals tested on day 2 were significantly more sensitive.

Because the above simple tests were all confounded by the remaining variables, the data were subjected to what is generally termed analysis of cross classified categorical data using the log linear model. The results of that analysis suggest that age is the only variable that has any significant impact on an individual's sensitivity to butanol odor.[3]

During the period October 1986 - April 1989, approximately 260 data sets, comparing the syringe dilution and olfactometer methods, were collected for sludge incinerator stack gas samples. A total of eight incinerators, with various types of wet scrubbers, are located at two of MWCC's treatment facilities. The data are presented in Figure 2 and the summary statistics in Table IV.

Discussion

Variability Characterization

If it is assumed that the above characterization of variability for butanol also applies to other odorants then confidence intervals for odor panel data can be established and a rational odor monitoring program can be implemented to minimize the probability of failing a standard because of poor testing reliability.

The OCU values obtained by both the syringe dilution and olfactometer method are estimates of the population median OCU. In the MWCC's case the population can be defined as the approximately two million residents of the Minneapolis-St. Paul metropolitan area. If it is assumed that odor panelists are chosen randomly from the population and that the distribution of OCU, for any given sample, is log normal with a standard deviation of 0.5 (log basis), then the confidence interval (CI) can be calculated for the log mean panel response as follows.

$$CI = \bar{X} \pm \frac{ts}{\sqrt{n}}$$

Where: $\bar{X}$ = the average of the logarithms of the dilution to threshold for each of n panel members, s = 0.5, and t = student t variate.[6]

Consider for example a report of 257 OCU for a panel of seven. In log form the 95% CI is determined as follows

$$CI = \log 257 \pm \frac{2.365 \cdot 0.5}{\sqrt{7}}$$

$$CI = 2.40993 \pm 0.44694 = 1.96299 ---> 2.85687$$

In arithmetic form the 95% CI is 92 to 719. If the sample in question was taken from a tall stack the MPCA limit would be 150 OCU and a violation would be reported. There is some uncertainty, however, if the "true" value is ≥ 150. The reported value may be high because the panel is biased with sensitive individuals. The uncertainty can be reduced by increasing the number of panelists (n). For example, if 257 OCU is reported for a panel of 20, the 95% CI is 150 to 440 OCU. In this later case both the reporting and regulating agencies are relatively certain a violation did occur.

Unfortunately the MWCC panel members are not chosen randomly from the population. Odor samples are routinely collected during morning hours and the analysis conducted in the afternoon. Although the MWCC pays panelists $6.10 per hour (for a minimum of 4 hours each test day), it is difficult to recruit younger people, males in particular. As of August, 1989 the active panel pool consists of 12 males with an age range of 18 to 74 and 18 females with an age range of 21 to 66. The average ages for males and females are essentially equal (48.5 and 48.8 respectively). It should be possible to normalize the panel by comparing their individual responses to butanol to that obtained from the population variability experiment. This has not been done to date, however, because it is not known if an individual's relative sensitivity to odors is independent of odor source/type. Limited MWCC data indicate that although an individual may be sensitive to butanol he/she may be insensitive to stack gas odors.

The question of random selection for panel members may be partially moot. An examination of 35 sets of data for stack gas analysis demonstrated that for each sample the distribution of the individual panelist's OCU values was log normal and the average standard deviation was 0.47. The observed variability of OCU values during normal odor panel tests is thus approximately equal to the estimate of population variability. The age of the panelists and the potential associated bias remains an issue however.

Method Comparison

The method comparison data demonstrate that there is no strong relationship between OCU values generated by the two methods (R^2=0.0024). It is interesting to note, however, that the olfactometer values are consistently higher, and, on average are approximately 4x the syringe dilution values. The comparisons presented in Figure 2 and Table IV are, however, confounded by the fact that more panel members were generally used with the olfactometer method because it is faster. The OCU values for the olfactometer method are being recalculated using only the responses from panelists that served for both methods. Preliminary analyses of these data (n=75) again yield a poor relationship between the two methods (R^2=0.06; however, the olfactometer, on average, yields OCU values 3.7x higher than the syringe dilution value.

The old ASTM definition of odor detection threshold was "the concentration of an odorous emission in air at which it is just barely sufficient to detect the presence of odorous materials in the air."[7] If this definition is accepted, it can be readily demonstrated that the values generated by the olfactometer have a substantial positive bias.[8] This bias is based on the relatively high probability that each panelist will make the "right" guess at dilutions below his/her threshold. For example, if an olfactometer using the typical dilutions of 7, 20, 60, 180, 520, and 1560 is used to test a sample with a "true" OCU value of 104 (geometric mean of 60 and 180) for an individual panel member, the expected reported value for that same individual will be approximately 300.

Although this systematic bias is high it does not, by itself, explain the larger differences summarized in Figure 2 and Table IV. The remaining bias, as well as the significant scatter, may be due to systematic and random errors associated with the syringe dilution technique. This uncertainty puts the MWCC in an awkward position. Although the olfactometer procedure appears to be the industry standard for estimating OCU values, the MWCC cannot petition the MPCA to adopt the procedure without also addressing the MPCA's numerical limits. Adopting the new analytical procedure without a concurrent increase in the limits by a factor of approximately 4x would result in certain failure of the MPCA regulation. Until such time that the scatter in Figure 2 can be adequately explained and a relationship developed between the OCU values generated by the two methods, the MWCC has no rational argument for increasing the limits.

Conclusions

Several experiments were conducted to characterize the variability, or uncertainty, associated with the determination of OCU values using the olfactometer procedure. The data collected from these experiments support numerous previous observations that the distribution of odor sensitivities in the population, in terms of OCU values, is log normally distributed with a standard deviation of approximately 0.5 on a log basis. Given the standard deviation and the panel size, confidence intervals can be calculated for reported OCU values; and, the panel size can be increased as necessary (at least theoretically) to minimize the probability of failing a regulatory limit because of normal test variability.

Based on the analyses of 266 stack gas samples there is no obvious relationship between the OCU values obtained using the syringe dilution and olfactometer test methods. The olfactometer procedure routinely yields larger OCU values and, on average, results in OCU values 4x those obtained via the syringe dilution method. Although it can be demonstrated that part of this bias is directly attributed to the triangle, force choice procedure, it remains to be demonstrated that a relationship exists between the two data sets. Until such time that this relationship is demonstrated, the MWCC cannot petition the MPCA to adopt the olfactometer procedure for odor compliance monitoring.

References

1.   W.H. Beyer, CRC Handbook of Tables for Probability and Statistics, 2nd ed. CRC Press, Inc., Boca Raton, FL, 1985, p. 282.

2.   A. Dravnieks, W. Schmidtsdorff, and M. Meilgaard, "Odor thresholds by forced-choice dynamic triangle olfactometer: Reproducibility and Methods of Calculation," Journal of Air Pollution Control Association. 36:900. (1986).

3.   R.C. Polta and M.N. Ho, Variability of Stack Gas Odor Evaluation Technical Memorandum No. 3, QC88-143. Metropolitan Waste Control Commission, St. Paul, MN, 1988.

4.   C.L. Koe and D. K. Brady, Quantification of Sewage Odors, Research Report No. CE 40, Dept. of Civil Eng., University of Queensland, St. Lucia, Australia, 1983.

5.   K.S. Brown, C.M. McClean, and R.R. Robinette, "The distribution of the sensitivity to chemical odors in man," Human Biology 40:456, 1986.

6.   J.K. Taylor, Quality Assurance of Chemical Measurement, Lewis Publishers, Inc., Chelsen, Michigan, 1987, pp. 27-28.

7.   ASTM, "Standard for the Measurement of Odor in Atmospheres (Dilution Method) - Designation D1391-78," 1984 Annual Book of ASTM Standards, American Society for Testing and Materials, Philadelphia, PA, 1984.

8.   R.C. Polta and R.L. Jacobson, Letter to A. Dravnieks, May 19, 1986.

Table I.   Results of 1st round robin.

Date	Lab	Sample	Test Order	n	Panel BET	Std. Dev.*
6/19/86	MWCC	A	1	7	99	0.43
		A	4	7	341	0.39
		B	2	7	86	0.70
		B	3	7	136	0.74
	Con	A	1	10	18	0.49
		A	3	10	18	0.49
		B	2	10	42	0.40
		B	4	10	35	0.55
6/26/86	MWCC	A	1	9	66	0.61
		A	2	9	175	0.83
		B	3	9	121	0.94
		B	4	9	149	0.39
	Con	A	2	10	20	0.22
		A	3	10	18	0.30
		B	1	10	20	0.46
		B	4	10	28	0.34

*log std dev of individual BET values

Table II.   Results of 2nd round robin.

Lab	Sample*	n	Panel BET	Std Dev**
Industry	B-1	10	195	0.87
	B-5	9	93	0.48
Consultant	B-2	8	30	0.42
	B-4	8	34	0.23
MWCC	B-3	6	110	0.42
	B-6	6	270	0.19
Industry	S-4	10	363	1.02
	S-5	10	269	0.73
Consultant	S-1	8	760	0.45
	S-2	8	590	0.48
MWCC	S-6	6	705	0.40
	S-3	6	705	0.40

 * B = butanol sample; S = stack sample
**log std dev of individual BET values

Table III. Results of simple hypothesis testing - α = 0.05.

Variable	Hypothesis	Status
Sex	mean male = mean female	accept
Smoke	mean smoke = mean nonsmoke	accept
day	mean day 1 = mean day 2	reject
age*	mean A = mean B	accept
	mean A = mean C	reject
	mean B = mean C	reject

*A (≤19); B (20-39); C(240)

Table IV. Method comparison summary statistics.

Parameter	Odor Test Method	
	Syringe Dilution	Olfactometer
Tests	266	266
Average panel size	5.4	7.5
Average OCU	92.8	382.1
Std. Dev. OCU	50.5	294.4
Max. OCU	408	3388
Min. OCU	19	76
Coefficient of Correlation	0.049	

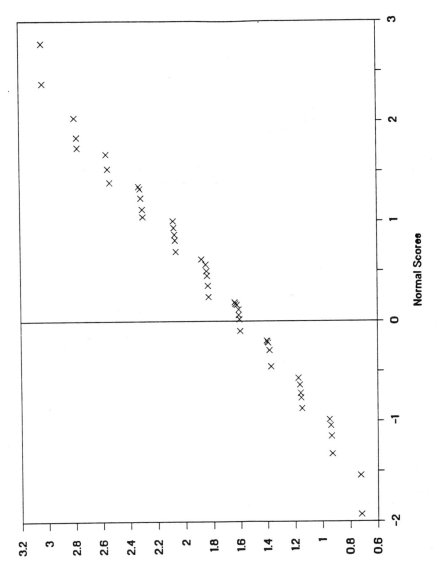

Figure 1.   Distribution of population odor sensitivity to butanol.

264

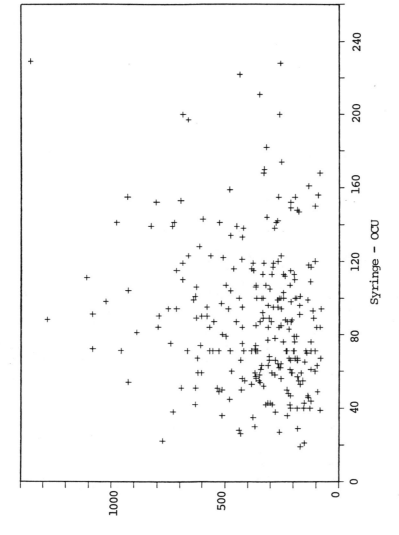

Figure 2.   Comparison of syringe and olfactometer methods.

265

AN APPROACH TO COMMUNITY ODOR SURVEYS

Kenneth E. Dowell
Stationary Source Environmental
  Control Office
Ford Motor Company
Dearborn, Michigan

An important element in addressing community odor complaints in the vicinity of manufacturing facilities is the objective definition of the problem. An analysis of agency and plant complaint records provides a preliminary indication of the problem but cannot be used solely as the basis of developing an odor abatement plan. A properly designed community odor survey documents actual community odor parameters. The effectiveness of any process and/or hardware improvements are assessed by periodically repeating the community odor survey and comparing results to some baseline evaluation. The odor parameters frequently considered are character, intensity, and areal extent of community odors.

Five trained observers survey the community twice daily for a two or four week period. The field manager assigns observer travel routes before each observation period. Current wind conditions provided from an on-site or nearby meteorological system dictate the area for study each period. Sixteen pre-determined route maps, one for each primary compass directions, are defined beforehand. At each of seven observation stops per observer, ten odor observations are taken at one minute intervals. Each observer notes the presence of an odor and, if present, the odor character and intensity.

Field observations of sampling location, odor intensity, and odor character are encoded for analysis. A typical four week study period (two observation periods per day, five observers) can yield over 14,000 observations. Actual wind speed and direction information supplements each survey. Odor statistics by location, frequency, intensity, and character provide a basis to develop an odor abatement program. Typical results from recent studies are discussed.

INTRODUCTION

Manufacturing facilities operate certain production processes that can occasionally contribute to community odors. Some examples include automotive painting and component casting operations. Community odor levels can become excessive due to control equipment failure, increased production levels, or increased biological activity in control systems. This paper delineates an approach used by Ford Motor Company to identify the source of odors around production facilities.

Addressing community odor complaints requires an objective definition of the character, intensity, frequency, and areal extent of odors. Without such definition, meaningful action cannot be expected. Corrective actions taken on the wrong source will not reduce community odor levels. Change will not occur until the responsible odor source is identified.

Agency and plant complaint records provide a preliminary insight on odor frequency and areal extent. Complaint information, though, can rarely be used alone to address odor character and intensity since the complainants cannot objectively describe the odor with enough detail to identify the source. Such sketchy information cannot be used as the sole basis of an odor abatement program.

Through the community odor survey approach described here, actual community odor parameters are documented in a consistent and systematic manner. Independent observers, trained to differentiate and report all odors by location, repeatedly cover the community downwind from the suspect source. Statistical analyses of the data form the basis for problem resolution. The effectiveness of any process modifications can be monitored by periodically repeating the survey process.

METHODS

Organization.

The personnel required to conduct a community odor survey and their roles include:

-- **Field team leader.** Initially, the field team leader qualifies, hires, trains, and orients a team of independent observers. The leader directs the field program and provides field support for the observers.

-- **Trained observers.** Five regular and one backup observers are trained and qualified. Screening is performed to assure that the team members can differentiate between distinct odors and can discern relative odor intensities using standardized butanol-in water samples. Potential observers with colds or sinus conditions are excluded.

-- **Data Reduction and Analysis.** An analyst or engineer reduces the data and prepares descriptive statistics, charts, and tables. This function can be performed by the consultant or sponsor.

Sampling Strategy and Procedures

Field observation periods are scheduled for times corresponding to normal production. Scheduled break and lunch times are avoided. For multi-shift operations, the sampling schedule should include all time periods such as morning, afternoon, evening, and nighttime.

Observer travel routes are assigned by the field manager just prior to each observation period based on current and recent past wind conditions. A 45 degree sampling sector is selected for observer deployment. Route maps are selected from a set of sixteen pre-determined routes, one for each primary compass direction.

Assigned routes typically start approximately two miles from the manufacturing location. Subsequent observation stops are located incrementally closer to the plant. The final observation point of each route is located as close as practical to plant property. Figure 1 depicts a typical observer route map. Note that the observers start away from the study focal point to minimize any potential olfactory fatigue or bias.

At each observation stop, ten odor observations are taken at one minute intervals. The observer then proceeds to the next closer sampling location.

Odor samples from key processes and locations characterized by distinct and characteristic odors are made available for observer reference both before and after each survey. Daily pre- and post-sampling odor orientation reinforces the skill to make subtle odor distinctions in the field.

Observer guidelines include:

i.   Perfume and deodorants are prohibited.

ii.   No spicy foods prior to the survey.

iii.   No food or drink other than water during the survey.

Wind observations are recorded prior to and during each survey for use in analyzing survey results.

Documentation

Observers are provided a clipboard, route map, and data sheet. The observer data sheet provides areas to record exact observation location, time, odor intensity and character. A comments section is provided for important notes regarding changing weather conditions or unusual activities in the area.

Odor intensities are reported using the following scale:

0	No odor	No odor perceived or an odor so weak that it cannot be readily characterized or described.
1	Slight	Identifiable odor, slight.
2	Moderate	Identifiable odor, moderate.

(continued on next page)

3	Strong	Identifiable odor, strong.
4	Extreme	Severe odor, one where the observer is compelled to leave the area.

All odors, whether plant-related or not, are recorded using this odor intensity scale.

TYPICAL RESULTS

Example statistics are presented here based on two actual studies conducted around one facility. One study was conducted in the winter. The second study was conducted the following summer after some process changes. The subject facility is an automotive assembly plant located in an urban area. The initial study documented the type and nature of plant-related odors in the community. The second study was performed to document the impact of the process modifications.

Community levels of three distinct odors were sought: spraybooth solvent, paint bake oven, and electrocoat. All spraybooths at the facility are exhausted through a single central exhaust stack at approximately 75 degrees F. Four paint bake ovens are individually incinerated and separately exhausted to atmosphere. The electrocoat tank is closed; tank ventilation air is incinerated with the system's oven incinerator.

Statistics presented in Table I summarize and compare odor detection frequency by odor characteristic for both surveys. Over 14,000 odor observations were collected during the first four week study using five observers, seven stops per observer route, and ten observations per stop. An additional 14,000 observations were made during the second survey. Upwind observations by the field manager supplemented the regular downwind observations.

Summary statistics shown in Table I suggest that odors, both plant-related and others, were detected for 32% (winter) and 40% (summer) of the observations. Plant-related odors were detected 14% of the total possible times or nearly 40% of the total odors detected. Given the 'plume chasing' nature of this study format, the long-term probability of detecting a plant-related odor at any one point around the plant is much less than 14% due to normal wind direction variations over time.

The number of spraybooth/solvent odor detections increased from 585 to 1134 between winter and summer. In the winter, the temperature contrast between spraybooth exhaust (75 deg F) and ambient air is frequently significant, adding to plume rise. Plume buoyancy is much less pronounced in the summer since plume and ambient temperature differences are minimal. Since the process and process rate did not change, the observed increase in spraybooth/solvent odors must be related to seasonal temperature differences.

The absolute number of oven-related odor observations decreased from 854 to 135 observations, a reflection of oven incinerator process modifications performed between the two test periods. As a percentage of plant-related odors, the decrease was from 47.1% to 7.0%. Since the oven incinerator exhausts are always heated well above ambient, seasonal temperature variations would exert small influence on plume dispersion characteristics.

The frequency of electrocoat odors in the community increased from 375 in winter to 654 in summer. Electrocoat odors were specified for approximately one third of the plant-related odors. Further investigation of the process was suggested by this information.

'Other' odor detections increased from 2867 in the winter to 3524 during the summer study. Characteristic winter odors included wood smoke from fireplaces. Grass and weed odors were commonly observed during the summer. Asphalt road construction odors were also occasionally noted. Traffic exhaust fumes from diesel engines were noted both seasons.

Downwind Distance Analysis

Figure 2 shows odor detection frequencies as a function of a normalized downwind distance from the plant. These data show that plant-related odors increased in frequency of detection with distance from the plant, especially solvent and electrocoat odors. The frequency of 'other' odors remained rather constant with downwind distance indicating a uniform distribution of 'other' odors.

Meteorology during Survey.

Average meteorology was calculated for each observation period. A comparison of the observed mean wind angle to the pre-survey wind angle showed a mean absolute angle difference of 28 degrees. Absolute wind angle differences were less than 45 degrees for 64 out of 80 observation periods. This suggests that the wind did not shift significantly during 80% of the survey periods.

CONCLUSIONS

The systematic study of odors downwind of a potential odor source yields information that allows:

a.   a definition of the extent of detectable plant-related odors; and,

b.   an assessment of the odor impact of multiple diverse processes.

Information from such studies provide a basis for an effective odor abatement program. The impact of process and/or hardware changes can also be assessed by periodically conducting community odor survey and comparing results to baseline statistics.

Table I.

Summary Statistics Comparison

Winter Observation Period
versus
Summer Observation Period

Number of Odor Detections (% of total)

Characterization Category	Winter	Summer
Electrocoat	375 (20.7%)	654 (34.0%)
Oven	854 (47.1%)	135 (7.0%)
Spraybooth/Solvent	585 (32.2%)	1134 (59.0%)
Total, plant-related	1814	1923
Other odors, not plant-related	2867	3524
Total number of observations with plant-related and other odors	4681	5447
Total observations made	14,580	13,730

Table II.

Community Odor Survey Summary Statistics

Number of Odor Detections by Intensity and Character
Summer Only

Intensity	ELECTROCOAT	SPRAYBOOTH/ SOLVENT	OVEN	OTHER	TOTAL
1	419	833	100	2953	4304
2	187	258	33	499	977
3	37	31	1	64	133
4	10	12	0	8	30
Total	654	1134	135	3524	5447

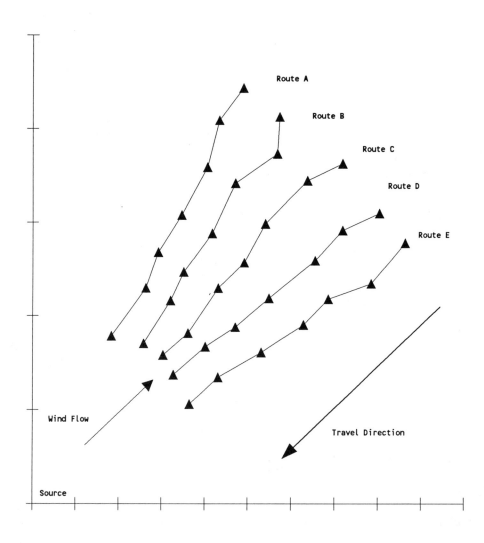

Figure 1   Observer sampling route example.

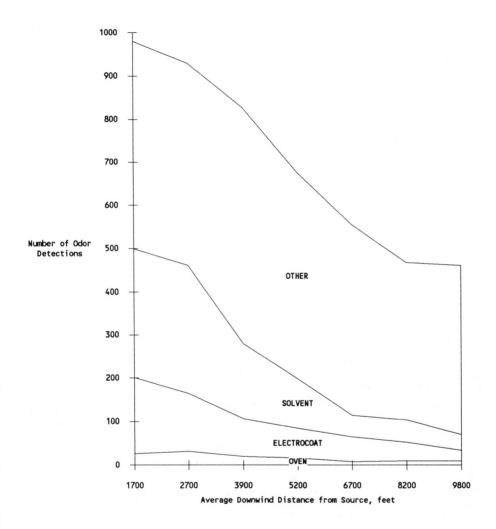

Figure 2   Community odor detection frequency and character as a function of
downwind distance.

ASTM Task Group E-18.04.25 on Sensory Thresholds

Morten C. Meilgaard, D.Sc.
Task Group Chairman
The Stroh Brewery Company
Detroit, Michigan

SUMMARY

The ASTM Task Group on Sensory Thresholds is looking for
collaboration with active users of odor threshold (ED_{50}) tests.
The Task Group's current proposals can still be modified; these
are (a) to reclassify ASTM Standard E679 as a rapid method,
mainly for use with small data sets of approx. 50-100 presentations; (b) to recommend that data sets of over approx. 300
presentations be treated as "research projects" which can be
handled by any of a number of methods such as French Standard
NF X 43-101; and (c) that intermediate data sets of approx.
100-300 presentations be the subject of a new ASTM Standard.
This will require that thresholds be determined for one individual at a time, using either a probability graph or iterative
curve fitting by computer; each individual threshold must be
reported; a group threshold can be calculated if the purpose of
the test so requires and only if the individual values form a
near-normal distribution.

INTRODUCTION

At the time of writing, the International Standards Organization in Geneva is working to standardize methods of determining sensory thresholds. This work is carried out in Subcommittee SC12 "Sensory Analysis" under Technical Committee TC34 "Agricultural Food Products". 39 countries are members of SC12; the US is a P-member (participating) which is the highest class. The US is represented by ANSI (American National Standards Institute) in New York and the actual work is done by ASTM Committee E-18 on Sensory Evaluation of Products and Materials. E-18 meets every 6 months and has 231 voting members. Our Task Group E-18.04.25 on Sensory Thresholds has 29 members. Its assignments are (a) to help prepare US submissions to the ISO on the subject of sensory thresholds; (b) to help revise ASTM Standard Practice E679-79 "Determination of Odor and Taste Thresholds by a Forced-Choice Ascending Concentration Series Method of Limits" (1) which is overdue for its routine 5-year revision, and (c) if a need is found to exist, to propose a new standard suitable for determining and calculating sensory thresholds for any substance in any medium. - A sister Task Group, E-18.04.21 on Atmospheric Odor Pollution, is in formation and will deal with items such as sample collection and the construction of olfactometers.

DEFINITION OF SENSORY THRESHOLD

Two basic possibilities exist:

(a) the stimulus value corresponding to a probability of perception of 0.5; or

(b) the stimulus value corresponding to 50% detection above chance,

under the conditions of the test. Definition (b) looks only at the data set at hand and arrives at the 50%-point by conventional interpolation while making a minimum number of assumptions. Definition (a), the one favored by most authorities, including this Task Group, makes the assumption that the data points belong to a normal (Gaussian) distribution.

Various symbols are in use for the dilution factor in olfactometry, e.g. Z, D, ED and K, with the threshold denoted as Z_{50}, D/T, ED_{50} and K_{50} respectively. The Task Group favors Z because it is the one used in E679, because it is specific to olfactometry, being named after H. Zwaardemaker, a Dutch pioneer in the field, and because it is less complex (when used in formulas and equations) than symbols consisting of more than one letter.

FRENCH STANDARD X 43-101

The ISO TC34/SC12 has before it a French and a German standard and also the E679 (see later); a Dutch standard is in preparation. The German Standard VDI 3881 "Olfactometry: Odor Threshold Determination" (2) is fairly general and allows many procedures. In contrast, the French Standard NF X 43-101 "Air Quality: Method of Measuring the Odor Intensity of Gaseous Effluents" (3) describes in detail a specific procedure aimed at air pollution testing. It applies to

"- in situ assessment of gaseous releases from industries"
"- checking devices for odor purification" and
"- laboratory inspection of samples of odorous gases"

Briefly, X 43-101 specifies the following procedure:

No. of Panelists

- 16 or more when a population sample is needed;
- 8 in most cases;
- 4 for comparative measurements.

Method of Presentation:

Three channels, two of which contain inodorous air (this type of triangle is called the Three-Alternative Forced Choice or 3-AFC Presentation).

Estimation of Individual Threshold

The individual receives a minimum of ten 3-AFC tests at each of three concentrations spaced $\sqrt{2}$ apart. The three concentrations are chosen by preliminary experiments using concentrations spaced 3-fold apart. Thus each individual receives approx. 3x10 + 8 = 38 presentations and an average group test using 8 panelists requires approx. 8x38 = 304 presentations. The individual threshold is calculated by plotting the data in a probability graph (Fig. 1).

Estimation of the Distribution of Thresholds in a Population

The panelist thresholds are sorted by rank i and plotted in a probability graph (Fig. 2), using as the ordinate the

"Rank Position" $F_i = 100 \, i/(n+1)$. For example, panelist no. 10 out of a panel of 20 would plot at $100 \times 10/(20+1) = 47.6\%$. If a straight line can be drawn through the points, the group is considered normally distributed with group threshold at the 50%-point and 1-σ confidence limits at the 16% and 84% points.

ASTM STANDARD METHOD E679

. This 1979 standard, written by the late Andrew Dravnieks, is at present widely used in North America in connection with the Illinois Institute of Technology's Dynamic Triangle Olfactometer. It is based on the 3-AFC method of presentation and the panelists (usually 8-16) receive only a single set of six 3-AFC presentations, i.e. 48-96 presentations per group threshold test. Concentrations are spaced 3-fold apart and all panelists receive the same six concentrations. Results are calculated as shown in Fig. 3. Briefly, a Best Estimate Threshold (BET) is calculated for each panelist as the geometric mean between the highest concentration missed, and the next higher concentration. Subsequently the Group BET is calculated as the geometric mean of the individual BET's. The instruction notes that the range of concentrations presented should be established by preliminary tests and should be such that the responses of a group of nine panelists distribute over the 3-4 middle concentrations. It is also noted that if the purpose of the test is to establish the sensitivity of the individual panelists (or the distribution of thresholds in a population), then panelists whose BET's fall near the extremes of the range of concentrations presented must be re-tested using a scale range extended to the right (resp. left) of the range as shown in Fig. 3.

DISCUSSION IN TASK GROUP E-18.04.25

<u>ASTM E-679</u>

The principal observation made by group members was that, while on the one hand E679 is a practical method allowing up to 16 panelists to complete testing of a sample brought to the laboratory in a Tedlar bag within 1-2 hours, it has on the other hand been subject to serious criticism for imprecision and bias (4,5,6). It obviously does not determine <u>the</u> threshold as defined above, but rather an approximate value somewhere near the threshold. The results if accepted at first attempt may on occasion be in error up to 3-fold or more in either direction. However, with some care in the training of panelists and in the preparation of the test range, and with additional testing of those panelists whose initial results fall near the edge of the range presented, results can be much improved without a significant increase in the amount of sample and time needed for a group test. Given that E679 uses only 50-100 presentations, it can be repeated twice or three times and thus provide a large increase in user confidence while still using less than half the amounts of sample and time required for one X 43-101 test. - It should also be noted that precision and accuracy with the IITRI Dynamic Triangle Olfactometer can be improved by raising the flow rate from 0.5 l/min to 1.5 - 3 l/min.

This method for most practical purposes requires 300-600 presentations and thus an amount of sample which cannot be transported to a laboratory (unless the odor happens to be very strong). At the time of writing, no member of the Task Group had attempted to use it. All agree that it is a satisfactory method, but members feel that it will be used too rarely, that it should be seen as falling into the group of "large projects" or "research projects" which would be undertaken mainly if a large investment is at stake, or for academic reasons. There was some criticism of the approach to determination of a group threshold, see below. - It was decided that the Task Group should not at this point attempt to cover the field above 300 presentations but should concentrate on developing a standard for the intermediate range of approx. 100-300 presentations per group test.

Calculation of Individual Threshold

The basic situation when a threshold is to be calculated is shown in the example in Fig. 4. The panelist was tested at 5 concentrations with a scale step factor of 2 and achieved results of 100% correct at the highest concentration, then 90%, 65%, 40% and 25%. By chance alone in the 3-AFC test one would obtain 33% correct, so the task is to fit a sigmoid curve, from 0% to 100% above chance, as shown here. One approach is to plot the data on probability paper as in Fig. 1; the data can then be fitted by eye, or they can be converted to Probits and fitted by least squares regression. The threshold is where the line crosses 50% above chance, or Probit 5. - One major source of bias with probability graphs is that they cannot handle results of 100% correct, nor 0% correct, and certainly not results of less than 0% correct above chance. An arbitrary choice must be made, such as to use 99.5% for 100%, 0.5% for 0%, and 0.1% or 0.5% for less than 0%; the bias introduced will be greater, the more such points are included.

A bias-free method of fitting a sigmoid is to use a computer program which can perform curve fitting by iterative regression. Programs using the normal curve of error do exist but are complex; essentially the same result for the threshold is obtained using the simple logistic model shown in Fig. 5 (according to A.W. MacRae (pers. comm.) it takes 5000 optimally located data points to distinguish empirically between normal and logistic models). Note that the computer program also calculates confidence limits.

Fig. 6 is an example of computer curve fitting for a single panelist. The Task Group proposes to allow calculation by probability graph or iterative regression using either the normal or the logistic model.

Calculation of Group Threshold

This is the most difficult aspect of threshold tests. In the simplest case, one could assume that the individual thresholds are log-normally distributed (Fig. 7, left hand histogram). In such a case, one may choose as the group threshold the Median, the Mode, or the Geometric Mean as the result would be the same, here $\log(Z) = 2.6$. However, in practice the assumption that thresholds are lognormally distributed is not justifiable. Populations exposed to odors are heterogeneous: the young and the old, the sick and the healthy, females and males, smokers and nonsmokers, the various ethnic groups, etc., etc. Large investigations (over 100 individuals) show that distributions can be skew (mid histogram) or even bimodal (right hand histogram) when part of a population is anosmic to the odor in question. Our problem is confounded by the fact that we have only 4-12 panelists; which measure of central tendency should we choose? An insoluble problem, hence the Task Group recommends that for data sets of 100-300 presentations, the report should simply list the individual thresholds and no attempt to calculate a group threshold should be made unless (a) this is needed because of the purpose of the investigation and (b) the data to all intents and purposes appear to be normally distributed when plotted as in Fig. 2.

The Task Group further recommends that the purpose for which thresholds are required should be clearly determined and included in the report. It is proposed to distinguish between the following four categories:

(a) Comparing an individual with a literature value, e.g. to diagnose anosmia or ageusia, to study sensitivity to pain, noise, odor, etc. This is the simplest category, requiring no more than 25-35 AFC presentations to the individual in question.

(b) Comparative measurement of a given stimulus, e.g. the efficiency of a pollution reduction device for air or water. Assuming that the nature of the stimulus does not change, the panel need only represent the affected population in the manner in which it reacts to a change in stimulus intensity. A selected panel of 3-5 will be adequate, and the group threshold can be calculated in any convenient manner, e.g. as the geometric mean. This is a case in which pooling of the raw data may be permitted, see below.

(c) A population threshold is required, e.g. the odor threshold of a population exposed to a given pollutant, or the flavor threshold of consumers of a beverage for a given contaminant. In this case, recourse must be had to the rules for sampling from a population (8,9) which require

- that the population be accurately defined and delimited,
- that the sample drawn is truly random, i.e. that every member of the population has a known chance of being selected,
- that knowledge of the degree of variation occurring within the population exists or can be acquired in the course of formulating the plan of sampling.

The experimenter clearly must calculate and tabulate the thresholds for each individual, i.e. the data can not be pooled. The Task Group recommends that the data be plotted as shown in Fig. 2. The variable of interest is the 50%-point, hence it is important to include panelists from all affected population groups, and it is less important to obtain accurate data from the most and least sensitive individuals.

(d) The Distribution of thresholds in the population is required, e.g. to determine which proportion of the population is affected by a given level of a pollutant. The requirements are those of (c) but in addition, accurate data are needed for the least and most sensitive panelists. For these, full testing using extension of the concentration range is a necessity, and repeat testing using a second or third sample from the same population is recommended.

Pooling of raw data for several or all of the panelists

A point much discussed in the Task Group is whether pooling of data as shown in Fig. 5 should be permitted. This graph represents the combined data from 25 panelists who each received one or more 3-AFC tests at each of 9 concentrations; the graph is based on 248 presentations. This procedure gives unambiguous and easily calculated numbers both for the group threshold and for its confidence limits, but the individual panelist thresholds are not calculated. The Task Group was tempted to permit this method of calculation but so far has decided, with 12 votes to 2, that we cannot permit it because it is based on an assumption that is not tenable, i.e. that the individual thresholds form a lognormal distribution. Comments are invited from AWMA members who use thresholds: should pooling of panelist data be permitted for certain applications, such as (b) above?

CONCLUSIONS

The Task Group proposes for discussion the following conclusions:

1. Standard Practice E679: Ascending Method of Limits

This procedure should be reclassified as a rapid method, suitable for data sets of approx. 50-100 3-AFC presentations. The text should be revised (a) to mention the New Standard for 100-300 presentations now being drafted, and also the option to choose a research method for more than 300 presentations; (b) to clearly indicate the conditions under which the results by the E679 procedure may be biased by a factor of up to three-fold and (c) to indicate the steps which may be taken to reduce such bias.

2. New Standard: Intermediate Method for Approx. 100-300 3-AFC Presentations

The new standard will contain the following: (a) the 3-AFC presentation is preferred, but other forced-choice methods can be permitted; (b) the experimenter must obtain each person's threshold; (c) whenever a panelist fails on all but the two highest concentrations presented, he or she must be tested at enough higher concentration steps to ensure a high proportion of correct responses at the two highest levels; (d) whenever a panelist is correct at all concentrations including the lowest, enough lower concentrations must be added to ensure that a near-chance response is obtained at the lowest level; (e) instructions for reporting should require that a table of the individual panelist's thresholds be included; and (f) calculation of a group threshold may be attempted only if the purpose of the test requires this and a straight line can be drawn through the points when plotted as in Fig. 2.

REFERENCES

1. American Society for Testing and Materials, E679 Standard Method for Determination of Odor and Taste Thresholds by a Forced-Choice Ascending Method of Limits. ASTM, 1916 Race St., Philadelphia, PA 19103, 1979.

2. Verein Deutscher Ingenieure, VDI 3881 Olfactometry: Odor Threshold Determination. Fundamentals. VDI, Beuth-Verlag GmbH, Burggrafenstr. 6, D-1000 Berlin 30, 1986.

3. Association Française de Normalisation, NF X 43-101 Air Quality: Method of Measuring the Odor Intensity of Gaseous Effluents. Determination of the Dilution Factor to Perception Threshold. AFNOR, Tour Europe - Cedex 7, 92080 Paris La Défense, 1986.

4. Polta, R., Odor Experience/Research at Metropolitan Waste Control Commission, Paper given at AWMA International Specialty Conference on Recent Developments and Current Practices in Odor Regulations, Controls and Technology, Detroit, October 1989 (to be published).

5. Morrison, G.R., Flavor Thresholds for Added Substances, J. Inst. Brewing 88:167, 1982; Measurement of Flavor Thresholds, J. Inst. Brewing 88:170, 1982.

6. Dravnieks, A., Schmidtsdorff, W. and Meilgaard, M.C., J. Air Pollution Control Assoc. 36:900, 1986.

7. SAS User's Guide, Statistics, Version 5 Edition. Cary, NC 27511-8000, 1985, pp. 565-606.

8. Snedecor, G.W. and Cochran, W.G., Statistical Methods, 6th ed., Iowa State University Press, Ames, IA 50010, 1967, Chapter 17.

9. American Society for Testing and Materials, E122 Standard Practice for Choice of Sample Size to Estimate the Average Quality of a Lot in Process. ASTM, 1916 Race St., Philadelphia, PA 19103, reapproved 1979.

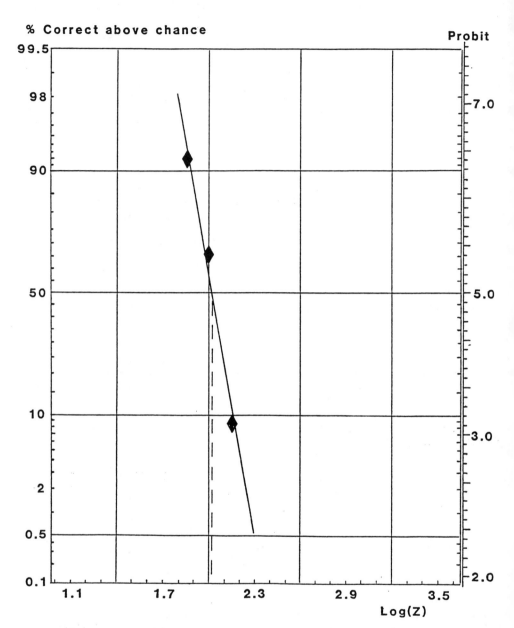

Fig. 1. NF X 43-101: Calculation of threshold for one individual tested 10 times at each of 3 Concentrations spaced $\sqrt{2}$ apart

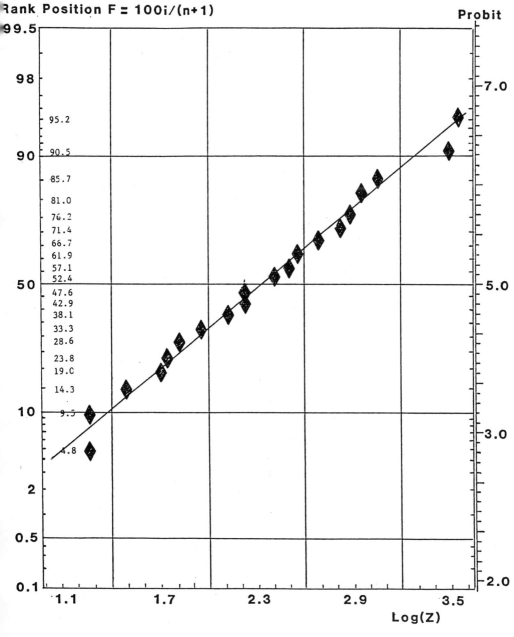

Fig. 2. NF X 43-101: Probability graph showing 20 panelists sorted by rank.

283

NOTE—This example has been selected to represent both extremes. Panelist 4 missed even at the highest concentration. Panelist 6 was correct even at the lowest concentration and continued to be correct at all subsequent higher concentrations.

Panel-ists	Judgments[A]						Best-Estimate Threshold (BET)	
	Dilution Factors							
	(concentrations increase →)						Value	log$_{10}$ of Value
	3645	1215	405	135	45	15		
1	0	+	+	0	+	+	78	1.89
2	+	0	+	+	+	+	701	2.85
3	0	+	0	0	+	+	78	1.89
4	0	0	0	0	+	0	9	0.94
5	+	0	0	+	+	+	234	2.37
6	+	+	+	+	+	+	6313	3.80
7	0	+	+	0	+	+	78	1.89
8	+	0	0	+	+	+	234	2.37
9	+	0	+	+	+	+	701	2.85

Group BET geometric mean Σ log$_{10}$ → 20.85
                                     209 ← 2.32
Standard deviation   0.81

[A] "0" indicates that the panelist selected the wrong sample of the set of three. "+" indicates that the panelist selected the correct sample.

Fig. 3. ASTM E679: Example of odor threshold calculation.

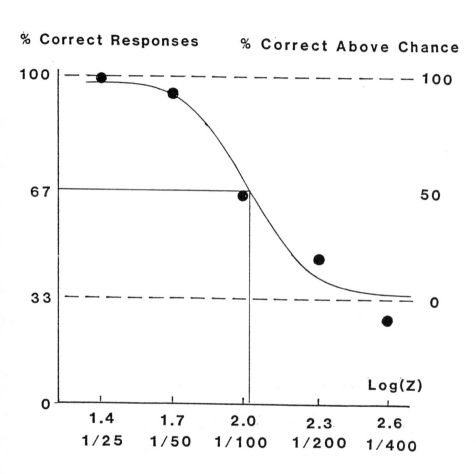

Logistic Regression Model Using SAS PROC NLIN

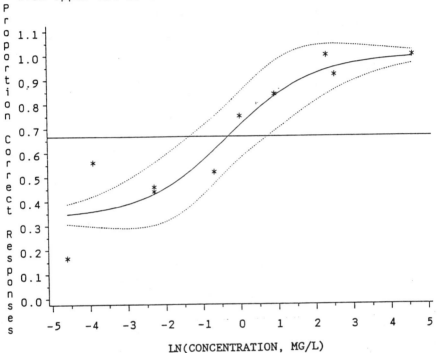

Fig. 5. Calculation of threshold of diesel oil in water (Howgate, private comm.) by iterative regression using a logistic model and SAS Proc Nlin (7). The logistic curve fitted is: $P = (Exp(LR) + 1/3)/(Exp(LR) + 1)$ where P = Probability of correct responses; $LR = A + B(ln \, concn., mg/l)$.

```
**  FILENAME:   ASTM_E18.SAS
**
**  PURPOSE:    Fit logistic models P = ( 1/3 + EXP(K) ) / ( 1 + EXP(K) ),
**                  where K = B( T - X ),
**                  P is the proportion of correct identifications,
**                  B is the slope,
**                  X is the actual ln(Z),
**                  and T is the ln(Z50).
**
**  INPUT:      In file, from Draft 4 of E-18.04.25 on Odor Thresholds
**
**  AUTHOR:     Tom Carr
**
**  DATE:       August 16, 1989

Title1 'Logistic Regression of Threshold Data Using SAS PROC NLIN';

data input;
  input judge X P;
cards;
1 1.1 .833
1 1.4 .667        3 1.1 1.00                    5 1.7 1.00
1 1.7 .500        3 1.4 1.00                    5 2.0 .667
1 2.0 .333        3 1.7 .333                    5 2.3 .667
1 2.3 .167        3 2.0 .333                    5 2.6 .000
1 2.6 .333        3 2.3 .500                    5 2.9 .500
2 1.1 1.00        3 2.5 .167                    6 1.7 1.00
2 1.4 .833        4 1.4 1.00                    6 2.0 .833
2 1.7 .667        4 1.7 .833                    6 2.3 .667
2 2.0 .333        4 2.0 .667                    6 2.6 1.00
2 2.3 .500        4 2.3 .333                    6 2.9 .833
2 2.6 .333        4 2.6 .333                    6 3.2 .500
run;                                            6 3.5 .333

proc sort data=input;
  by judge;
run;

proc nlin method=dud data=input;
  by judge;
    parms B=5 T=1.5;
      K  = B*(T - X);
      E  = EXP(K);
      N  = (1/3 + E);
      D  = (1 + E);
    model P = N/D;
  title2 'Logistic Regression Models';
run;
```

<pre>
            Logistic Regression of Threshold Data Using SAS PROC NLIN
                            Logistic Regression Models

                                    JUDGE=2

                     NON-LINEAR LEAST SQUARES ITERATIVE PHASE

              DEPENDENT VARIABLE: P   METHOD: DUD

       ITERATION   B   T   RESIDUAL SS

          -3   5.000000000   1.500000000   0.06384440003
          -2   5.500000000   1.500000000   0.065067331922
          -1   5.000000000   1.650000000   0.033640046758
           0   5.000000000   1.650000000   0.033640046758
           1   5.368411447   1.660518332   0.033416406696
           2   5.334687430   1.655343936   0.033374507108
           3   5.337596664   1.655396561   0.033374352545
           4   5.355260969   1.654833159   0.033373158371
           5   5.356049417   1.654783791   0.03337315305
           6   5.356113845   1.654786895   0.033373151824
           7   5.358886637   1.654847471   0.033373128743
           8   5.359359792   1.654832621   0.033373128384
           9   5.359276837   1.654838461   0.033373128265

       NOTE: CONVERGENCE CRITERION MET.
</pre>

Plot of Logistic Regression Model for Judge=4

PREDICTED VALUES:

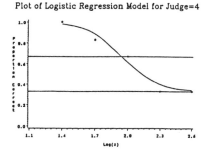

P	C	X
.967	.95	1.482
.933	.90	1.603
.833	.75	1.768
.667	.50	1.943
.500	.25	2.133
.400	.10	2.297
.367	.05	2.412

Fig. 6. Calculation of individual threshold by iterative regression using a
logistic model and SAS Proc Nlin (7).

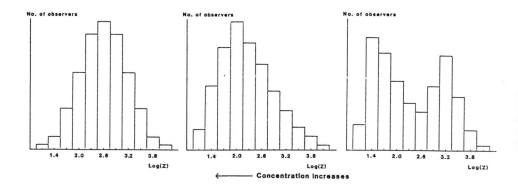

<u>Fig. 7</u>. Three hypothetical histograms showing the distribution of individual thresholds in a population.

ODOR MODELING - WHY AND HOW

Richard A. Duffee, Martha A. O'Brien and Ned Ostojic
Odor Science & Engineering, Inc.
57 Fishfry Street
Hartford, CT 06120.

Abstract

Complaints of ambient malodors are associated primarily with the intensity and frequency of perceived odors. Odor perception, however, is a near instantaneous reaction not a time-averaged response. Thus, an effective odor dispersion model should be able to predict peak (instantaneous) odor intensities and their frequency of occurrence at any given receptor. In this paper the predictions of a fluctuating plume puff odor model are compared to those produced by the EPA's Industrial Source Complex Short Term (ISCST) Gaussian dispersion model and the EPA's INPUFF 2.3 Gaussian integrated puff model for a theoretical single source application. In addition the results of the fluctuating plume puff odor model and the ISCST model are compared with actual odor observations for two situations: (1) multiple vent rooftop releases with severe building wake entrainment; and (2) multiple ground level area sources.

For the theoretical single source test case the puff odor model results ranged from 12 to 65 times higher than the ISCST model depending on stability class and from 5 to 8 times higher than the INPUFF model for five second averaging time with Pasquill-Gifford class D stability. For the multiple vent rooftop releases the puff odor model predictions coincided with observations, while the ISCST model results were 10 to 20 times less than observed values. For the ground level area source situation the puff model again agreed with observations and the ISCST model results were 7 times lower than observations with Pasquill-Gifford class E stability. Adjustments to the ISCST values were made based on published peak-to-mean ratios or power-law relationship between two averaging times, but the resulting values still underpredicted observed odor levels. As a result, the conclusion was drawn that a fluctuating plume puff model must be used with valid odor emission rate data and odor concentration/ intensity/ complaint probability relations to resolve actual and prevent potential future community odor problems.

INTRODUCTION

The purpose of odor dispersion modeling, whether in diagnostic or prospective applications, is to determine the maximum allowable odor emission rate for each existing or potential source that will maintain ambient odors below complaint levels at downwind receptors. Before an appropriate modeling procedure can be selected, one should consider the nature of odor complaints and the characteristics of potentially available models.

Complaints of malodors result primarily from the perceived intensity of the odor sensation and the frequency of occurrence. (Odor character is also a factor, but is reflected in the odor intensity evaluation. People judge the intensity of odors they consider unpleasant higher than those they consider as normal to their environment, e.g. mown grass or cooking odors, or as intentionally pleasant, e.g. air fresheners or perfumes). In addition, odor perception is a near instantaneous reaction not a time averaged phenomenon. Thus, effective odor models should yield essentially instantaneous values of odor intensity and their frequency of occurrence in a given time period.

All Gaussian plume dispersion models, however, are based on mass emission rates and predict concentrations, i.e. mass per unit volume, at specified receptors. Accordingly, to conduct odor dispersion modeling effectively, it is certainly desirable, if not necessary, first to establish the relation between odor concentration and perceived odor intensity (Steven's Law) and then the relation between odor intensity and complaint probability. Odor concentration typically is expressed as the dilution-to-threshold ratio (D/T) of the odorous emission. Since it is a ratio it is dimensionless; it is independent of volumes and does not describe the intensity of the perceived odor. The odor emission rate of a source is the product of the D/T ratio and the volume flow rate. Although this product has the units of volume flow rate, e.g. m^3/s, it can be considered as the mass emission rate of odor. When this product is used as input to a dispersion model in place of the mass emission rate, the model results will also be dimensionless, i.e. the D/T ratio.

For existing sources, i.e. diagnostic applications, direct measurement of odor emission rates and related ambient odors is reasonably straightforward although requiring considerable care and technique. For new construction purposes, i.e. prospective applications, it is necessary to derive odor emission factors from similar processes or facilities or to develop them from bench scale tests. It is tempting in these situations to use published odor thresholds of specific odorants as a means of deriving odor emission rates for input to a model. One should strive mightily to resist such temptation for the following reasons. Few odorous emissions are attributable to a single compound. Published odor thresholds for a single compound can vary by as much as five orders of magnitude. Even when published odor threshold values are subjected to careful evaluation of the procedure used to develop the threshold value, the resulting "acceptable" values still vary by two orders of magnitude.[1] The range of olfactory sensitivity in people of "normal" acuity ranges over four orders of magnitude. Finally, there no correlation between odor threshold values and annoyance.

MODEL COMPARISONS

As indicated previously, odor perception is an essentially instantaneous reaction. Typical Gaussian plume models, such as the ISCST model, yield values consistent with 10-minute to one-hour averages. Högström[2] developed a method for predicting odor frequencies from a point source on the basis of a fluctuating-plume dispersion model. It is used to give odor frequencies around a point when the odor threshold (dilution ratio) of the material emitted is determined by sensory methods. A fluctuating plume puff model was developed by Bowne at TRC which incorporated the puff dispersion coefficients developed by Högström, as described by Murray et al[3,4]. This model predicts the number of occurrences of specified odor dilution ratios during a specified period, such as 1 h. A variant of this model has been developed by the staff of Odor Science & Engineering, Inc. (OS&E). This model has an averaging time of approximately two seconds and predicts the contribution of each source to the instantaneous (two second) peak odor and hourly average value at a given receptor as well as the frequency of occurrence of specified dilution ratios.

In order to determine the relative output values of different models currently used in evaluating odor problems, predictions of the fluctuating plume puff odor model were compared to the predictions of the EPA's ISCST model and the EPA's INPUFF (version 2.3) integrated puff model using the same odor input data and meteorological conditions. A simple test case was used for this comparison. The physical parameters of the assumed source of a single isolated stack were:

    Stack height = 5 m
    Stack diameter = 0.91 m
    Stack gas flow rate = 6.66 m^3/s
    Stack gas velocity = 10.16 mps
    Stack gas temperature = 300°K
    Stack gas odor concentration = 2000 D/T

Meteorological conditions assumed are summarized in Table I.

TABLE I. METEOROLOGICAL CONDITIONS FOR COMPARISON TEST

Stability Class P-G	Wind Speed m/s	Temp $^{\circ}$K	Temp Gradient Puff $^{\circ}$C/m	ISCST $^{\circ}$K/m	Mixing Ht ISC & INPUFF
A	2.0	298	-0.02	0	5000
B	2.0	298	-0.02	0	5000
C	3.0	298	-0.02	0	5000
D	4.0	298	0.00	0	5000
E	2.0	298	0.01	0.02	5000

Each model requires some specific inputs. The puff odor model requires specification of the surface roughness length, terrain (urban, suburban, or rural as defined by Bowne), building height and building width used in calculating wake effect. For this exercise, a roughness length of 0.75 m

291

was used for all stabilities with an urban terrain specified. Building
height and width were set at zero since the source is an isolated stack.

The ISCST model specifies an emission rate input of g/s, which yields
an output in ug/m^3. Accordingly, the odor emission rate for this model
was calculated as $(2000 \text{ D/T})(6.66 \text{ m}^3/\text{s})(10^{-6}) = 0.01332 \text{ m}^3/\text{s}$. This
model also requires specification of a wind profile exponent for each P-G
stability class. A value of 0.07 was used for stabilities A and B; a value
of 0.10 was used for class C; 0.15 for class D; and 0.35 for class E.
Building height and width are also used in the wake effect algorithm in
this model. For this analysis these were set at zero.

The INPUFF 2.3 model also utilizes a mass emission rate of g/s but the
output is given in terms of g/m^3 instead of ug/m^3 as in the ISCST
model. Accordingly, an odor emission rate of $(2000 \text{ D/T})(6.66 \text{ m}^3/\text{s}) =$
13,200 m^3/s was used. This model also requires specification of a wind
profile exponent. A value of 0.150 was used for stability class D, the
only class used for this model. The model asks for specification of the
initial plume (puff) sigmas (termed R and Z). Based on the stack diameter
these were calculated as 0.2 m for R (lateral) and 2.5 m for Z. Finally, a
simulation period of six minutes was specified with an averaging time of
five seconds beginning at five minutes after release of the initial puff.
A default option was chosen for puff release rate and puff combination
criterion. The model calculated a puff release rate of 2.5s and a puff
combination criterion of 0.500 sigmas.

Three receptors were located directly downwind from the stack (plume
centerline) at distances of 200 m, 500 m, and 1,000 m. The outputs of the
three models are summarized in Table II.

TABLE II. COMPARISON OF MODEL PREDICTIONS - SINGLE SOURCE

STABILITY	RECEPTOR DISTANCE KM	PUFF MODEL PEAK	PUFF MODEL 1H AVG	ISC 1H AVG	INPUFF 5S AVG
A	0.200	15.85	1.57	0.61	–
	0.500	3.59	0.39	0.10	–
	1.000	1.30	0.14	0.02	–
B	0.200	15.85	1.98	0.61	–
	0.500	3.59	0.50	0.10	–
	1.000	1.30	0.19	0.02	–
C	0.200	9.44	4.37	0.78	–
	0.500	2.46	1.01	0.14	–
	1.000	0.90	0.34	0.04	–
D	0.200	22.39	7.01	1.14	2.87
	0.500	5.87	2.33	0.40	1.19
	1.000	2.23	0.93	0.11	0.43
E	0.200	3.63	0.16	3.36	–
	0.500	14.54	1.79	1.18	–
	1.000	5.46	1.42	0.42	–

These results clearly show that, except for the values calculated for the closest receptor under stable (P-G class E) conditions, the ISC model yields values ranging from 12 to 65 times less than the peak values of the puff odor model. The INPUFF values for 5s averaging time were 5 to 8 times less than the puff odor model results for the one neutral stability condition (P-G class D) compared.

The next step in this evaluation was to see how the model predictions compare with actual odor observations using measured odor emission rates in real situations. The INPUFF model was not included in this further analysis because it does not contain any algorithm for building wake effects nor any treatment of terrain effects except through the wind field. Two cases were evaluated. The first was a manufacturing plant which had 12 short stacks atop a gable roof. All of the stacks were below the peak height of the roof. Four of the stacks had a measured flow rate of 6,000 ACFM (2.826 m^3/s) whereas the other eight stacks had a volume flow rate of 4,000 ACFM (1.884 m^3/s). All of the stacks had a measured odor concentration of 175 D/T. Odor monitoring downwind of the facility resulted in observed odor values, as measured by Scentometer, of between 7 and 15 D/T at a distance of approximately 100 m from the stacks; between 2 and 7 D/T at a distance of approximately 300 m; between 1 and 2 D/T at a distance of near 400 m; and threshold or below at a distance of 550 m. These observations were made with southerly winds of 7 mps, overcast skies and cool temperatures (P-G stability class D).

For the model comparison eight receptors were located downwind as shown on Figure 1. The stacks were positioned on the plant roof at their specific locations. The physical parameters used were:

```
Stacks 1,2,3 and 4 --height = 7.9 m
                    diam.   = 0.7 m
        stack gas temp      = 293°K
     stack gas velocity     = 12.6 mps
       building height       = 11.0 m
       building width        = 34.0 m
Stacks 5 - 12 ------height = 7.9 m
              diam  = 0.6 m
    stack gas velocity = 10.2 mps
```

The meteorological parameters for the puff odor model were wind speed of 7 mps, wind direction of 180 degrees, roughness of 0.750 m, a temperature gradient of 0.00°C/m, and a suburban terrain. The same parameters were used for the ISC model. For this model the mixing height was specified as 5,000 m, the wind profile exponent was 0.15, and the potential temperature gradient was 0.00°K. The odor emission rates were calculated as 0.0004946 m^3/s for stacks 1 through 4 and 0.0003297 m^3/s for stacks 5 through 12.

The results are summarized in Table III. Obviously, the peak odor values from the puff odor model agree very well with the observed odor levels. The ISCST model results, one hour average values, are 10 to 20 times lower than the observed odor levels.

TABLE III. MODELED AND OBSERVED ODORS COMPARISON - MULTIPLE POINT SOURCE

| RECEPTOR | PUFF ODOR MODEL | | ISCST | OBSERVED |
	PEAK D/T	1H AVG D/T	1H AVG D/T	D/T
1,2	6-19	1.03-1.75	0.47	7-15
3,4	2.1-2.9	1.08-1.45	0.30	2-7
5,6	2.0	0.95-1.12	0.16	1-2
7,8	1.4	0.75-0.81	0.08	T

The second comparative analysis was done for a situation with multiple area sources located at essentially ground level. These sources were aerated static compost piles separated by an active mix area. The area of the active mix zone was determined to be 115.5 m^2 with an average height of 3.0 m. The odor level of emissions from this area was measured as 750 D/T with a vertical velocity of 0.5 mps. The active compost piles covered an area of 464 m^2. The odor of the emissions from the piles had an average value of 287 D/T. Average height of the piles was also 3.0 m. With the puff odor model the area sources were modeled as large point sources by converting the areas to equivalent diameters and calculating the volume flow rate from the area and measured vertical velocity. In this scheme, the diameter of the active mix area was 12.2 m and the volume flow rate was determined to be 58.75 m^3/s. The diameter of the active compost area was determined to be 24.3 m and the volume flow rate was calculated as 5.28 m^3/s.

With the ISCST model the sources were modeled as area sources. In this case the emission rate specified by the model is in terms of $g/s/m^2$. Accordingly the equivalent odor emission rate for the active mix area was determined as follows: (750 D/T) (58.875 m^3/s) (10^{-6}) (1/115.5 m^2) = 0.0003823. The calculation of the odor emission rate from the active compost area was done in a similar manner:
(287 D/T) (5.28 m^3/s) (10^{-6}) (1/464 m^2) = 0.0000033.

The mixing height value for the ISCST model was again 5,000 m, the potential temperature gradient was 0.02°K, and the wind profile exponent 0.300. For the puff model a roughness length of 0.750 m was again used with suburban terrain. Ambient temperature was set at 300°K for both models.

The models were run with three receptors located directly downwind at distances of 0.8 km, 1.6 km and 2.4 km for E stability conditions at a wind speed of 2.0 mps. During community odor monitoring conducted by OS&E staff downwind of these sources, compost odors were noted under P-G E stability conditions with wind speeds of 2 mps at downwind distances between 1.6 and 2.4 km. They were measured as 8 D/T by Scentometer (4-hole model) which means the actual D/T ratio was greater than 8 D/T but less than 32 D/T. The mixing height value for the ISCST model was again 5,000 m, the potential temperature gradient was 0.02°K, and the wind profile exponent 0.300. For the puff model a roughness length of 0.750 m was again used with suburban terrain. Ambient temperature was set at 300°K for both models.

The results of this model comparison with observed odors are summarized on Table IV. Again the puff odor model peak values agree well with observed odors while the ISCST 1h average values are 6 to 7 times less than the observed odor levels.

TABLE IV. COMPARISON OF MODELED AND OBSERVED ODORS - MULTIPLE AREA SOURCES

RECEPTOR KM	PUFF ODOR MODEL		ISCST	OBSERVED
	PEAK D/T	1H AVG D/T	1H AVG D/T	D/T
0.8	34.53	4.13	5.67	-
1.6	14.11	2.89	2.01	8-32
2.4	8.55	2.02	1.14	8-32

CONCLUSION

A power-law relationship that depends on time and relates the ratio of one averaging time to another has been utilized by various agencies to convert from short-term (3-min) values to hourly average. This technique used the form:

$$X_1 = X_s(t_s/t_1)^p \qquad \text{(Eq. 1)}$$

where X_1 = the concentration for the longer averaging time
X_s = the concentration for the longer averaging time
t_s = the shorter averaging time, h
t_1 = the longer averaging time, h
p = the power law exponent

Power law exponents for sources with stacks less than 30 m high are given by stability class as:

Stability Class	p
A	0.5
B	0.5
C	0.333
D	0.2
E	0.167
F	0.167

If this relation is applied to the values for the ISCST model shown in Tables III and IV to convert them from 1h values to 3-min averages, the resulting values will be lower than the observed odor values by a factor of 10 for the D stability situation and by a factor of 5 for the E stability case. Thus, effective odor dispersion modeling requires the use of a fluctuating plume puff model. The validity of the results of such modeling, of course, is totally dependent on the accuracy of the input odor data.

REFERENCES

1.) Duffee, R.A., J.E. Hooper. Odor thresholds for chemicals with established occupational health standards. Prepared for the American Industrial Hygiene Association, 1988.

2.) Högström, U. A method for predicting odour frequencies from a point source. Atmos. Environ. 6:103-121, 1972.

3.) Murray, D.R., S.S. Cha, and N.E. Bowne. Use of a fluctuating plume puff model for prediction of the impact of odourous emissions. Preprint 78-68.6. Presented at the 71st Annual Meeting. Air Pollution control Association, Houston, Texas, June 1978.

4.) Duffee, R.A., D.R. Murray. Ambient odor modeling to determine control requirements. Technical Paper 83-37.6. Presented at Annual Meeting of the Air Pollution Control Association, 1983.

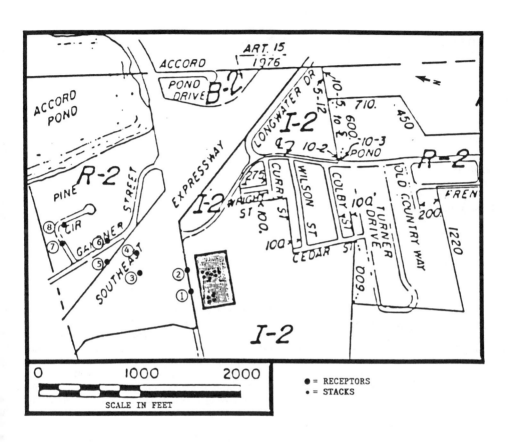

FIGURE 1. MODELING RECEPTOR LOCATIONS (1-12)
(SOUTHERLY WIND)

A MOBILE ODOR LABORATORY FOR CONTINUOUS AMBIENT
MEASUREMENT OF REDUCED SULFUR GAS

Charles H. Padgett
US EPA Region IV
IPA to Jacksonville
Jacksonville, Florida

The City of Jacksonville, Florida, Bio-Environmental Services Division (BESD) ha developed a unique ambient odor monitor-on-wheels, which provides in situ measureme of aggregate reduced sulfur (ARS) gases. Equipped with two sulfur dioxide (SO_2) analyzer and a thermal oxidation furnace, the mobile odor unit is used to determine ambient odc levels around five major ARS sources, as well as wetlands, water and sewage treatme plants, landfills and various odor episodes determined by a spurt of citizenry complaint Aggregate reduced sulfur gas concentration is measured by subtracting the concentratic of sulfur dioxide from the concentration of total oxidizable sulfur in ambient air. The AR monitoring system is housed in a converted 1975 Chevrolet 30 step van. The odor measure ment system rides on a super-single water bed which acts as a cushion for the sensitiv equipment. An Onan 6500 watts generator supplies onboard power for the odor equipme while the unit operates in the mobile mode. The mobile unit plugs into a 220 volt AC outle while operated in the stationary mode. To prevent power loss while switching from o mode of operation to another, a SOLA uninterruptible power supply (UPS) is used. The UI also conditions the onboard voltage and protects the odor monitoring equipment from powe surges.

The mobile odor unit was put into operation in October 1988 and has proven to be ve successful. Whether monitoring while parked or on-the-roll, BESD's unique odor lab is tru an odor monitor-on-wheels.

INTRODUCTION

In April 1987 a new Mayor was elected on a platform to rid Jacksonville of objectionable odors. Mayor Thomas L. Hazouri moved quickly to expand the air pollution control staff and to increase the agency's resources. Reduced sulfur gases emitted from two kraft pulp mills, two terpene chemical plants and a municipal sewage treatment plant caused Jacksonville's major odor problem. The key to eliminating odor nuisances was to develop an ambient odor standard, and then develop industry-specific emission limits designed to achieve compliance with the ambient standard. Using an ambient odor standard in this manner is similar to the manner in which criteria pollutant ambient standards are used. A monitoring method had to be developed to measure odorous compounds concurrent with development of the ambient odor standard.

The Jacksonville BESD attempted to take whole air grab samples using various containers: Summa polish canisters, tedlar bags, stainless steel teflonlined one liter cylinders and other types of containers. These samples were returned to the agency's laboratory for GC analysis. None of the grab sample containers proved to be completely successful. All of the containers tested showed a partial sample loss.

Another approach to measuring odorous compounds was tried using an SO_2 continuous monitor coupled to a thermal oxidizer and an SO_2 selectable scrubber. SO_2 would be removed selectively from the sample gas and reduced sulfur compound passing through the scrubber would be thermally oxidized to SO_2. The scrubber was a heated, temperature controlled molecular sieve material and effectively removed SO_2. However, the scrubber also removed hydrogen sulfide (H_2S) which is one of the four compounds of interest. Because H_2S was absorbed in the SO_2 scrubber, this method was abandoned.

Because SO_2 would also be measured along with the thermally oxidized ARS compounds, the amount of SO_2 in the sample gas had to be known. One way to measure the amount of SO_2 in the sample gas was using a second SO_2 monitor. Establishing five to seven ARS monitoring stations using two SO_2 monitors, thermal oxidizer, and support equipment would cost the agency several hundred thousand dollars. Funds were not available to establish such a fixed ARS monitoring network. However, since the agency had a low mileage step van (1975 Chevrolet with 58,000 miles) available, it was suggested that the step van be used as a mobile odor measurement laboratory. Realizing monetary savings and the ability to cover a large part of Duval County, the staff developed a unique ambient odor monitor-on-wheels.

The mobile ARS monitor consists of a step van, two Thermo Electron Model 43A SO_2 continuous monitors, a CD Nova 101 Thermal Oxidizer, a SumX data acquisition system, a Tandy 1400 LT Computer, an Environics Model 200 Computer gas blender, a water bed, a Coleman roof-top air conditioner, an Onan 6500 watts generator, a SOLA UPS, and miscellaneous hardware.

This method provides a measurement of the concentration of aggregate reduced sulfur (ARS) in ambient air for determining compliance with the City of Jacksonville ARS ambient standard of 55 ppbv. The method is applicable to the measurement of ambient ARS concentration using an averaging period of 3 minutes.

PRINCIPLE OF OPERATION AND DETECTION LIMITS

The aggregate reduced sulfur (ARS) continuous monitor consists of a thermal oxidatio furnace and two SO_2 automated reference or equivalent analyzers. A thermal oxidize converts ARS compounds to SO_2.

The sample gas stream is first split into two equal channels using a teflon union tee One channel is analyzed directly in a SO_2 automated reference method analyzer for SO content. The second channel is directed through a quartz tube housed within a high tem perature ceramic oven. The quartz oven chamber is designed to provide retention, a maximum flow rate (1.5 liter/min.), well in excess of the recommended minimum (0.1 sec for oxidation.

For ARS applications, a temperature range between 800 and 950°C is used. At lowe retention times or lower temperatures, dimethyl sulfide (DMS) and dimethyl disulfide (DMDS are not oxidized. If the temperature is too high, SO_2 will be oxidized to SO_3.

After the ARS compounds have been oxidized to SO_2, the cumulative SO_2 is the monitored by the second SO_2 automated reference method analyzer. The SO_2 measure in the second channel is the sum of the SO_2 ambient gas concentration and the SO_2 con verted from ambient ARS gases as a result of oxidation in the thermal oxidation furnace The difference between the ambient SO_2 concentration monitored in channel one, and th cumulative SO_2 concentration monitored in channel 2, is ambient ARS.

Thermo Electron Model 43A Pulsed Fluorescence SO_2 Analyzer is an Environments Protection Agency (EPA) approved SO_2 analyzer which can be operated on a 0 to 100 pp range, the lower detectable limit is 1.0 ppb.

The sample gas is drawn into the Model 43A measurement chamber. Pulsed UV ligh passes through a reflection mode optical filter system to the Model 43A measuremen chamber, where it excites SO_2 molecules. As these molecules return to the ground stat they emit a characteristic fluorescence with intensity linearly proportional to the concen tration of SO_2 molecules in the sample. The fluoresced light then passes through a secon filter to illuminate the surface of a photomultiplier tube (PMT).

Electronic amplification of the output of the PMT provides an analog signal for a dat acquisition system. Long-term span stability is maintained by a second UV detector whic senses UV lamp intensity and maintains a constant UV excitation level by a closed loo feedback control of the UV lamp power supply. Long-term zero stability is assured by th reflective filter optics, which results in a very low level of scattered light.

ASSEMBLY

BESD had on-hand a low mileage 1975 Chevrolet step van. A local van conversio company installed carpet, panels with insulation on both walls and ceiling, and a roof-to Coleman air conditioner. Also, a divider wall was added to separate the driver from th equipment.

The agency next purchased an Onan 6500 watts AC generator which was installe according to staff specifications.

Using a water bed in the step van proved very important. The equipment had to ride on a good cushion. Rubber mounts, foam rubber, and other such devices do not provide adequate insulation from vibration for a unit operating in a mobile mode. A board $2\frac{1}{2}$ by $4\frac{1}{2}$ feet in size was placed on the water bed and the SO_2 monitors and ancillary equipment were placed on the board.

After installing the water bed, the equipment compartment of the step van was wired to a 220 volt breaker box. The breaker box has a 220 volt electric range top pigtail that plugs into a 220 volt outlet. The step van can be powered directly by the onboard Onan generator, or by being connected to an external source. This enables the step van to be used either in a mobile mode or a stationary mode. At the breaker box, the 220 volts is split to provide 110 Vac to the equipment.

After the electrical circuits were tested and proven satisfactory, two Thermo Electron 43A EPA designate equivalent SO_2 analyzers, a SumX 444 DAS, an Environics Model 200 computerized gas standard calibrator, and a Tandy Radio Shack Model 1400LT computer were installed and calibrated, as applicable, according to the manufacturer's specification. After the equipment was installed, the van was ready for field testing.

FIELD TEST RESULTS

It was first necessary to determine how this equipment operates using the onboard power generator, and how well the instruments ride on the water bed.

While operating the ARS monitoring system in the mobile mode, the 6.5 kilowatts generator supplied all electrical power. The Coleman air conditioner and thermal oxidizer drew the major load. The air conditioner has a running load of 1100 watts and, in the heating mode, 1500 watts. The thermal oxidizer electrical demand is about 500 watts after the unit has reached it set point temperature, 850° C, and stabilized. The rest of the equipment in the system drew less than 500 watts. With all the equipment turned on, there was more than 3.0 kilowatts of power in reserve. With 3.0 kilowatts of reserved power, no major problem was anticipated; however, there was a hidden problem in the data acquisition system (DAS).

The air conditioner cycling on and off caused changes in the load on the Onan generator. Although the generator has an automatic choke, it cannot adjust the RPM's on the generator rapidly enough so as to maintain a clean 120 volts 60 hz current. The actual fluctuation in the voltage was from 85 volts to 135 volts depending on the electrical load and on the amount of acceleration/deceleration needed to adjust for the load.

If the DAS output averaging time was a minute or longer, such short term voltage fluctuation could not be detected. When displaying instantaneous readings from the DAS, negative concentrations would be displayed on the computer screen once or twice every minute or so. For example, five second readings like 5, 5, 5, 6, -90, 5, 5, ... would appear on the screen. The false negative readings would be included in an average time of interest, and of course, bias the results low.

The problem with the voltage fluctuation had to be solved not only because the data results would be biased, but all of the monitoring equipment was being operated by the Onan generator. It was obvious that the problem was not in the wiring of the monitoring system since voltage fluctuation only occurred in the mobile mode. In order to solve the voltage problem in the mobile mode, the line voltage from the Onan generator had to be conditioned and cleaned up before it got to the monitoring equipment. A SOLA uninterruptible power supply (UPS) solved the voltage problem. The SOLA UPS power supply provides power that is completely isolated from the utility power system. The UPS contains an inverter that is operational 100% of the time that the unit is in normal mode. When there is power available from the utility line, the inverter draws power from the unit's battery charger; when there is no power from the utility line, the UPS draws power from its battery. The battery is always kept in a charging or charged state by the battery charger when utility air conditioner is available.

Prior to installing the SOLA UPS, it was necessary to turn the ARS monitoring system off while switching from external utility power to onboard utility power. With the UPS operational 100% of the time, no power interruption occurs when switching from stationary to mobile mode or vice versa.

The use of a water bed as a cushion was a unique and innovative idea. When electronic equipment is shipped from the vendor to the user, the vendor may place packaging material inside the instrument chassis to prevent electronic PC boards from vibrating or dislodging from their sockets. If such equipment was operating while in a mobile mode, a PC board insecurely held in its sockets could short-out or fall against another PC board. Damage could be bad enough to require replacing an instrument. To insure the integrity of the equipment operating in the mobile mode required using a good cushion system to prevent shock to the equipment while driving over railroad tracks, hitting pot holes, driving on the interstate roads, stopping unexpectedly, accelerating rapidly, parking on embankments with major slopes and generally subjecting the equipment to any condition imaginable while riding in the back of a step van.

The water bed has been able to handle all of these conditions. All the equipment sits unsecured on a $2\frac{1}{2}$ by 4 feet board on the water bed. The equipment is neither tied down to the bed nor secured to the wall of the step van. The water bed acts as a self-leveler or self-adjuster to forces acting on it. When the operator has to stop the vehicle quickly the equipment wants to slide forward because of the quick deceleration. The equipment is prevented from moving because the water moves to the front of the bed lifting the front of the bed upward. This self-leveling of the water bed occurs continuously no matter what motion the step van is in. Also, the leveling of the equipment occurs when the step van is parked on an embankment, or on a curb. The success of the ARS mobile monitoring system can be attributed directly to two components of the system: The SOLA UPS and the water bed.

ARS MEASUREMENT RESULTS

After the mobile ARS monitoring system was put into operation in October 1988, several monitoring studies have been conducted to collect ambient ARS data. One study measured ARS values around sewage treatment plants, potable water aeration plants, landfills and salt marshes. During the ARS rule development, several commenters expressed a concern that these sources were major contributors to Jacksonville's odor problem. ARS concentrations in the range of 30 to 80 ppb for 3 minute averages have been measured around sewage treatment plants and up to 50 ppb for 3 minutes averages around potable water treatment plants. No detectable amounts of ARS have yet been recorded adjacent to Jacksonville's three major landfills. Sulfur emissions from salt marshes are currently being investigated. The initial indication is that sulfur emission is higher on incoming tide rather than outgoing or slack low tide.

The mobile ARS monitoring system was used in conjunction with a study to quantify the relative proportions of the four major ARS compounds present in the air around the major industrial facilities, with emphasis on emissions from terpene chemical plants. The mobile ARS monitoring van was used to determine if significant amounts of ARS were present n ambient air around a particular facility and if so, a grab sample was taken and returned o the lab for analysis. The results of the study showed that about 95% of the reduced sulfur compounds in the air around the terpene chemical plants was DMS. Other organic sulfur compounds such as carbonyl sulfide (COS), carbon disulfide (CS_2), methyl mercaptan (MeSH), and DMDS, did not occur in concentrations great enough to be major contributors to malodors n the air.

In addition to these and other studies being done with the ARS mobile unit, its primary unction is to collect ambient ARS data around the five major sources. It is used in a hunt and search manner to locate prevalent ARS odors. ARS odors are detected either by the operator smelling them or by reviewing the data after driving the ARS van around the source. Once odors are detected, the van is parked and operated in a stationary mode for at least an hour or until the ARS levels drop to less than 10 ppb.

A summary of the two highest ARS recordings collected downwind of each of the five major odor sources for FY 1988/89 can be found in Table A.

In addition to the ARS equipment, a Varian 3400 gas chromatograph (GC) has been installed in the ARS van to measure individual sulfur species. A second SOLA UPS has also been installed in the van.

COST

The total cost of an ARS monitoring system is about seventy-five thousand dollars. The cost to BESD was about forty-five thousand dollars since BESD already had on-hand a van (A new step van cost about thirty thousand dollars). Is it worth it? BESD thinks so; the agency is in the process of assembling a second unit.

CONCLUSION

To date, the mobile ARS van has proven to be an effective monitoring tool. Several months of data have been collected thus far mainly downwind of the five major ARS sources. Average ARS values downwind of Jacksonville's two terpene chemical plants range from 8 to 29 ppb. ARS values average from 17 to 19 ppb for Jacksonville's two pulp plants. The major sewage treatment plant in town averages approximately 21 ppb. Data above 0 ppb is most easily collected downwind of a facility within 1/4 mile of the facility. The highest valves (such as those in Table A) are mostly obtained during the fall and winter months when weather conditions are more conducive to ambient odors.

The success of the mobile ARS van can be attributed to two major components: the SOLA UPS and the water bed. This method is applicable for averaging periods of 3 minutes or longer. The mobile ARS monitoring system can be used in a variety of places, such as salt marshes, to quantify ARS levels. The mobile ARS monitoring system can be used as a fixed-site, semi-fixed, semi-mobile, or completely mobile monitor station. Since the mobile odor step van was put into operation about a year ago, it has proven successful; so much so that BESD is putting together a second unit. With the addition of a GC operating in conjunction with the ARS monitoring equipment, BESD's unique mobile odor measurement laboratory is truly an odor monitor-on-wheels.

303

TABLE A

FY 1989/90 ARS Data Summary

Two Highest Recordings at Each Source

Source	ARS (ppb) 3–min. avg.	Month
Buckman Sewage Treatment Plant	95.6	November 1988
	52.0	July 1989
Union Camp Corporation	65.3	February 1989
	63.9	September 1989
SCM Glidco Organics	82.1	March 1989
	27.0	March 1989
Jefferson Smurfit Corporation	81.8	April 1989
	71.1	August 1989
Seminole Kraft Corporation	82.9	September 1989
	65.2	March 1989

MEASUREMENT OF ODOR EMISSIONS
FROM MUNICIPAL SEWAGE SOURCES

William H. Prokop, Prokop Enviro Consulting, Deerfield, IL
Michael G. Ruby, Envirometrics, Seattle, WA

A sampling and odor sensory panel evaluation program was conducted for the Municipality of Metropolitan Seattle, WA. Samples were collected of the odor emissions from 28 different locations in the wastewater treatment processes for three different sewage treatment plants and also from 28 different locations in the sewage collection system. The odor samples were evaluated by the forced-choice, triangle dynamic olfactometer technique and also were analyzed for hydrogen sulfide concentration.

The odor sensory results, expressed as ED_{50} values, obtained at the sewage treatment plants are discussed and compared with similar data obtained previously. The ED_{50} values obtained at different lift pump stations of the sewage collection system provided a measure of the odor removal efficiency achieved by various odor control systems in operation. The ED_{50} sensory values and the H_2S concentrations measured analytically resulted in a relatively poor correlation suggesting that odorous compounds other than H_2S could have been present in significant quantities.

A method of estimating the area odor emission rate, $V_SA(ED_{50})$, is described. It consists of calculating the evaporation rate of water vapor escaping from the water's surface. Comparisons are made of the area odor emission rate for various area type emissions from the treatment process which vary in the degree of turbulence at the surface.

INTRODUCTION

A previous technical article[1] describes odor sensory measurements which were
conducted during early October of 1986 to obtain odor dilution to threshold
values for emissions from a municipal sewage treatment plant in Renton, WA.
The forced-choice triangle dynamic olfactometer technique[2,3] was used to
evaluate the samples taken from different stages of the treatment process.
This article also discussed the use of gas chromatograph/mass spectrometer
(GC/MS) analyses which were used in an attempt to identify and quantify the
specific compounds contributing to the odor emissions. Unfortunately, the
GC/MS results identified a large percentage of chemical compounds but could
not account for the odor sensory values obtained.

This study describes a more comprehensive sampling and odor evaluation
program which was conducted for the Municipality of Metropolitan Seattle, WA.
Samples were collected of the odor emissions from 28 different locations in
the treatment process for three different sewage treatment plants (STP) and
also from 28 different locations in the sewage collection system consisting
primarily of lift pump stations.

The three treatment plants include Richmond Beach, West Point and Renton.
Richmond Beach is a smaller plant that includes only primary treatment and
performs anaerobic sludge digestion. West Point is a larger plant which
includes activated sludge treatment and also anaerobic sludge digestion.
Renton includes activated sludge treatment and during 1987 was in the process
of installing anaerobic sludge digestion. The sampling program conducted for
the sewage collection system included eleven specific lift pump stations

which have odor control systems: either activated carbon adsorption units or sodium hypochlorite scrubbers.

ODOR SENSORY MEASUREMENT

The sampling program was conducted during the first two weeks of September 1987 under generally warm, dry weather conditions. Two types of samples were collected: point sources and area sources. An example of a point source would be the duct discharge from a carbon adsorption column or a hypochlorite wet scrubber. In this case, the sample was obtained by a stainless steel probe inserted into the duct. For an area source such as an aeration basin or clarifier in a STP, a styrofoam box was used as a hood and fitted with a stainless steel probe to capture the odor released from the surface of the water. The bottom surface of the hood was suspended approximately 1 to 2 inches above the water's surface. The area of water surface covered by the hood was approximately 2 sq. ft.

Samples were collected in 20-liter Tedlar bags located inside a semi-rigid container which was evacuated by means of a pump. The vacuum created inside the container allowed the sample to be drawn into the Tedlar bag. The sampling procedure included an initial half filling of the bag with an odor sample and then exhausting it in order to equilibrate the inside surface of the bag. Teflon tubing was used to connect the stainless steel probe with the Tedlar bag.

The odor samples were evaluated within four hours of collection using the IITRI forced-choice, triangle dynamic olfactometer.[2,3] Two separate panels of 9 persons each were used in these odor sensory evaluations to determine the ED_{50} dilution ratio, a measure of the number of dilutions to the detection threshold for a panel where 50 percent detect and 50 percent do not detect odor.

The members of both panels were screened to determine whether a prospective panelist was anosmic (unable to detect odor) by use of the butanol reference intensity scale. Only one potential panelist was found to be anosmic and was disqualified. All members of both panels were over 18 and under 50 years of age. Most of the panel evaluations were conducted with 9 panelists, however, during one or two days a panel was operated with only 7 or 8 panelists.

The two panels were calibrated to determine their relative detection sensitivities by the presentation of two samples of hydrogen sulfide at 64 and 650 ppb, respectively. For the 64 ppb sample, the estimated ED_{50} dilution ratio was 70 for panel no. 1 and 75 for panel no. 2. For the 650 ppb sample, the ED_{50} ratio was 2150 for panel no. 1 and 3100 for panel no. 2.

The H_2S concentrations were measured using a Jerome Model 621 Hydrogen Sulfide Analyzer. The use of this instrument above concentrations of 500 ppb is considered to be questionable due to the need for the dilution module attachment. This is discussed under analytical measurements.

A detection threshold of 0.9 ppb was obtained for H_2S for the 64 ppb calibration sample. H_2S has a detection threshold of 1.1 ppb according to Thiele.[4] This value has been critiqued and judged to be acceptable.[5]

The ED_{50} dilution ratios are shown in Table I of the Appendix for those samples collected at the sewage treatment plants (STP) and in Table II for those samples collected at the sewage lift pump stations.

ANALYTICAL MEASUREMENTS

The odor emission samples were also tested for hydrogen sulfide, total mercaptans and ammonia. Sewage samples were measured for dissolved oxygen and sulfides.

Hydrogen sulfide samples were usually collected at the same time and from the same sampling location as the odor samples. Hydrogen sulfide and total mercaptans were measured using a Jerome Model 621 H_2S Analyzer. The 621 uses a thin gold film sensor. This permits sensitivity to 1 ppb, which is greater than most other portable H_2S monitors. The system is calibrated in the range of 1-500 ppb H_2S. Hydrogen sulfide concentrations greater than 500 ppb are measured by attaching calibrated dilution modules to proportionally reduce the concentrations and to extend the range up to 50 ppm. The resultant readings are multiplied by the dilution factor to determine the H_2S concentration.

The Jerome 621 actually measures a "total" reduced sulfer, consisting primarily of H_2S and the lower mercaptans. A copper sulfate scrubber is used to separate the H_2S from other reduced sulfur compounds, such as mercaptans. When used without the copper sulfate scrubber, the 621 measures the mercaptans plus hydrogen sulfide. With the scrubber, it measures mercaptans only. Hydrogen sulfide is determined by subtraction and adjustment for the differential sensitivity of the instrument with the scrubber in place. The use of the scrubber creates an uncertainty in the results. Also, the use of the dilution module attachment to the Jerome 621 to extend the range above 500 ppb is a further complication which could result in inaccurate readings.

The Jerome 621 was calibrated in accordance with the manufacturer's recommendations and checked in the field against a calibration gas standard to ensure accuracy. The manufacturer's reported accuracy is $\pm$ 3% at 100 ppb H_2S. The manufacturer's reported precision is 3% relative standard deviation at 100 ppb H_2S.

The H_2S and total mercaptan concentrations in ppb are shown in Tables I and II for the various odor samples which were collected.

The ammonia concentrations in the air were measured using a colorimetric method involving the use of Nessler reagent. The samples were collected by drawing air through a sodium hydroxide solution with midget impingers. Total air flow was determined by using previously calibrated air pumps. This method is capable of detecting 13 ppm of ammonia. The odor detection threshold of ammonia is reported to be 17 ppm.[6] This value has been critiqued and judged to be acceptable.[5]

DISCUSSION OF RESULTS

Odor sensory results of replicate samples taken one immediately after the other showed good reproductibility. For example, two samples taken on 9/8 at the primary clarifier of the Richmond Beach STP both had an ED_{50} of 50 (see

Table I). Similarly, two samples taken on 9/2 at the outlet of the scrubber
at the North Mercer pump station both had an ED_{50} of 40 (see Table II). Two
samples taken on 9/8 at the South Mercer Wetwell had ED_{50} values of 3500 and
3200. Two samples taken on 9/3 at the Kenmore Wetwell had ED_{50} values of
4600 and 3800.

The ED_{50} values obtained for the three STPs normally follow a pattern where
the higher values coincide with the sewer influent and the first stages of
the treatment process: bar screens, grit chambers and primary clarifiers.
The lower values obtained occur toward the end of the treatment process:
activated sludge aeration tanks and secondary clarifiers. The highest values
were obtained from the sludge anaerobic digestors and from the handling of
this type of sludge.

Substantial variations in ED_{50} values occurred at similar stages of the
treatment process for the three STPs. These are primarily due to differences
in the physical facilities and their operation. For example, the sewage
received at the Richmond Beach STP is less odorous than that received at
either Renton or West Point because its connecting sewer system has a much
lower residence time. This also applies to the hydraulic flow through the
Richmond Beach STP, having less holdup than at Renton and West Point.
Similarly, the influent sewer ED_{50} value at Renton is much greater than that
obtained at West Point because the access at Renton is at the bottom of a
deep, enclosed concrete structure whereas the sampling point at West Point is
located in a large, open wetwell. Also, the bar screen room at Renton had a
much lower ED_{50} than the bar screen room at West Point because only two bar

screen units were being operated at Renton at a reduced flow with less turbulence.

The ED_{50} values obtained at the Renton STP for the sampling program conducted on 9/10/87 may be compared with those obtained for this plant during three days of sampling conducted on 9/30, 10/1 and 2 of 1986.[1] The samples in 1986 were collected in thick-walled, polyethylene bags by means of a peristaltic pump and were evaluated by the same odor sensory method of measurement approximately 16 hours after collection. This comparison is shown in Table III.

Table III - Comparison of ED_{50} Values at Renton STP

Sampling Location	1987	1986-1	1986-2	1986-3
Influent Sewer	3180	----	----	----
Bar Screen Room	245	----	----	----
Morning Glory Distributor	----	813	1120	1050
Grit Chambers	830	----	----	----
Primary Clarifier	150	71	94	118
Grit Hopper	100	62	59	61
Aeration Tank (1st pass)	130	----	----	94
Aeration Tank (2nd pass)	38	29	31	----
Secondary Clarifier	40	62	39	48
Act. Sludge Mixing Box	160	52	88	88

The ED_{50} value for the grit chambers in 1987 relates to the ED_{50} values for the morning glory distributor in 1986. This comparison of values obtained during 1986 and 1987 is relatively consistent throughout the treatment process at Renton. The generally lower values obtained in 1986 may be readily attributed to the cooler weather conditions. It seems reasonable to conclude that the use of different sampling methods and containers did not significantly affect the odor sensory results.

The odor results obtained at the lift pump stations also provided a comparison of inlet and outlet values for the odor control units operated at some of these pump stations. This comparison is shown in Table IV which includes ED_{50} values and H_2S concentrations.

Table IV - Comparison of Measurements Across Odor Control Units

Sampling Location	ED_{50} Values Inlet	Outlet	%	H_2S Conc. - ppb Inlet	Outlet	%
Barton	50	130	---	29	0.5	98
Murray	110	30	72	712	0	100
53rd Street	690	90	87	7383	15428	---
63rd Street	430	80	81	2407	1	100
Medina	270	40	85	12	0	100
Sunset	270	50	81	132	0	100
Heathfield	510	50	90	5700	2	100
North Mercer	160	40	75	336	2	99
South Mercer	3350	15	100	8462	0	100
Sweyolocken	560	80	86	4535	0	100
Yarrow Bay	70	170	---	0	0	---

Note: All odor control units are impregnated activated carbon adsorption units except for the hypochlorite wet scrubbers at North Mercer and Yarrow Bay.

In general, the carbon adsorption units were effective in removing a significant amount of odor and the residual odor emitted to the atmosphere was not considered to be excessive. These emissions ranged from 115 to 1600 CFM air flow. Regarding the 53rd Street pumping station, a subsequent investigation discovered the carbon bed was only half full. Further, the outlet sampling locations for the odor (ED_{50}) evaluation and for the Jerome 621 reading were not the same. The sample for H_2S was obtained at a location where the air flow was completely by passed which could account for the relatively high H_2S reading. At Yarrow Bay, the inlet odor level to a

hypochlorite scrubber was quite low and the obvious odor from the
hypochlorite solution resulted in a higher ED_{50} value in the outlet.

The H_2S concentrations measured by the Jerome Model 621 instrument are highly
variable when compared with the ED_{50} odor sensory results in Tables I and II.
This seems to be particularly true at H_2S levels exceeding 500 ppb where a
calibrated dilution module must be added to extend the range of detection.
Although the H_2S levels generally follow the same upward or downward trend as
the ED_{50} values, the mathematical correlation between ED_{50} and H_2S is
relatively poor (including the correlation with the logarithm of ED_{50}). No
significant improvement in the correlation was obtained when total mercaptans
and H_2S (adjusted for odor threshold) were compared to ED_{50}. It seems likely
that other odorous compounds are present which are not measured by the Jerome
Model 621.

Ammonia was not detected in the odor emissions at any of the sampling sites.
In only one case, the measured value approached the detection threshold level
of 17 ppm. In all other cases, the ammonia concentration was below this
detection threshold level.

APPLICATION OF ODOR SENSORY DATA

Odor emissions from wastewater treatment plants include both point and area
sources. For example, certain categories of the treatment process
(screening, grit removal and sludge processing) often are enclosed within a
building which is equipped with an air ventilation system. These are point

sources. Area sources include the primary clarifier, aeration basin and secondary clarifier.

As indicated in the section on sampling, the point sources are sampled by means of a probe inserted into a duct or discharge from an exhaust fan. The sample is evaluated to obtain the ED_{50} value. Also, the volumetric air flow rate for an odor emission is readily measured in cu ft per min. The quotient, ED_{50} x CFM, is known as the odor emission rate and provides the basis for conducting atmospheric dispersion modeling to estimate the downwind ED_{50} at ground level.

As indicated in the section on sampling, the area sources are sampled by using a plastic box which is inverted above the water to capture the odor released from its surface. The sample is evaluated to obtain the ED_{50} value. However, in the case of the area source, the volumetric air flow rate in CFM is not readily available. For area sources, the following formula represents an area odor emission rate:

$$OER_A = V_S A (ED_{50})$$

where
OER_A	-	area odor emission rate, CFM
V_S	-	surface escape velocity, ft/min
A	-	source area, sq ft
(ED_{50})	-	odor dilution to threshold ratio, dimensionless

The surface escape velocity relates to the evaporation rate of water plus the volatilization rate of those gases which are soluble in water. Based on known data obtained from a sludge processing operation, the combined evaporation of water and volatilization of ammonia resulted in an emission consisting of 95 percent water vapor and 5 percent ammonia by volume.

Ammonia was the predominant odor present at a concentration of 727 ppm. Ammonia has a relatively high odor threshold level of 17 ppm compared to those for the sulfides and amines (0.1 to 100 ppb) often found in wastewater. As indicated on page 4, an H_2S concentration of only 650 ppb resulted in two panel ED_{50} values of 2150 and 3100 which are relatively high odor levels for wastewaters found in STPs (see Table I in the Appendix and Table IV on page 9). As a result, it may be assumed that the volatilization rate of the odorous compounds from the waste-water is relatively insignificant compared to the evaporation rate of water. This assumption is based on typical concentration levels of odorous compounds found in domestic sewage and may be incorrect for higher concentrations in industrial wastewater.

Accordingly, it may be assumed that the volumetric air flow, V_SA, is directly related to the evaporation rate of water from the water's surface and V_S represents the surface escape velocity from the water's surface. V_S is dependent upon wind speed, water temperature, ambient air temperature and humidity, and the turbulence existing at the interface between the ambient air and water. It should be noted that this upward velocity should be related to the ambient air actually receiving the escaping water vapors. For quiescent water, this velocity V_S can be determined using a particular correlation that has been confirmed by actual water evaporation data.

Eagleson[7] describes the evaporation process of water vapor escaping from the surface of a lake or similar body of water. It discusses both the diffusion method and the energy-balance method being applied to calculate these rates of evaporation from a water's surface. For the energy-balance method, an equation and a graphical solution of this equation is provided by Eagleson in Figure 12-5 on page 222. The evaporation rate in inches of water per day is calculated by knowing the ambient air temperature, its dewpoint temperature, the wind velocity and the net solar radiation in langleys per day.

An example of estimating V_S is provided, based on an odor emission study performed in 1989 at a municipal sewage treatment plant (STP) in Michigan. This study included an odor sensory panel evaluation of samples taken from point and area sources within the treatment process to determine the ED_{50} values.

The sampling was conducted generally under the following conditions:

        ambient air temperature - 85 F and 50% relative humidity
        water temperature   - 70 F
        wind velocity   - 5 mi/hr
        radiation   - 300 langleys per day[8]

Based on this data and the energy-balance method of calculation, the evaporation rate for the more quiescent surface category of treatment, such as primary and secondary clarifiers, was estimated to be 0.16 inches per day. Converting an evaporation rate of inches per day to the volume of water vapor escaping vertically from one sq ft of area surface yields the water vapor velocity, W_S, in ft/min. For a specified air temperature and % humidity (and atmospheric pressure), the specific volume of air including the water vapor is obtained from psychrometric charts for air-water vapor data expressed in cu ft per pound of bone dry air. This also applies to the specific volume of water vapor contained in one pound of bone dry air at the specified temperature and humidity. Therefore, the surface escape velocity for the ambient air, V_S, can be calculated as follows:

$$V_S = W_S \frac{\text{(sp. volume of air)}}{\text{(sp. volume of water vapor)}}$$

In the above case, the evaporation rate of 0.16 inches per day is converted to 0.013 ft/min, known as W_S. If air at 85 F is saturated with water vapor, the upward air flow velocity, V_S, is 0.3 ft/min. If air at 85 F is saturated to a 50% relative humidity, then the upward air flow velocity, V_S, is 0.6 ft/min.

The energy-balance method of calculation was also checked by reviewing data on evaporation from pans and lakes.[9] The data obtained from this reference varied no more than 25 percent when compared with evaporation rates obtained by the energy-balance method of calculation as determined using the graphical solution to the equation provided by Eagelson.[7]

The aeration basin is another category of water surface evaporation which is more turbulent than that for the clarifier. It was estimated that 34,000 CFM of air for aeration was passing upward through 67,500 sq ft of surface for 6 aeration tanks (3 passes per tank) that were in operation during the odor study. Therefore, the upward air flow was approximately 0.5 CFM per sq ft of surface or 0.5 ft/min. It was assumed that this air flow leaving the aerated water at 79 F was saturated with water vapor at this temperature. Adjusting for the different saturation humidities (100% of relative humidity) at 79 F and 85 F, it was estimated that the resulting upward air velocity was 1.2 ft/min at an air temperature of 85 F and 50% relative humidity.

The rate of evaporation from a quiescent surface is less than that from a turbulent surface. Further, the evaporation rate will tend to increase as the degree of turbulence increases. For example, a circular shaped clarifier has three separate zones of turbulence: 1) the center ring where the entering wastewater has a "moderate" turbulence, 2) the main part of the clarifier where the wastewater moves radially from the center ring out to the weir overflow. This is considered to be a "quiescent" zone of turbulence, and 3) the outer concentric ring which accumulates the weir overflow and discharges from the clarifier. This is considered to be a "highly" turbulent zone.

The degree of turbulence is directly related to the surface velocity of the water in contact with the ambient air in a direction that is parallel to the water's surface. For the municipal STP in Michigan, the following surface velocities were calculated. In the main part of the primary clarifier just outboard of the center ring, the surface velocity was estimated to be 1.6 ft/min. In contrast, in the outer concentric ring at a distance of 5 ft from the discharge point from the clarifier, the surface velocity was estimated to be 193 ft/min. For the grit removal tank, the surface velocity was estimated to be 39 ft/min which is considered to be "moderately" turbulent.

Table V illustrates a comparison between quiescent, moderately and highly turbulent surfaces in the treatment process at the Michigan STP. It is particularly interesting to compare the ED_{50} values obtained for the different categories of turbulence. For example, the quiescent region of the primary clarifier yielded an ED_{50} of only 95 compared to 1,745 obtained in the highly turbulent zone of the clarifier.

Table V - Comparison of CFM Rates and ED_{50} Values at STP

Source Description	Degree of Turbulence	V_S ft/min	Area sq ft	Emission CFM	ED_{50}	Emission Rate 10^6
grit removal tank	moderate	1.2	676	810	1,320	1.07
primary tanks (main)	quiescent	0.6	9,400	5,640	95	0.54
primary tanks (effl)	highly	2.4	500	1,200	1,745	2.10
aeration tanks	moderate	1.2	67,500	81,000	66	5.35

CONCLUSIONS

A comprehensive sampling and odor sensory panel evaluation program was conducted to determine the odor dilution to threshold values of the emissions from three STPs and from 15 lift pump stations in the sewage collection system of Seattle, WA. H_2S concentrations were also measured at the same odor sampling locations but they did not correlate satisfactorily with the ED_{50} values. The odor sensory results appear to be more reliable than the H_2S values. In future odor studies on municipal sewage sources, more reliable analytical tools are needed to characterize the odorous compounds or classes of compounds which may be present.

The selection of the odor sampling sites for characterizing the treatment process of municipal or industrial STPs is an important consideration that may strongly influence the odor sensory results obtained. The turbulence of the wastewater stream, its exposure to or isolation from the atmosphere and its relative location with respect to the influent flow are important factors to consider. Also, the physical characteristics of the STP facility and its method of operation should be fully understood.

A method of calculating the volumetric emission rate, V_SA, for an area source is proposed. It is based on estimating the evaporation rate of water from a quiescent surface. This rate of evaporation is dependent upon the degree of turbulence at the water's surface. It is recommended that this concept be investigated further to establish whether it is a valid assumption to ignore the rate of volatization of those odorous compounds which are present in the wastewater treated by a municipal STP.

REFERENCES

1. M. Ruby, W. Prokop and D. Kalman, "Measurement of Odor Emissions from a Sewage Treatment Plant," Paper 87-75A.4. Presented at the 80th Annual Meeting of Air Poll. Cont. Assn., New York, June 1987.

2. "Standard Practice for the Determination of Odor and Taste Threshold by the Forced-Choice Ascending Concentration Series Method of Limits," (ASTM E-679), Am. Soc. Test. Mater., Philadelphia, PA, 1979.

3. A. Dravnieks and W. Prokop, "Source Emission Odor Measurement by a Dynamic Forced-Choice Triangle Olfactometer," _Journal of APCA_, Vol. 25, p. 28, 1975.

4. V. Thiele, "Experimentelle Untersuchungen zur Ermittlung Eines Geruchsschwellenwertes fur Schwefelwasserstoff, _Staub_, _39_, 159-160, 1982.

5. "Odor Thresholds for Chemicals with Established Occupational Health Standards," Am. Ind. Hygiene Assoc., Akron, OH, 1989.

6. K. Nishida, M. Yama Kawa and T. Honda, "Experimental Investigations on the Combined Actions of Components Mixed in Odorous Gas," _Mem. Fac. Eng._, _Kyoto Univ._, _41_, part 4, 552-565, 1979.

7. P.S. Eagleson, _Dynamic Hydrology_, pp. 221-223, McGraw-Hill, 1970.

8. I.F. Hand, "Weekly Mean Values of Daily Total Solar and Sky Radiation," Technical Paper No. 11, U.S. Weather Bureau, Washington, D.C., 1949.

9. M.A. Kohler, T.J. Nordensen, and W.E. Fox, "Evaporation from Pans and Lakes," Research Paper No. 38, Hydrolic Services Division, U.S. Weather Bureau, Washington, D.C. (May 1955).

Table I - Sewage Treatment Plant Odor Data

Date	Time	Sample Location	Odor ED_{50}	H_2S ppb	Mercaptan ppb
RICHMOND BEACH TREATMENT PLANT					
9/8	10:35a	Influent Sewer	320	81	8
9/8				86	0
9/8	10:30a	Bar Screen Room	120	53	0
9/8	9:40a	Grit Chambers	100	42	4
9/8	10:00a		240		
9/8	9:15a	Primary Clarifier	50	20	0
9/8	9:30a		50	21	0
9/8	9:05a	Grit Hopper	170	17	3
9/8	9:30a	Anaerobic Digester	2400	200	5
9/8	10:30a	Sludge Loading	9600	34464	36
RENTON TREATMENT PLANT					
9/10	12:40p	Influent Sewer	3180	13772	28
9/10				6079	21
9/10	11:45a	Bar Screen Room	245	188	8
9/10				183	9
9/10	11:20a	Grit Chambers	830	1333	23
9/10	11:00a			1639	28
9/10	11:00a	Primary Clarifier	150	435	15
9/10				394	16
9/10	11:20	Grit Hopper	100	21	20
9/10	10:15a	Aeration Tank (1st pass)	130	30	12
9/10				12	25
9/10	10:00a	Aeration Tank (2nd pass)	38	0	6
9/10				0	7
9/10	9:00a	Secondary Clarifier	40	0	5
9/10				0	5
9/10	9:30a	RAS Mixing Box	160	13	15
9/10				14	17

a - a.m.
p - p.m.

Table I - Sewage Treatment Plant Odor Data (cont'd)

Date	Time	Sample Location	Odor ED$_{50}$	H$_2$S ppb	Mercaptan ppb
WEST POINT TREATMENT PLANT					
9/4	2:00p	Influent Sewer	530	23675	31
9/9	2:15p		640	465	38
9/4	1:25p	Bar Screen Room	1500	2000	0
9/9	1:50p		3100	1274	9
9/4	12:00a	Grit Chambers	475	442	25
9/9	12:15p		380	458	0
9/4	12:30p	Primary Clarifier	190	2	20
9/9	12:50p		160	1	4
9/4	1:15p	Grit Hopper	250	109	16
9/9	11:25a		140	11	2
9/4	9:15a	Pilot Plant TF	520	5837	39
9/9	9:00a		500	12698	2
9/4	9:40a	Pilot Plant AB	100	0	8
9/9	9:00a		40	4	7
9/4	9:30a	C3 Aeration Basin	130	132	13
9/9	10:15a		40	1	0
9/4	10:08a	C3 Secondary Clarifier	110	0	23
9/9	9:45a		30	3	3
9/4	10:45a	Dewatering Room	240	49	25
9/9	11:00a		150	33	0
9/4	11:00a	Dewatering Sludge Hopper	1140	2	0
9/9	11:30a		3100	0	2
9/4	11:40a	Anaerobic Digestor		466	21
9/9	11:40a		3670	112	0

a - a.m.
p - p.m.

Table II - Sewage Lift Pump Station Odor Data

Date	Time	Sample Location	Odor ED_{50}	H_2S ppb	Mercaptan ppb
9/1	9:30a	Barton Wetwell	50	29	4
9/8	3:15p	Barton Adsorber Outlet	170	0	2
9/8	3:30p		90	1	1
9/1	11:30a	Murray Wetwell	110	712	12
9/1	11:00a	Murray Adsorber Outlet	30	0	0
9/2	3:00p	53 rd St Wetwell	690	7266	24
9/2	3:00p			7501	28
9/2	2:45p	53 rd St Adsorber Outlet	90	15428	62
9/2	3:45p	63 rd St Wetwell	430	2407	35
9/8	4:30p	63 rd St Adsorber Outlet	80	1	1
9/8	4:30p		80	1	1
9/3	11:40a	Hollywood Wetwell	420	2667	33
9/3	11:40a	Hollywood Effluent	180	72	15
9/2	12:00a	Medina Wetwell	270	12	0
9/2	12:00a	Medina Adsorber Outlet	40	0	0
9/3	9:00a	Sunset Wetwell	270	132	35
9/3	9:00a	Sunset Adsorber Outlet	50	0	1
9/3	10:00a	Heathfield Wetwell	510	5700	31
9/3	9:30a	Heathfield Adsorber Outlet	50	2	0
9/2	8:30a	N Mercer Wetwell	160	336	28
9/2	8:30a	N Mercer Scrubber Outlet	40	2	8
9/2	9:00a		40		
9/1	2:00p	S Mercer Wetwell		8462	16
9/8	2:30p		3500	2984	16
9/8	3:00p		3200	3059	11
9/1	2:00p	S Mercer Adsorber Outlet	15	0	0
9/1	3:00p		15		
9/3	3:30p	Matthews Wetwell	2600	461	24
9/11	12:55p		1280	69	12
9/11	12:55p			67	9
9/3	5:20p	Matthews North Portal	4250	11103	55
9/11	2:30p		2530	120	37
9/11				118	23
9/3	2:30p	Kenmore Wetwell	4600	3530	31
9/3	3:00p		3800		
9/11	3:00p		2370	294	19
9/11				286	19
9/2	10:30a	Sweyolocken FM Discharge	560	4535	15
9/2		Sweyolocken Adsorber Outlet	80	0	0
9/11	10:35a	Yarrow Bay Wetwell	70	0	14
9/11	11:00a	Yarrow Bay Scrubber Outlet	170	0	6
9/10	2:00p	Interurban Wetwell	480	16252	47
9/10				13860	40

a - a.m.
p - p.m.

V. Odor Emission and Control Technologies at Waste Water Treatment Plants

ODOR CONTROL SYSTEMS FROM AN
ENGINEERING FIRM'S VIEWPOINT

Perry L. Schafer, Managing Engineer
Brown and Caldwell Consulting Engineers
Sacramento, California

Abstract

Odor control at wastewater facilities includes a wide variety of methods. These methods include the following: source control; limit the production of odor (within sewers and treatment plants); treat odors in the liquid phase; cover, ventilate, and treat foul air; and, enhance dispersion of released odorants. This session of the Specialty Conference is involved primarily with only one of these methods: the treatment of foul air in various types of systems. Owners and engineers need to be aware of how this method is integrated with the other methods of odor control so that the entire odor problem is solved cost-effectively.

Engineers need a variety of data to design a foul air treatment system. The engineer needs to match the type of system with the problem so that maximum benefit is obtained. To do this, the engineer needs a host of information about the odorants to be treated, performance of foul air treatment systems, process parameters, and materials of construction, as well as costs. Once process selection is made, the details of design are refined so that a cost-effective treatment system is constructed. Monitoring of all systems are necessary to insure that performance meets expectations.

Introduction

Odor control for wastewater treatment facilities is a complex subject. Odor can be produced and emitted by almost every process unit within a plant. And often the worst odors originate in the wastewater collection system. Thus, these odorants are brought into the headworks of the plant on a near-continuous basis. The odor from most wastewater facilities is a mixture of many specific odorants, involving hundreds of compounds in some instances.

Odor control work, more often than not, involves other issues such as corrosion control and worker safety. Ventilation systems must be designed for safety whenever structures are enclosed for odor control. These other issues are critical when dealing with hydrogen sulfide. It is often mandatory that all three issues (odor, corrosion and safety) be handled together as a group because failure to assess these issues in a coordinated manner can lead to control systems that may not be ultimately desirable.

It is also possible that volatile organic compounds (VOCs) or toxic air compounds emitted from wastewater facilities will become an issue that should be handled together with odor control. If health authorities demand that VOCs be controlled from publicly owned treatment works (POTWs) foul air treatment systems will likely be called upon in many cases to perform this role. Therefore, it will likely be important in the future to determine the performance of foul air treatment systems in removing VOCs.

Odor Control in Wastewater Facilities

There are various methods used to control odor from wastewater facilities. The methods can be grouped into the following categories:

1. Source Control. In many cases, odorants can be minimized from wastewater facilities by controlling the characteristics of the wastes discharged to the sewers. This could include direct discharge of sulfide by an industrial firm, or discharge of a specific organic material that has a strong odor. It can also involve the discharge of slugs of biochemical oxygen demand (BOD) or suspended solids or large pH swings that affect odorant production and off-gassing within sewers.

2. Limit Production of Odor. In almost all cases the specific design of the collection and interceptor system and treatment plant has a major influence on production or generation of odorants. For instance, slow-moving wastewater caused by too flat a slope for the sewer, will allow solids to settle, forming deposits and sediments within the pipes. In some situations these deposits can generate large quantities of sulfide and other reduced sulfur compounds. Also, the recycle of anaerobic sidestreams from solids processing systems can cause significant odor emissions if not handled properly. The proper design and operation of sewers and treatment facilities is crucial to minimize the production of odorous compounds.

3. Treat Odors in the Liquid Phase. In many cases, perhaps most cases, it is more cost-effective to treat odorants while they are in the liquid phase, rather than waiting to let them move into the gas phase. Also, from a

corrosion and safety standpoint, it is better to avoid the off-gassing of many compounds including hydrogen sulfide. Treating odors in the liquid phase usually involves adding a chemical. There are chemicals that oxidize odorants, precipitate odorants, or otherwise treat or remove the odorous compounds before they have a chance to move into the gas phase.

4. Cover, Ventilate, and Treat Foul Air. If the above control measures fail to eliminate the problem, process units at treatment facilities will need to be covered, the gas space under the cover ventilated, and the foul air withdrawn for treatment in a treatment device. This category of control is one of the least desirable for wastewater agencies since whenever a piece of equipment or a process unit is covered, it increases the complexity of operation and maintenance and may, therefore, lower the performance or reliability of performance. If it is necessary to cover, ventilate, and treat foul air, the project should be engineered carefully to minimize the problems for operation and maintenance and to insure adequate safety.

5. Enhanced Dispersion. There are various methods that can be used to enhance the dispersion of odors from wastewater facilities and, therefore, reduce the odor concentrations reaching nearby residents. The method of ventilation and location of ventilation points can be important. Properly designed and located ventilation stacks can enhance dispersion. Even the plant layout and the way buffer areas adjacent to the plant are used can be critical to minimizing odor complaints from downwind residents. Wastewater agencies should be cognizant of the possibilities with this category and use it to its maximum because it is likely to be cost-effective.

This session of the Specialty Conference is concentrating on the gas phase treatment of odorous compounds. Therefore, category 4 listed above, is the one to be considered in detail here.

What Engineers Need to Know

Engineering firms need to know a great deal of information to select and design a foul air odor control system. In particular they need to know how all the above-described five categories can be used to solve a problem, and how category 4 interacts with the others.

Engineers need to know the expected odorant concentrations as well as the degree of variation and the peak concentrations of odorants. Gas flow rates also need to be determined, and these should be carefully evaluated to insure adequate capture of odorous compounds and maximum safety of personnel.

The following describes many of the factors important in selecting a foul air treatment process train and conducting the specific design of the equipment.

1. Performance. The engineer must insure that adequate performance can be obtained from the treatment system. The owner is demanding performance primarily because the plant neighbors need to be assured that the system will work.

2. Reliability of Performance. The reliability of performance is obviously important to the engineer and the owner. Reliability depends on many factors including the materials used, type of equipment used, ability to adjust to varying concentrations of odorous compounds, downtime for maintenance, and possibility of process upset.

3. Sound, Technical Basis for Treatment Process. The engineer needs to know how the system works if he is to recommend it to an owner. There must be a sound description of how odorants are treated and processes used to reduce the odor level.

4. Operating Installations. It is important to know that the process is working at other similar installations. The names and phone numbers of contact people at these installations are important so that confirmation can be made on performance of the process. In some cases, site visits are needed to confirm operation and performance details.

5. Recommended Design/Operating Parameters. Previous experience should be used to recommend specific design and operating parameters including chemical consumption and utilities requirements.

6. Size and Equipment Configuration. With increasingly tight space availability at wastewater treatment plants, size of equipment can be a problem. Also, the configuration of the system is important, especially such issues as whether the system should be operated at atmospheric pressure or at slightly positive or negative pressure.

7. Maintainability Issues. The engineer should determine the equipment warranties, availability of spare parts, and the availability of operation and maintenance information.

8. Instrumentation for Trouble-Shooting. The engineer needs to be convinced that adequate instrumentation exists to find and solve a problem easily. Also important are the recommendations on monitoring the system to insure that it is performing adequately.

9. Materials of Construction. The engineer is concerned that the facilities serve the owner for many years and that materials of construction are appropriate for the specific conditions they will be subjected to.

In addition to these non-cost factors, engineers need to know the costs of the systems, both capital and operation and maintenance costs. Engineers need to balance the costs of a system with the non-cost factors so that the selected system meets the needs of the owner at minimum overall cost.

Consider Nuisance Risk Assessment

Due to the variables involved in odor emissions, foul air treatment process performance, and meteorological variation, the engineer and owner should consider an approach using nuisance risk assessment. Appendix A describes this concept. Undertaking nuisance risk assessment for odor control could result in the requirement that a foul air treatment system meet 95 percent total odor removal (ED_{50}, for instance) 90 percent of the time. Or the specification could be written around a particular odorant such as hydrogen sulfide, dimethyl sulfide, or ammonia, and would involve certain percentages of samples that must achieve a specified level of treatment.

Nuisance risk assessment results in an evaluation of the amount of time that odor at or below a specified level will occur at a certain location. This reduces the problem to one which can technically be evaluated and solved and offers the public the commitment of specified levels of control that can be documented and monitored.

Summary

This paper discusses the methods of controlling odor at wastewater facilities as well as the factors that are important in selecting and designing a foul air treatment system. Foul air treatment systems are an integral part of many wastewater plants and are expected to become even more critical in the 1990s. These systems need to work reliably and adequately to make the neighborhoods adjacent to our wastewater facilities livable and enjoyable.

APPENDIX A

VARIABILITY OF ODOR AT WASTEWATER FACILITIES

The variability of a number of factors leads to widely varying odor concentrations at the fencelines of wastewater and other waste processing facilities. These factors can be summarized as follows and are supported by the three figures on the following pages:

1. Variability in Source Emissions. An example is provided showing one to two orders of magnitude variation in surface odor emission rates from sludge lagoons.

2. Variability in Foul Air Treatment System Performance. The example shows that performance of foul air treatment systems can be excellent and still provide widely fluctuating odors in the treated gas stream. This foul air was from a pilot test of vacuum deodorization of anaerobically digested sludge.

3. Variability in Meteorological Factors. The graph shows about two orders of magnitude variation in the atmospheric dispersion coefficient at the Sacramento plant.

When these variations are coupled together, there is obviously widely changing odor levels at downwind locations. Since zero odor downwind is not feasible, the situation almost demands statistical analysis. Downwind residents can then be informed as to the small percent of time that certain levels of odor are likely to occur. Treating this in terms of "nuisance risk assessment" can be helpful.

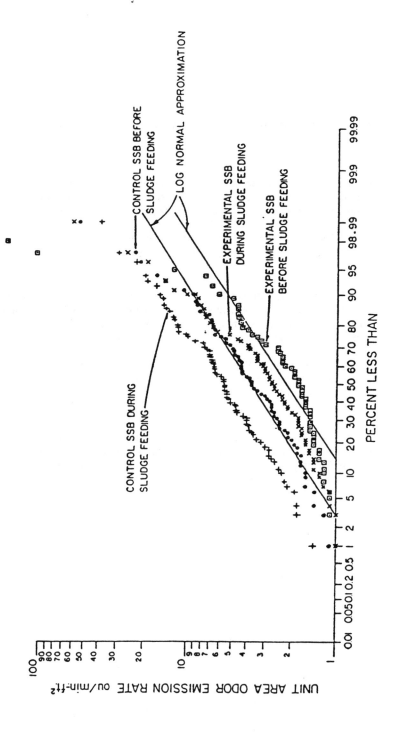

EXAMPLE OF VARIATION IN EMISSION RATES
(Sacramento Sludge Lagoons)

EXAMPLE OF CARBON ADSORPTION ODOR SCRUBBER
PERFORMANCE VARIATION
(SACRAMENTO VACUUM DEODORIZATION EXPERIMENT)

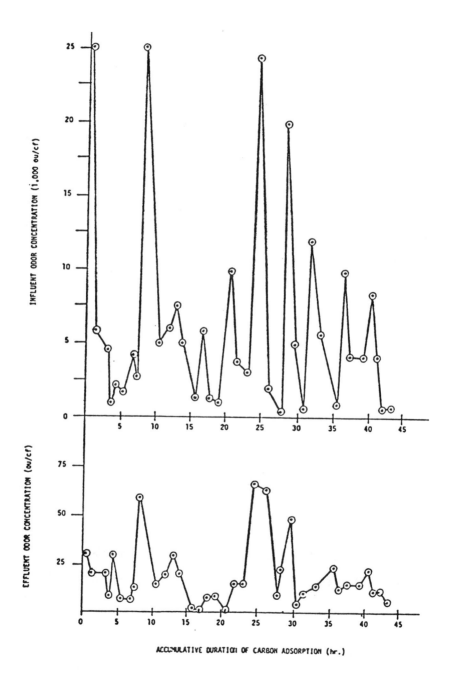

ACCUMULATIVE DURATION OF CARBON ADSORPTION (hr.)

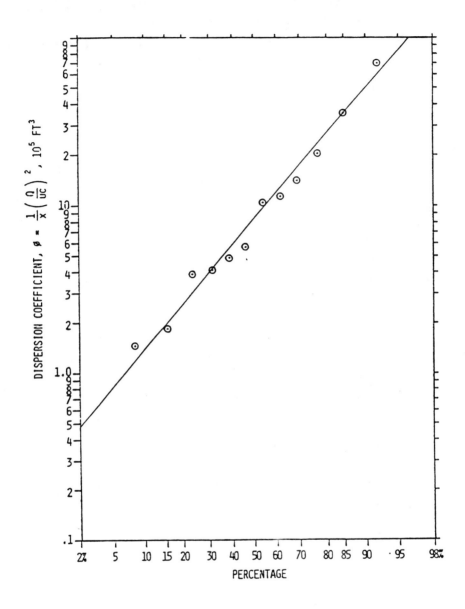

EXAMPLE OF METEOROLOGICAL VARIATION

(Frequency Distribution Dispersion Coefficient–Sacramento Plant)

334

MIST SCRUBBING TECHNOLOGY

Harold J. Rafson
President
QUAD ENVIRONMENTAL TECHNOLOGIES CORPORATION

Mist scrubbing technology, invented and patented by QUAD in 1975, is described and compared to alternate wet scrubbing technologies. The QUAD Chemtact™ System offers more contact, reaction time, and more reactive chemical solution than traditional designs, resulting in higher efficiencies of odor removal.

Applications of the QUAD Chemtact™ System to numerous wastewater applications is described. Designing an odor control system, and the options in design are discussed. Effectiveness of treatment of VOC's is commented upon.

I want to express what a pleasure it is for me to participate in such an excellent seminar arranged by AWMA. And specifically this session which was scheduled by Dr. Kruse, that has such an outstanding group of knowledgeable people making presentations.

Being one of the first speakers on the subject of odor control equipment gives me the opportunity to make some preliminary comments. First, I urge you that these presentations should not be looked upon as a bunch of odor control equipment salesmen all competing for the sale, and all praising the merits of our systems, and damning the other technologies. Rather it is for you to understand the merits, and demerits, of each system, and to find where they are best applied. As you will see from my and my colleagues presentations the technologies are all different, and you do not have apples and apples comparisons, you have apples and oranges comparisons, and it is necessary to evaluate your situation to find out whether you need an apple, an orange, a little bit of each (since several of the systems are complementary), or something else altogether. The point that I think you can be assured of, and I think I can speak for all of us, is that we all will be pleased to work with you to evaluate that question to determine the best solution for your needs.

There is another preliminary comment that I'd like to make. And that is that the environmental regulations are becoming more stringent, and more rigidly enforced. It is not the same ball game. And the same solutions that might have been adequate, or borderline, will be found wanting in the new regulatory climate. You have heard from speakers in the previous days that removals of gaseous chemical contaminants will need to be much more complete. An older technology that removed 75% of the odors, or 50% will not be adequate. Buying odor control equipment, at the lowest price, just to show the regulator that you were doing something in your designs and construction, whether it works in practice or not, and then throwing up your hands when it doesn't work, will not be adequate. Buying odor control equipment, of any design, and then not maintaining it or not operating it at all, will not be adequate. And considering gaseous emission control in just the limited sense of those odors which we can smell, will not be adequate, since there is an urgent concern about volatile organic chemicals, and hazardous and toxic emissions. It has been too long that these issues were not addressed, or were swept under the rug. The spotlight is on gaseous emissions, and we, as an industry supplying you with equipment, and you and the designers and users of that equipment, have to do a better job.

I'll tell you about QUAd, and what QUAD is doing. And I'll tell you about QUAD's technology, for those who are not familiar with it. QUAD's technology has been in use for about 15 years now, and I still find it necessary to explain how the QUAD system works, and why it is more effective at removing odors and gaseous chemical contaminants than any other gas-liquid contacting system – than any chemical reaction scrubbing system, and doing it at lower operating costs.

The explanation is simple. And it all gets back to the mass-transfer equation - and no scrubber can get away from that. It states that the Efficiency of Removal is a function of Contact (and that is visualized as Surface, or Number of Droplets) times Reaction Time (the time the gas and liquid are in contact) times the Reactivity of the Solution. If you compare any scrubber, all you have to do is look at how its design approaches each of these design parameters.

Conventional designs, whether conventional spray scrubber, or packed columns, have large droplets, or films of liquid. Let me give you an example: a typical droplet is about 1,000 microns; a QUAD droplet is 10 microns. For comparison the thickness of a human hair is approximately 100 microns. So you see that the droplets QUAD makes are very small indeed. And why? Because, for the same amount of liquid, comparing these two different droplets, you will obtain 10,000 times the Surface, in a QUAD system, and 1,000,000 times the number of droplets. Well, with that kind of increase in contact, we don't use the same amount of liquid, we use less, but we still have enormously better contact.

And these droplets, which are like London fog, do not fall like rain drops. Therefore in a conventional spray scrubber, or a packed column, the liquid falls through the air and the contact times are in hundredths, or tenths of a second. QUAD designs for longer reaction time - usually between 3 and 60 seconds. And with this longer reaction time, with this longer time of contact for mass transfer, it is obvious again how higher efficiencies of removal are achieved.

And now we come to the last part of the equation, the reactivity of the solution, and here QUAD's further advantage is not as immediately apparent. You could ask why since the QUAD system uses
the same chemicals as other systems, should it also have better reactivity? Let me explain. The contact we achieve is so great, that we can react all the chemical. If you can react all the chemical, why do you have to recirculate? You don't. Therefore, you can feed just the chemical you need (and a little more for completeness) – and, and this is the big and, you can control the amount of chemical being fed by monitoring continuously the effluent liquid stream, and the exhaust gas stream. And the response time is almost immediate, usually less than a minute. This allows feeding of the chemical needed, and not wasting chemical. It means having enough chemical when you need it. And less when you don't.

All recirculating systems however, don't work that way. first, the contact surface and time is inadequate, and all chemical could not be reacted, so the solution had to be recirculated. This means several things - first, the recirculating liquid builds up dissolved and reacted compounds in it, and then revolatilizes, then into the air, recontaminating the air. It is attempting to get clean air; by washing it with a dirty solution. What is the necessary correction for that? to add chemical to refreshen the solution. This creates a waste stream of chemical which is a great operating cost. Please, whenever you evaluate chemical scrubbers, don't overlook the wasted chemicals. Secondly, this recirculating liquid is a pot of liquid, with very slow response time to varying inlet odors.

It is my observation that most odor sources are very variable. This means that chemical concentrations must be maintained higher. This means more wastage. This means poorer diffusion into the liquid.

Again what does QUAD do? We have fresh chemical. Controlled to the need. Excellent contact. Long reaction time. Efficient usage of chemicals, and low wastage, and the bottom line, high effectiveness of odor removal.

QUAD has about 150 systems in operation for as much as 15 years. In wastewater, about 120 systems in operation for up to 10 years. And wide ranges of applications.

The odor panel testing method is the most useful, since it answers the question most directly, "to what level have odors been reduced?"

Further the QUAD odor control system is practical, and low cost to operate. QUAD builds quality equipment that will seldom break down. We refuse to underdesign, and we refuse to build cheap equipment that will give you maintenance headaches. We might cost a little more up front - we will cost you far less in the long run - we will run continuously and get rid of odors effectively. Every manufacturer has to make a decision of where they want to be in the market place. Our decision is to make equipment that works, and continues to work. Ask our clients.

Why are QUAD's operating costs low? Because of less chemical usage, because of lower pressure drop and power usage. We do have an added cost of power to atomize. We urge you to make life cycle cost that is present worth comparisons. That is the only way to compare the varied kinds of equipment offered to you. Some equipment may have low capital cost, possibly low installation costs, and be high in maintenance, in replacement of components, in operating cost. Operating cost is the most significant factor in the equation, and you have to compare all systems using all factors.

What chemicals does QUAD use? That depends upon what is to be reacted, since QUAD can use any chemical. Usually in odor control it is an oxidant such as sodium hypochlorite (other oxidants can be used), a base, sodium hydroxide (other bases could be used) or an acid, sulfuric acid (other acids can be used).

Let me get back to my comparative comments again. And this gets back to the question of what you are actually trying to accomplish in your design.

If you have lots of H_2S, then the regenerable reactant you will hear about later, will probably be your low chemical operating cost choice. But that system is specific to H_2S and does not treat odors, and must be supplemented by a mist system, or packed column to get rid of odors.

If you have low concentrations of odors, and low air flow volumes, a carbon system may be your choice, since it will have a long life. Or, if you must remove a wide range of compounds from high to very low levels you may need a mist system or packed column followed by a carbon system.

There are many options, and a highly skilled industry here to serve you.

Let us talk a little bit further about some of the problem areas.

There are some odor and emission generating sources that are not easy to treat, that are not straight forward, and that take real care in design. I'd like to mention two of them, and they will also serve to illustrate the problems that we have worked on, and continue to work on, and ways to their solution.

Let us begin with compost odors. If anyone underestimates the problem of odors generated in composting they just don't know what they are talking about. The odors generated are intense, variable, of a complex mixture, and extremely persistent. QUAD has a system that has been operating successfully now for about 40,000 hours, or 5 years of continuous operation. The owner has done a good job of making refinements in the operation of the system, as it is adapted to the specific needs. In this case, the inlet odors are hot, coming undiluted from a downwardly drawn, compost piles. The ammonia level is high ranging up to 800 ppm NH_3. It has been found desirable to eliminate the ammonia before further treatment, and at WSSC this is done with a first stage acid mist and spray chamber. Then the odors are oxidized with sodium hypochlorite. And this is followed with a final wash step. Analyses both of GC/MS and odor panel testing have shown QUAD Chemtact™ system to be very effective.

Sludge absorbs large quantities of VOC's, and the composting process volatilizes them. We have done GC/MS analyses of many points of a waste treatment plant, and it is remarkable the variety and quantity of VOC's carried with wastewater and sludge. These are volatilized throughout the aerobic waste treatment. And it seems there still is enough left over to be absorbed by the sludge. Here is a typical analysis of how effective QUAD was in treating these compounds.

QUAD is continuing a series of studies of VOC removals at waste treatment facilities, and we expect to have much more information in the coming months. If there is anyone in the audience with similar concerns, and would want to work with us, or if you want me to keep you apprised of developments, please let me have your card at the next break.

SEWAGE ODOR CONTROL USING
ALKALI IMPREGNATED ACTIVATED CARBON

BY D. A. BISCAN,
RESEARCH & DEVELOPMENT
& J. 'L. RIZZO,
COMMERCIAL DEVELOPMENT
CALGON CARBON CORPORATION
PITTSBURGH, PA 15230

Introduction

Historically, most municipal sewage treatment plants were
located on the outskirts of built-up areas where any
undesirable odors were discharged directly into the air.
Today, those same areas may be densely populated and odors
are of greater concern.

Plant managers and their consultants are now focusing
attention on these odors, particularly those that emanate
from points where sewage is agitated or where sludge
accumulates such as pump stations, head works, trickling
filters, clarifiers, digesters and at sludge handling and
storage areas.

Specifically, these sewage odors consist of a complex
mixture of hydrogen sulfide (rotten egg odor), mercaptans,
and other organic compounds. The threshold level at which
some of these compounds are detected is as follows:

Odor	Threshold (ppb) In Air
Hydrogen sulfide	0.47
Indole	1.00
Skatole	1.20
Methyl Mercaptan	1.10
Methyl Amine	21.00

To be successful in minimizing these odors, the treatment
technology must not only reduce these compounds to below
threshold levels but it must be cost effective and
reliable.

For more than a decade, Calgon Carbon Corporation has been actively involved in the application of an activated carbon process for sewage odor control. In the late 1970's, Calgon was issued a patent on the use of an impregnated carbon that was specifically developed for controlling the aforementioned types of odors. The activated carbon itself was manufactured from bituminous coal which was then impregnated with sodium hydroxide. It is known as Type IVP activated carbon and can adsorb up to 24% by weight or hydrogen sulfide (compared to about 6% by weight hydrogen sulfide on a similar, but unimpregnated activated carbon).

During the past 12 years, Calgon Carbon has installed over 800 adsorbers of various sizes containing Type IVP carbon for the control of sewage plant type odors. This presentation will discuss the reaction mechanics, the key design parameters and an improved and more cost-effective process [1] for the in situ regeneration of the carbon when it is exhausted. In situ regeneration can result in reduced operating costs.

Reaction Mechanism

The odor control technology with an alkali impregnated carbon combines adsorption of organic odorants as well as the catalytic oxidation of hydrogen sulfide and mercaptans. The adsorptive function removes organic compounds like Skatole, Indole and methyl amine. Simultaneously, hydrogen sulfide is oxidized in the presence of moisture and activated carbon to form elemental sulfur and sulfates. In the case of methyl, ethyl and isopropyl mercaptans, the organic sulfide is oxidized to a disulfide which is less odorous and more readily adsorbed by the Granular Activated Carbon, this results in a higher removal capacity. The overall reactions are as follows:

1. $2H_2S + O_2 \xrightarrow{\text{NaOH}}{H_2O} 2H_2O + 2S$

2. $4RSH + O_2 \xrightarrow{\text{NaOH}} 2RSSR + 2H_2O$

Another sulfur compound (SO_2) is also a specific pollutant often regulated by air discharge standards. Sulfur dioxide (SO_2) is also controlled by Alkali impregnated Activated Carbon. These carbons are able to control sulfur dioxide emissions by relying on their ability to act as a catalyst. In air, they catalyze the conversion of SO_2 to sulfur trioxide which then reacts with the Alkali to form a salt.

Performance Characteristics

The following represent some of the important performance characteristics of Caustic impregnated activated carbons.

1. The hydrogen sulfide test method employed to determine the carbon capacity is Test Method No. 41R. Using this test, the hydrogen sulfide capacity reported on a weight percent basis for IVP carbon is 24%.

(1) Regeneration is covered by U.S. Patent No. 4,558,022

2. No significant effect has been observed on H_2S capacity over a range of influent concentrations. We have concluded that the saturation capacity for H_2S is relatively independent of H_2S concentration.

3. We have determined that pre-humidification of IVP with humid air (85% R.H.) has a negligible affect on H_2S capacity. Moreover, inlet gas with a high moisture level does not deteriorate H_2S capacity. However, interstitial water can have a negative affect on H_2S removal due to mass transfer and solubilization of the alkali impregnant.

4. With a fixed adsorber length, the H_2S capacity can be affected by the contact time. For example, H_2S capacity for IVP is approximately 24% at 1.4 seconds contact time. This capacity decreases with contact times less than 1.4 seconds. We generally utilize a conservative contact time of 1.8 seconds for design purposes.

5. Carbon bed depth is defined by the gas velocity and contact time. The present design is based on 1.8 seconds minimum contact time, i.e.

Velocity ft./min.	Bed Depth–ft.
50	1.5
75	2.25

6. A strongly adsorbed organic compound on the impregnated carbon will occupy pore volume and reduce the H_2S capacity. Tests have shown that a 15% w/w toluene loading on IVP carbon reduced the H_2S capacity by 33% of its virgin capacity. Therefore, a sewage odor stream should be characterized for H_2S and organic compound levels prior to the system being designed. If the vapor stream has high concentrations of organic compounds, and a relatively low H_2S level, then the stream could be treated with an unimpregnated activated carbon.

Air flow should be determined based on adequate ventilation of the area in question. We have found the following to be adequate ventilation rates, but they may have to be modified for specific cases:

Area	(Air changes)hr.)
Normal treatment building	6
Chlorine room	60
Sulfur dioxide room	60
Comminutor room	30
Rough screening room	30
Vacuum filter room	10
Motor control center (room)	4–6
Lavatory	30

Note: $CFM = V\ C_s \times 1/60$

V = Room Volume
CS = Air change per hour

The following table shows typical hydrogen sulfide
levels measured at various areas in a typical sewage
treatment plant. It also shows typical air
changes/hour:

Area	Typical H_2S Levels, ppm	Air Change per Hour
Pump Stations	0-50	5-10
Lift Stations	0-50	5-10
Primary Screening Chamber	50-500	15-30
Grit Chambers	50-500	15-30
Primary Settling Tanks	50-500	5-10
Wet Wells (Pump pits)	5-100	1-5
Clarifiers	0-10	1-5
Sludge Holding Tanks	100-1000	5-10
Sludge Conveying Systems	0-10	(Hood Design)
Vacuum Filter Area	0-50	5-10
Control Rooms	0-5	4-6

With the above information, gas volume flow rates can
be established.

Design Considerations for Odor Control Systems

The critical design considerations include:

o   Selecting the proper air flow/

o   Maintaining a high bed efficiency from both an odor
    removal and carbon use standpoint/

o   Minimizing power requirements/

o   Estimating carbon usage/

Since the capital cost of an odor control system increases
in proportion to the air flow, it is recommended that
odors be treated as close to the source as possible to
minimize the air volume. To contain the odors, a slight
negative pressure must be maintained on the controlled
area.

The carbon bed must be deep enough to assure complete odor
elimination, and have some excess capacity to handle
concentration excursions. That portion of the carbon bed
in which adsorption and catalysis occurs is called the
mass transfer zone (MTZ).

As this mass transfer wavefront moves through the carbon
bed, some contaminants will begin to appear in the
effluent. When this occurs, it means that the mass
transfer zone has progressed through the carbon bed depth
and that the carbon is approaching exhaustion.

Based on laboratory and field testing, we have learned
that the minimum mass transfer zone length, for effective

odor control is usually about six inches. Deeper beds
enable more of the column to be saturated and are more
efficient. If a 12 inch bed is used, for example, the
service life will be about 75% of the total theoretical
available capacity. When a 36 inch bed is utilized, 92%
bed efficiency can be expected. Three foot deep beds are
usually more cost effective from a carbon use standpoint,
and they provide excess capacity if an odor excursion
occurs.

Power requirements for a system using a 4x6 mesh IVP
carbon are directly proportional to the pressure drop and
air flow. Pressure drop will increase with face velocity,
carbon bed depth, and will be related to carbon mesh size.
For example, Type IVP 4x6 mesh carbon has at least a 25%
lower pressure drop than a conventional 4x10 mesh vapor
phase carbon. This reduced pressure drop results in a
reduction in power (operating cost savings).

Caustic impregnated carbons can be regenerated in-situ,
most adsorbers are of FRP construction so the materials of
construction can handle various acid gases, moist air and
the regeneration chemicals (caustic). The adsorbers
should have multiple access ports to permit safe filling
and emptying of the carbon. The adsorber should also
include an adequate carbon support system to ensure even
air distribution and adequate carbon support. Grounding
rods should be used to dissipate any possible static
charges which could build up in the FRP vessel. The air
blowers are generally of FRP construction as well.
A unique feature of the systems that we supply relates to
the selection of the dampers. We use self-sealing dampers
to assure essentially complete air shut-off during periods
of system shutdown. The use of this type of damper is
important, since heat excursions can occur under
conditions of high temperature, high humidity, and low air
flows that could occur during a system shutdown if non
self-sealing dampers are used.

We have recently developed a unique Hydrogen Sulfide
Monitor which can be easily installed on one of the carbon
sample ports to measure the H_2S concentration at that
point in the carbon bed. This type monitor is inexpensive
and permits rapid detection of H_2S breakthrough. A
monitor of this type should be specified for all systems.

We offer a complete line of units to handle flows in a
single bed up to 10,000 cfm. For higher flows multiple
single beds or dual beds can be employed. The following
represent physical adsorber sizes and the flows that can
be handled by each:

Vessel Diameter(ft.)	CFM
3 SB	350-530
4 SB	625-940
5 SB	980-1470
6 SB	1410-2120
6 DB	2830-4240
8 SB	2510-3770
D DB	5030-7540

344

```
       10 SB                    3925-5890
       10 DB                    7850-11780
       12 SB                    5650-8475
       12 DB                    11310-16960
```

Note: SB Means Single Bed
       DB Means Dual Bed

Regeneration

When H$_2$S is detected in the effluent, the carbon is
considered to be spent (exhausted). It must, therefore,
be replaced or regenerated with an alkali (NaOH in the
case of Type IVP carbon). Our calculations show that it
is more economical to replace the carbon in units 6 feet
diameter and less. Conversely, it is more economical to
regenerate units greater than 6 feet in diameter.

The H$_2$S capacity of the carbon can be restored in place by
using a suitable chemical regeneration procedure. This is
accomplished by contacting the carbon with a two-stage
sodium hydroxide solution followed by a water rinse.

Specifically, the carbon is chemically regenerated with a
NaOH solution by completely covering, and soaking the bed
in caustic for about 24 hours with the 50% solution, and
about six (6) hours with a 15% solution. Then the carbon
bed is drained and soaked with water for about one-half
(1/2) hour.

This procedure will result in organic compound levels
(water soluble) being preferentially reduced by the 50%
caustic soak while the sulfur levels are reduced by the
15% caustic soak.

In most sewage odor control applications, chemical
regeneration will increase the useful life of the carbon;
thus making it more cost effective. Most clients are able
to regenerate the carbon three to five times before the
carbon has to be replaced. Although the chemical
regeneration does not return the carbon to virgin H$_2$S
capacity, it usually returns the carbon to the 60-90% of
virgin capacity.

Examples of Calgon Carbon Odor Control Systems Proven
Effective in Field Use:

San Diego, CA

In 1981, five 12 ft. diameter dual bed adsorbers were
installed to control odors at a large pumping station. In
addition, twelve, 12 ft. diameter dual beds were installed
to handle odors from a variety of sources at a new
treatment plant. Of interest is the fact that six 8'
diameter caustic scrubbers were installed ahead of the
adsorbers to reduce the high H$_2$S concentration (~100 ppm).
Since the caustic scrubbers do not handle all of the H$_2$S,
the carbon continues to efficiently handle an influent
concentration of 5-10 ppm H$_2$S. This system has been
successfully regenerated several times and has been in
operation for more than six years.

The twelve (12) 12 foot dual beds treat a total six volume of 162,000 CFM. These dual bed units are equipped with a unique carbon bed support, which allows for uniform distribution of air through the carbon bed.

Topeka, KS

In 1982, five(5) 12 foot diameter dual bed adsorbers were installed to control odors associated with primary clarifiers. Three(3) additional 8 foot diameter single bed adsorbers were installed at pump stations in 1985 to control odor in the wet wells. The twelve(12) foot diameter dual bed units treat a total air volume of 70,000 CFM and all of the units have been successfully regenerated. The 8 foot diameter single bed units are an upflow design, and contain the unique carbon bed support which allow for uniform air distribution.

Conclusion

During the past twelve years adsorption systems using IVP carbon have been applied to effectively remove odors from municipal sewage plants. These systems are easy to install and require virtually no maintenance. Moreover, chemical regeneration of the spent carbon has proven to be an effective method of reducing operating costs. By using the key design parameters presented in this paper, an efficient and economical odor control system can be designed.

PACKED TOWERS IN MUNICIPAL
ODOR CONTROL APPLICATIONS
PRINCIPLE OF OPERATION

Richard J. Kruse,P.E.
Duall Division, Met-Pro Corporation

There are numerous odor causing chemicals generated, principally, by bacterial action in waters containing high organic content. Generally those compounds formed under anaerobic conditions are the most offensive to man.*

Virtually any unit operation where the water comes into contact with the air can become a source of odor at Wastewater Treatment Plants.*

The successful resolution of odor complaints requires a careful analysis of the problem. A systematic approach involves:

1. Identification of the sources of odor
2. Identification of the chemical composition of the odor
3. Quantification of the intensity and determination of the degree of control that must be achieved to eliminate or reduce complaints
4. Selection of a method of controlling the odors.

Identification of the odor source may be simple in a small plant, perhaps the influent station or an anaerobic digester may generate the only obvious odors. A large treatment complex may have numerous odor sources. Only a careful odor survey combined with dispersion modeling will provide sufficient information to determine the required control strategy.

Numerous analysis of odorous gases from a variety of municipal sources have shown the presence of between one and two hundred organic molecules. Virtually every family of organic compounds has been found including organic acids, aldehydes, ketones, ethers along with mercaptans, amines and halogenated hydrocarbons.

*For the types of odors, mechanisms of odor production and sources of odor release, the reader should consult any of numerous available texts.

347

The majority of obnoxious odors are generated as the result of bacterial action under anaerobic conditions, that is, the formation of molecules not fully oxidized. The most commonly reported gas is hydrogen sulfide, the result of anaerobic decomposition of inorganic sulfates and organic sulfur. Hydrogen sulfide has a reported odor detection threshold of between 1 and 10 parts per billion.

Control of odor emissions may include chemical addition to the waste water or process modifications in an effort to prevent anaerobic conditions in the system itself. More commonly, the offending odor sources are covered and a ventilation system delivers these off-gases to an odor removal system.

There are numerous odor control systems available, including activated carbon, incinerators, biological towers, and wet scrubbers. Each system has advantages and disadvantages, and only a careful analysis of the problem will prevent an incorrect selection of equipment.

This paper describes the use of and the mechanisms of mass transfer in packed columns.

Wet scrubbers are the most commonly specified control device in Wastewater Treatment Plants. There are two differing approaches to wet scrubbers, one method disperses a small quantity of finely atomized droplets into the gas stream and provides a 15 second to 30 second contact time for exchange of odorous gases to the liquid droplets.

The other method utilizes absorption columns filled with a "packing" material. The purpose of the packing is to provide a retention time for the liquid, create a large liquid surface area and allow thorough contact between the gas and the liquid. Packing is available in a variety of shapes, sizes and materials.

Generally, packed absorption columns consist of a vertical cylindrical tower with appropriate supports to hold the packing material, a liquid distributor above the packing to provide even water distribution. Air enters the bottom of the tower and exits at the top after demisting. Water falls countercurrently under the influence of gravity, where the flow is impeded and redistributed as it spreads around, across and through the packing surfaces. See Fig. 1.

The scrubbing solution falls into a collection tank (sump) where chemical adjustments are made and the solution is recirculated to the top of the scrubber.

The major criteria in the design of the absorption column is selecting the correct chemistry and providing optimum conditions for transfer of the odorous molecules from the air stream into the liquid for removal.

Since the odorous gases have been released from waste water, the recapture of these odors by another volume of water requires that a method of enhancing the gas solubility be employed.

Hydrogen sulfide, mercaptans, amines and many other odors are readily absorbed, and oxidized in a wet scrubber when the recirculating solution contains the correct chemicals at sufficient a concentration.

A brief discussion of principals and mechanisms that govern mass transfer processes is provided so that the reader can better understand the mechanism.

This presentation uses hydrogen sulfide as an example gas. Hydrogen sulfide demonstrates several traits which may be applied to many other odors, with obvious modifications.

Equilibrium

The principal of equilibrium is an important concept in designing any absorption column. The solubility, or equilibrium distribution of a gas in water is expressed as a function of a Henry's Law Constant (H). $Y_i = Hx_i$, where Y_i and X_1 refer to the mole fraction of the gas in the gas and liquid respectively. Namely, the Henry's Law Constant expresses the distribution ratio of the compound between the liquid and vapor.

The gas Hydrogen Sulfide has a Henry's Law Constant of 350, thus

$$Y_i = 350 \ X_i, \quad \text{or} \qquad \frac{Y_i}{X_1} = 350$$

and states that the gas phase concentration will always contain 350 times the liquid phase. Is it any wonder then, that waters containing even a few parts per billion of H_2S produce a "big stink"? In familiar terms it states that when the water contains 5 mg/l of dissolved H_2S, the vapor will contain about 900 ppm v/v of H_2S when equilibrated.

A system at equilibrium has no **NET** transfer of molecules between the two phases. If a system, initially at equilibrium is adjusted in one phase, then there will be a transfer between the phases until equilibrium is again established.

Diffusion (Mass Transfer)

When a gas containing, say, 100 ppm of H_2S is brought into contact with a liquid initially free of H_2S, the H_2S will migrate to the water surface, dissolve in the liquid and migrate through the liquid until the liquid attains the same concentration throughout. The process will continue until the ratio of concentrations between gas and liquid established equals 350 (the Henry's Law Constant in this case).

The theoretical **minimum** quantity of pure water required to remove 100 ppm of H_2S from a 1000 cfm air stream is nearly 2,000 g.p.m.!

Although the solubility of H_2S in pure water is relatively low, fortunately H_2S ionizes in water to form a non–volatile salt (the HS^- ion). Addition of sodium hypochlorite or other oxidants will convert the H_2S to elemental sulfur So or sulfate SO_4^{-2}. The addition of chemicals provides a "sink" for the H_2S absorbed in the liquid by converting the gas to a non–volatile salt.

The degree of ionization of H_2S is dependent upon the available caustic in the absorbing solution. The ionization also follows a distribution law, namely

$$H_2S + OH^- \rightleftharpoons HS^- + H_3O^+ \quad \text{then} \quad \frac{HS^- + H_3O^+}{H_2S} = k_1$$

By a control of the solution pH, the concentration of H_2S may be almost totally depleted, thus providing a continuous "sink" for absorption from the gas phase.

A second step occurs in the ionization at still higher caustic values

$$HS^- + OH^- \rightleftharpoons S^= + H_3O^+ \quad \text{then} \quad \frac{S^= + H_3O^+}{HS^-} = k_2$$

The second ionization constant becomes significant at pH values greater than pH 11.

The constant removal of the H_2S dissolved in the liquid is the most important mechanism occurring within the column. In contrast to the earlier example, where water alone was used, at 2,000 gpm only 5–10 gpm of solution containing a caustic or an oxidant would achieve the same removal. The solution could be "reused" provided the chemical balance is maintained.

The following simplified equations describe the basic driving forces that determine the rate of absorption.*

<u>Solubility and Distribution Coefficients</u>

Equilibrium
Expressions

1. $Y \text{ (gas)} = H_1 \text{ x } X \text{ liq}$ $Y/X = H_1 = 350 \text{ (for } H_2S)$

Ionization of dissolved gas (Salt Formation)

2. $[H_2S]_L \rightleftharpoons [H_3O^+] + [HS^-]$ $K_1 = \dfrac{[H_3O^+][HS^-]}{[H_2S]_L}$

Rate Processes

Rate
Expressions

3. $\dfrac{d\,(Y)}{dt} = K_{Ga} \text{ } (Y - Y^*) \text{ x } A$ Gas/Liquid

4. $\dfrac{d\,(X_1)}{dt} = k_1 \text{ } (OH^-) \text{ } X_1$

Equation 1 merely states the equilibrium condition.

Equation 2 states the distribution between dissolved H_2S and the inorganic sulfide salts.

Equation 3 states that the rate of disappearance of H_2S from the gas stream (mass transfer) is proportional to the surface area, and proportional to the concentration difference at the gas and liquid interface. KGa is defined as an overall rate constant – determined empirically.

Equation 4 states that the disappearance of the dissolved H_2S is proportional to the available hydroxyl ion concentration. <u>These are simplified representations of the overall system but serve to describe the phenomena.</u>

*Dissociation and ionization principals are readily found in any basic chemistry textbook.

In short, the mass transfer occurs at the greatest speed when a large surface area is available and the largest driving force is maintained. A large driving force is maintained when the H_2S is removed from the liquid interface at a fast rate. The fastest removal is obtained when excess chemical is available to convert H_2S to dissolved salt, and when the liquid surface is constantly renewed with unsaturated (H_2S) solution. The packed column excels in gas/liquid mass transfer applications. The user can recirculate moderate quantities of liquid providing large surface areas. At the same time a great excess of reactant is always available, providing the necessary "sink" for removal of the gases.

Y = gas phase concentration; X = liquid phase concentration of H_2S (as dissolved gas); H_3O^+ = hydrogen ion concentration; subscript L refers to the liquid phase; A = interfacial area of liquid/gas; H_1 = Henry's Law Constant; K_G = generalized mass transfer co-efficient; k = reaction rate coefficient; k_1 and k_2 = ionization constants.

FIGURE 1 depicts a typical packed column in countercurrent operation.

Gas enters the bottom of the absorber and exits at the top. Liquid is distributed over the top of the packing and falls by gravity over and around the packing surfaces, exiting at the bottom.

FIGURE 2 shows the mass balance around the overall column.

FIGURE 3 depicts the mechanisms that are occurring in the mass transfer process itself.

The H_2S migrates to the water surface by a process called diffusion. The migration occurs at a rate that is proportional to the concentration difference. Refer to Fig. 3. The molecules diffuse through a thin stagnant gas film of length Δ G, pass into the liquid and diffuse into the bulk liquid through a film of length L. Diffusion always occurs from a region of high concentration to region of lower concentration. Upon absorption into the liquid surface the H_2S migrates through the liquid to a region of lower concentration, at a rate proportional to the concentration difference. Diffusion of H_2S in water occurs at a rate less than 10^{-5} as its speed in air.

In Fig. 2 the lines labeled A depict a system wherein the liquid is partially saturated with dissolved gas, that is the liquid eventually will become saturated with the gas. The driving force for transfer is $(Y_A - HX_A)$. It is obvious that with increasing concentration in the bulk liquid the driving force will approach zero and absorption will cease.

In Case B, the H_2S $(X_B)_L$ is at a low concentration and the driving force is $(Y_B - HX_B)$, considerably larger than in Case A. In Case B, the H_2S could be removed almost entirely from the gas stream, providing that sufficient water of the indicated capacity is available.

It is obvious that diffusion through the film and through the liquid potentially limit the overall rate of transfer. Turbulence in the liquid mixes the solution, bringing the dissolved gas, into contact with fresh reactant, simultaneously renewing the liquid surface.

The packing material promotes turbulence in both the liquid and vapor phase.

If the chemistry is maintained such that dissolved H_2S is instantaneously removed upon absorption, then the dissolved gas concentration (X_1) essentially becomes zero. For this case the equilibrium condition would be such that $Y_i = 0$. The rate expression reduces to only those terms describing the gas phase mass transfer to the liquid.

In order to absorb all the contaminant sufficient neutralizing chemicals must be present, and sufficient surface area available to accomplish the mass transfer from the gas phase to the liquid within the residence time provided.

In summary, packed absorbers provide mass transfer in a compact space and provide operating flexibility over a wide range of inlet loadings, in a minimum of plan area.

Some Rules of Thumb

*Residence times between 1 and 3 seconds can achieve greater than 99% removal of H_2S in a properly designed packed column. Ammonia, mercaptans, amines and many other odorous gases can be similarly absorbed by selection of the proper chemistry, and knowledge of reaction kinetics. The same mass transfer principals will apply, however oxidation kinetics, solubilities and overall rates will vary tremendously! These differences are readily accommodated with the proper change in liquid rates, chemical strength hold-up time etc.

*Residence times refer to the ratio of volumetric gas flow to packing volume and times stated are for properly designed packed bed absorption column.

Typical gas velocities range from 400 to 500 ft/minute of tower cross section when liquid rates are in the range of 3 to 10 gallons per minute per square foot of cross section. Higher gas or liquid rates will increase gas pressure drop excessively, or cause flooding.**

Gas side pressure drops will rarely exceed 3" w.c. for the absorber.

Theoretically there is no upper limit to the total gas flow rate in a single column, however, shipping limitations generally limit vessel diameter to 12 ft. and smaller in some areas. Parallel columns are used when high volumetric flow rates are needed.

The packed absorber can handle wide fluctuations in H_2S concentrations merely by adjusting the strength of the recirculating solution. Turn down capabilities are almost infinite, without loss of efficiency. Chemical consumption will vary according to system demands, there is little penalty for recirculating excess chemicals. The ability to provide excess chemical in the scrubber solution offers the purchaser the assurance that the desired odor removal will be completed in the minimal space and over fluctuating loads.

Multi-Stage Systems

Occasionally two and three stage scrubber systems are used to control complex odors. Multi-stage operations generally employ a different chemistry in each of the stages. Compost operations may generate not only mercaptans, and hydrogen sulfide, but also large quantities of ammonia and amines. A two stage scrubber would employ an acid solution in the first stage to remove the ammonia and amines. The second stage would use caustic and an oxidant to remove the remaining odors. Processes that generate large amounts of H_2S may benefit from a first stage scrubber employing only caustic, followed by a second stage scrubber using caustic/hypochlorite. The first stage operates at the least cost utilizing the more expensive chlorine only in the second stage for the remaining compounds.

**Flooding occurs when the gas velocity for a given liquid rate exceeds a critical value and the system will surge and pulsate causing high and low pressure swings.

Successful odor control installations using packed columns include, numerous WWTP processes, municipal and industrial anaerobic digesters, cheese manufacturing wastes, fish frying, food processing, oil reclamation, sludge pelletizing, brewery, flavor/fragrance manufacture, refineries, pulp and paper industry, and chemical manufacturing. Specific odor causing compounds include iso-propyl alcohol, methanol, ethanol, acetic acid, phenols, creosols, methyl and ethyl acrylates, ethylene and propylene oxides, styrene, amines, ammonia, mercaptans, asphalt, formaldehyde, arsine, hydrogen selenide, phosphine, chlorine, and bromine.

A complete odor control system consists of collector (hoods, covers) duct system, packed bed absorbers, blower, chemical control system, pumps. Monitoring and alarm functions in the control panel are best furnished by the equipment supplier. The total system must be properly engineered to ensure successful integration of all the components.

Maintenance

Maintenance on scrubbers will vary from location to location. The most frequently encountered problem is the gradual fouling of the packing surface. The surface fouling may result from improper operation at too high a pH with resultant carbonate deposition. Excessive water hardness can increase the rate of fouling by deposition of calcium, manganese or magnesium salts. Generally a routine acid wash will restore the scrubber to like new condition.

Bacterial fouling will not occur in systems recirculating oxidants, but may occur in systems operating between pH 3 and pH 10. Bacterial populations are readily inhibited by the addition of hypochlorite on a regular schedule.

Performance failures of packed absorption columns are found to be caused by (in order of occurrence):

1. Improper operation of chemical additions, generally due to lack of knowledge by operators.
2. Improper maintenance, either due to poor scheduling or lack of instruction.
3. Insufficient tower sizing resulting in less than required mass transfer capabilities.

4. Poor liquid or gas distribution resulting in poor gas/liquid contact.
5. Poor fabrication and/or components resulting in structural failures or excessive failures of pumps, seals, chemical feed systems.

The economical resolution of any odor problem requires an experienced engineering approach at all stages of the project.

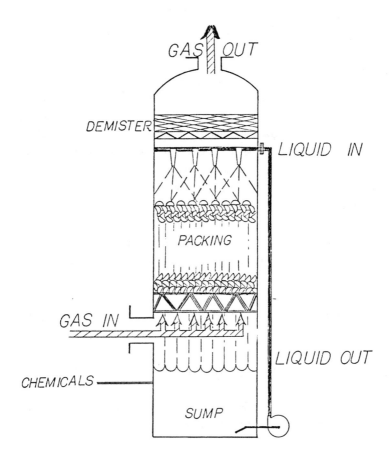

FIGURE 1

356

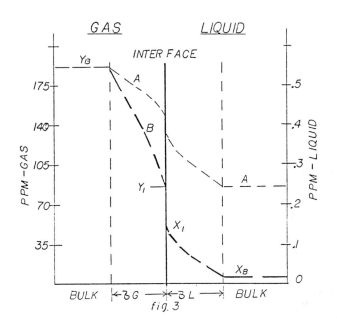

fig. 3

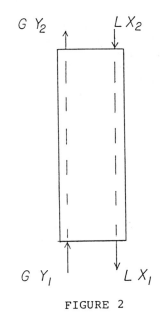

G = lb. Moles/hr. Gas

L = lb. Moles/hr. Liquid

Y = Mole Fraction Gas

X = Mole Fraction Liquid

H = Henry's Law Constant

FIGURE 2

$$G (Y_1 - Y_2) = L (X_1 - X_2)$$

357

SAN ANTONIO BALANCES

SOLUTIONS TO PLANT AND SEWER ODORS

Freylon B. Coffey, P.E.
Permits Division
Texas Air Control Board

Dr. Jeffrey Lauria
Richard J. Pope
Malcolm Pirnie, Inc.
White Plains, New York

Pursuant to recommendations of the 201 Facilities Plant Report, the City of San Antonio in 1987 replaced an aging wastewater treatment plant with a new 83 million gallons per day treatment plant. The new plant incorporated several odor control measures in an effort to comply with community expectations. After start-up, persistent citizen complaints indicated that further odor controls would be necessary. With the aid of a consultant and participation by the Texas Air Control Board, the City of San Antonio conducted a comprehensive study that resulted in a long range plan for effectively controlling odors from the several sources involved with the new plant.

Introduction

Texas communities increasingly are complaining about odors attributed to the collection of wastewater and/or operation of wastewater treatment plants (WWTPs). The rise in number of complaints is the result of public awareness regarding environmental issues, in particular over airborne contaminants, coupled with the reduction of effective buffer zones surrounding WWTPs.

Historically, planners located treatment plants far away from urban environs and residential housing, and developers were unlikely to build nearby. However, with the recent expansion of cities and the lack of available land, areas around WWTPs that once served to buffer the public's exposure to the facility are now being developed. Today, it is not uncommon for residents to share a common property boundary with WWTPs. Citizens therefore are more likely to perceive and be affected by odors.

Case Study

The City of San Antonio (City) has received persistent complaints about odors from residents who live around the Dos Rios Wastewater Treatment Plant and the associated sewer collection system.

This paper presents a case history of how the City, working with the Texas Air Control Board (TACB), and the engineering firm of Malcolm Pirnie, Inc. (Pirnie), conducted an odor control study, found the odor sources and proposed solutions.

Background

Pursuant to the recommendations of a 201 Facilities Plan Report, the City in 1987 replaced the aging Rilling Road Wastewater Treatment Plant with the new Dos Rios Wastewater Treatment Facility. The Dos Rios Plant was designed to provide advanced wastewater treatment for an average flow of 83 million gallons per day. The advanced treatment process consisted of secondary treatment with conventional air activated sludge followed by separate, second stage nitrification, effluent filters for final solids removal and chlorination before being discharged to the Medina River.

To minimize sewer line construction, the Facility Plan recommended that the wastewater continue to be collected at the headworks area of the Rilling Road facility. From this central location, a single 90-inch diameter, plastic lined, sewer pipe was constructed to convey the wastewater to the Dos Rios facility. This 90-inch diameter sewer pipe, referred to as the Rilling

Transfer Line (RTL), combines the wastewater received from up to seven interceptors. The interceptors range in size from 30 to 84 inches in diameter and collect from a service area of approximately 125 square miles in the south San Antonio region. The wastewater flow mainly consists of domestic sewage with varying input from commercial, industrial and/or hospital facilities.

To reduce the length of the RTL, a direct route between Rilling Road and the Dos Rios Plant was selected. This precluded using existing road right-of-ways to place the pipe. As a result, resident property easements along nearly the full length of the RTL were necessary. This meant that the RTL, with associated accessways and vents, was predominantly located on residential property.

Odor Control Design

Based on the City's experience with incidences of odorous gas releases and resident odor complaints resulting from the collection of wastewater and operation of the Rilling Road Plant, odor control systems were an integral part of the RTL and Dos Rios designs. Specific odor control systems included:

- a wet chemical scrubber to treat the odorous gases collected from the airspace of the RTL (Rabel Road Odor Control Unit - RROCU);

- a wet chemical scrubber designed to control the release of odors that are collected from under the covered plant influent channels, Parshall Flume and grit tanks (headworks scrubber);

- a wet chemical scrubber designed to treat odors collected from the enclosed sulfide extended aeration/oxidation tanks (preaeration scrubber);

- waste anaerobic gas flares.

Despite the operation of these odor control systems, the City received numerous odor complaints from the local residents. The RROCU was perceived as the main problem, with less of an impact attributed to the treatment plant. To reduce the odors released by the RROCU, the City modified the wet chemical scrubber system by changing the chemical used to scrub the odors from chlorine to a caustic solution. In addition, the City added an activated carbon unit to the exhaust from the wet scrubber. Even with the additional odor control provided by the polishing carbon unit, the RROCU was still unable to provide continuous odor control for the strong odors collected from the RTL. At the same time, the City continued to receive complaints which were attributed to the operation of the RROCU.

As a result of continuing odor complaints, the City retained an outside consultant, (Pirnie) to conduct an in-depth odor situation assessment and to develop a recommended odor control plan for the RTL, RROCU and Dos Rios Plant.

Developing an odor control plan is difficult because a comprehensive solution is required to solve the problem completely. There is no single odor control solution that when implemented would fix the City's problems of odor perception by local residents. In this case, implementing one solution might tend to relocate the problem. For example, shutting down the ROCCU would solve the odor problems for those residents who live nearby. However, the build-up of odorous gas in the airspace of the sewer could then vent at other locations along the RTL and the upstream collection system interceptors. Because of the complexity of this project, attributed to the interaction of the collection system and Dos Rios, the recommended odor control plan must provide for a total solution, consisting of many applicable, effective and interrelated odor control actions.

Plan Objectives

To develop an effective plan that would address the concerns of the City and residents, a set of study objectives were prepared at the start of the project. The study objectives were:

- Conduct comprehensive odor sampling of the Dos Rios Plant.

- Identify wastewater treatment plant odor sources.

- Identify off-site odor impacts of the identified sources.

- Collect and analyze air samples from the Dos Rios Plant for the presence of volatile organic compounds (VOC).

- Conduct comprehensive odor sampling and evaluate the odor potential of the 90-inch RTL.

- Evaluate the existing RROCU, and the headworks and preaeration wet scrubbers performance, and develop a long-term operating strategy.

- Develop a recommended long range plan for the RTL, RROCU and the Dos Rios Plant.

Project Approach

To achieve these objectives, Pirnie and the City followed a "rational approach." This approach has successfully been used to solve odor problems at over 40 wastewater treatment plants nationwide. The steps involved in this approach are defined below.

The first step involved listening to the community affected by the odors and the operators of the City facilities about the history of odors. The second step consists of independently identifying the source(s) of odors through comprehensive on-site investigations. This second step included evaluating each identified odor source to determine its relative strength and off-site impact potential so that all nuisance odor sources could be ranked. Ranking the odor sources provides guidance in focusing the plant's control efforts to the areas where the biggest benefit could be accomplished. The third step included long-term plan odor control selection, design and implementation, together with operator training. The fourth step consisted of providing for citizen involvement throughout the previous three steps.

Project Implementation

At the start of the project, a technical advisory panel (TAP) was formed. The TAP was made up of representatives from the City, the TACB, local residences and Pirnie odor specialists. Formation of the TAP served to bring all these parties together to discuss the odor problems and proposed solutions, and to provide a means for communicating the progress of the odor study as the work was being performed. From a public relations standpoint, this "open-door" policy demonstrated to the public that the City was committed to providing odor control to reduce the off-site impact of wastewater related odors.

Sampling

Pirnie conducted intensive on-site and off-site sampling at the Dos Rios Plant, RROCU and the collection system up to Rilling Road. The activities included a determination of total sulfides in wastewater and hydrogen sulfide (H_2S) in air, an evaluation of VOCs and existing odor control systems, an assessment of the qualitative odor strength of air samples, and fence line and neighborhood odor quality surveys.

The results of the intensive sampling and analysis of odors are presented in the following paragraphs.

Sulfides in Wastewater

Sulfides, typically present in wastewater in dissolved form or as metallic sulfides in solid form, are a good indicator in assessing the potential of the waste stream as an odor source. In general, concentrations of 0.3 mg/l or less do not represent a significant odor impact potential.

The highest concentrations of total sulfide in wastewater were measured at the plant headworks and the collection system, 3.8 and 3.5 mg/l, respectively. Average concentrations at these two locations fell between 1.0 and 2.0 mg/l. Concentrations of total sulfide in wastewater averaged less than 0.2 mg/l throughout the rest of the plant.

Gaseous Hydrogen Sulfide

H_2S is an important indicator of wastewater odor impact since it is almost always present when sulfides in wastewater are observed. As well as having a characteristically unpleasant (rotten egg) odor, H_2S also has one of the lowest odor threshold values of those compounds associated with domestic wastewater. According to odor threshold studies published in literature and confirmed by field experience, typical levels of detection of H_2S are 1 to 10 parts per billion by volume in air.

The highest plant H_2S concentrations were associated with the anaerobic digesters, ranging from 90 to 465 parts per million (ppm) by volume in air, with an average of 290 ppm. High levels of H_2S were also recorded within covered Rilling Road junction chambers and within the RTL. Average concentrations of H_2S ranged from 30 to 380 ppm in the vapor space in the pipeline.

An evaluation of the H_2S removal efficiency of the odor control systems demonstrated that none of the three (RROCU, headworks and preaeration systems) were consistently achieving the control of H_2S for which they were designed. The observed H_2S removal efficiencies ranged up to 99+%, although on average the efficiencies were less than 90%. These removal efficiencies suggested that optimization of the existing systems was necessary.

Odor Strength

In many instances, the foul air perceived is made up of many odorous components. Rather than measure each odorous component individually, Pirnie used a modified field odor panel procedure to assess the overall strength of air samples collected at all unit processes. This saved the City the time and money involved with conducting elaborate compound identification procedures. The modified field odor panel consists of a group of field odor specialists that qualitatively evaluate various dilutions of an air sample to determine if an odor is present. The results of the procedure represent the number of times an air sample must be diluted before the average person can identify the presence of a detectable odor. The values recorded are used to compare the strengths of individual unit processes and assess the potential impact in the community.

The results of the odor strength evaluation also identified the influent, headworks and preliminary treatments areas of the plant as having the highest potential for being perceived off-site. In addition, the odor strength results indicated that higher removal efficiencies were necessary by the odor control facilities.

Odor Control Surveys

To supplement the data collected on-site, odor quality plant fence line and neighborhood surveys were conducted by field personnel. The purpose of the odor quality surveys was to establish a consistent set of closely related observations which would indicate whether odors perceived at the fence line and the neighborhood could be attributed to the Dos Rios Plant, RROCU and/or RTL odor sources.

Several plant unit processes were perceived at the fence line during the odor quality surveys, including anaerobic digesters, headworks, grit tanks, primary settling tanks, headworks and preaeration wet scrubbers and the belt filter presses. In addition, the odor sources were perceived during the neighborhood odor surveys. These included anaerobic digesters, RROCU, headworks and preaeration wet scrubbers, plant headworks and grit tanks.

Odor Source Ranking

The results of the on-site investigations indicated that the RROCU exhaust was the source of the majority of the residential odor complaints. In addition, several sources were identified at the Dos Rios WWTP.

To focus the City efforts on those areas of the Dos Rios Plant that have the highest potential for off-site odor impact, a ranking analysis of the key odor sources identified during the intensive on-site investigations was conducted. The results of the analysis are:

- anaerobic digesters (highest off-site odor potential)
- headworks
- grit tanks
- headworks scrubbers
- preaeration scrubber
- sludge dewatering operations - primary settling tanks
- polymer storage
- preaeration tanks
- sludge thickening
- activated sludge processes
- cascade outfall (least off-site odor potential)

Based on the results of the ranking analysis and observations in the field, Pirnie estimated off-site odor impact potential for all the plant sources. The results attribute 50% of the estimated off-site odor impact due to the anaerobic digesters of the Dos Rios Plant; 35% with the integrated headworks, grit tanks and scrubber systems; 5% each with the belt filter press and primary settling tanks and 5% for the remaining sources. This analysis is useful in that it helps direct the City priorities to the odor sources where they are most needed and estimates the percent reduction in odor impact that can be achieved.

Recommended Long Range Goals

As presented above, measurements and observations indicated that the RROCU as well as several Dos Rios Treatment Plant processes can contribute to off-site odors. Based on the results of the comprehensive, three-month odor impact assessment, the recommended elements of the long range plan were developed. Implementation of the long range plan is expected to provide effective control of off-site odors. The recommended plan addresses the RROCU and Dos Rios Plant.

RROCU

For this study, an intensive investigation was undertaken to determine the effectiveness, usefulness and performance characteristics of the RROCU. Sewer odor control performance was of particular interest since word-of-mouth reports indicated that odor improvements were noticed miles upstream in the sewers north of Rilling Road which might be associated with operation of the RROCU.

In addition, to site odor investigations at the RROCU and in and along the 90-inch diameter RTL, Pirnie, together with the City, conducted a comprehensive investigation of the South Rilling Road collection system. These investigations included detailed sewer mapping, odorous substance measurements, residence and business location inspections and personal interviews with area residents.

The ROCCU and the South Rilling Road collection system odor investigations produced major findings which drove several important long range plan decisions. Those major findings are summarized below:

a. It was concluded that costs to upgrade the installed wet scrubber/ carbon system to consistently achieve the desired level of odor control at the RROCU location would be prohibitive.

b. Odor benefits at the Dos Rios Facility do not appear to be associated with operation of the RROCU.

c. The system requires large operations and maintenance investments.

d. Pirnie estimated that a 1 to 1.5 million dollar modification plan would be necessary to upgrade the RROCU to best achieve reliability and performance, with operating costs approaching $400,000 to $500,000 annually.

e. With regard to improving the overall community odor situation, a higher efficiency, more comprehensive control strategy appeared justified at the source.

f. The recommendation to shut down the RROCU is unique and highly site-specific. The wet scrubber/carbon adsorption combination on a technical basis is proven in several other locations as cost effective and the technology of choice for controlling odors due to H_2S.

Based on the findings presented above, Pirnie recommended that the City ultimately cease operation of the RROCU. In the meantime, the City should retain the RROCU odor control equipment in standby condition until the local sewer odor program is firmly in place and thoroughly reviewed. Subsequently, the City should plan on re-using the RROCU equipment at other locations.

Pirnie recommended that the City evaluate odor control in the South Rilling Road Collection System in place of the RROCU. Accordingly, the plan for local odor control in the South Rilling Road Collection System should include a balance of the following procedures:

a. Installation of sewer traps on affected residences and commercial buildings.

b. Raising of residential roof vents, as necessary.

c. Replacement of residential sanitary facility seals and traps.

d. Investigate and repair of possible cracked yard sewers.

e. Sealing of abandoned piping which is connected to potential odor
    sources, such as abandoned septic tanks.

f. Discontinued use of the 84-inch diameter stormwater overflow
    interceptor; except as intended for storm relief.

g. Sealing of vents on the 84-inch stormwater overflow interceptor, subject
    to hydraulic investigations.

h. Better gas sealing of all affected sewer injunction chambers, especially
    along the 90-inch RTL and at Rilling Road.

i. Selectively seal and use carbon inserts at manholes on Ashley Road.

j. Implement chemical addition for odor reduction in the wastewater at the
    Yett Street and Ashley Road areas.

k. Move septage dumping from Rilling Road to another location with lower
    odor impact.

l. Continue dialogue with affected community regarding actions implemented
    to reduce odor.

It is important to note that the balance of procedures for optimum sewer
odor control must often be evaluated through full-scale trials. In addition,
laboratory tests should be conducted for chemical treatment effectiveness
prior to commencing full-scale tests.

Dos Rios Wastewater Treatment Plant

Today, we know that the effective wastewater odor control transcends
operating equipment and includes a balance among effective technology,
standby systems, well-trained operators, attentive maintenance procedures,
serious odor monitoring and feedback programs and a commitment by the owner
to make it work. The recommended plan for the Dos Rios facility emphasizes
upgrading and optimizing odor control equipment along with effective
training, operating and odor monitoring programs. The following items
outline the main components of the recommended long range plan.

a. Provide an additional wet scrubber to function as a second, polishing
    stage for the existing headworks and preaeration scrubbers and as a 100%
    back-up unit for both operating existing units.

b. Implement laboratory and full-scale carbon adsorption testing for the
    Dos Rios Plant wet scrubber finished gas. If timely, relocate the RROCU
    carbon absorber to the Dos Rios site. Prepare a study protocol with
    full advisement of the TACB, and participate with the TACB in performing
    the tests. Focus on removal performance of VOC, bed service life,
    regeneration effectiveness and costs.

c. Cover primary settling tank effluent channels.

d. Add potassium permanganate or hydrogen peroxide to belt filter press
    feed sludge to reduce odors. Also, investigate changing to a less
    odorous dewatering polymer and implement a less turbulent washwater
    conveyance.

366

e. Seal all digester overflow boxes.f.Eliminate digester gas release to atmosphere via gas distribution management and the addition of another waste gas flare.

g. Minimize the use of chlorine at the outfall in keeping with NPDES permit disinfection criteria.

h. Develop and implement a plant-wide and community odor monitoring program.

i. Establish an odor operations group with sole responsibility for all odor control activities for the Dos Rios site.

The recommended long range plan for the RROCU and the Dos Rios Plant has been incorporated into an agreement between the TACB and the City. Currently, the City is initiating implementation of the recommended odor control actions.

BIOLOGICAL DEGRADATION SYSTEM FOR ODOR CONTROL

C. Richard Neff, P.E.
KBN Engineering and Applied Sciences, Inc.
Gainesville, Florida

The treatment of the odor causing compounds (e.g. hydrogen sulfide, organo-sulfur compounds, volatile organic compounds, etc.) generated in manufacturing processes or waste treatment systems can be a difficult and expensive process. Conventional treatment methods (including wet scrubbing, activated carbon sorption, direct flame incineration, catalytic oxidation, etc.) can provide some degree of control but have limitations in controlling variable air streams and may result in the generation of a byproduct which needs treatment and/or disposal.

An alternative treatment method is the aerobic biological degradation of the odor compounds in air streams using biofilters. The biofilter treatment process has been used sporadically in the US since the mid-1950s, while its use in Europe is extensive. A reason for the limited use of the process in the US is the traditional biofilter design which uses either perforated pipe or pressure chamber air distribution. Both system designs operate with low air to filter volume loading rates and as a result require relatively large area requirements for acceptable filter operation.

This limitation of the biofilter design has been overcome by the development of an alternative filter bed design. The key to the design is a patented filter bed system which provides uniform air distribution to the filter media, thus greatly increasing air to filter volume loading rates while preserving high odor compound removal rates (approaching 99%, depending on the compound). An example of the this new biofilter effectiveness can be seen at Gainesville (Florida) Regional Utilities Kanapaha Waste Water Treatment Plant where H_2S concentrations are reduced by 99.9%.

INTRODUCTION

The processes associated with waste water treatment, composting, pulp and paper production, and food processing often result in the production of odorous waste gases. In some cases, the odorous gases create primarily a nuisance condition; but in many other cases (especially those involving H_2S) a health hazard can be present.

The traditional odor control technologies can be segregated into either physical or chemical treatment processes. The physical treatment processes involve either sorption of the odor compound to another media (e.g., activated carbon) or ultra-violet radiation. Chemical treatment is accomplished by oxidation of the odor compounds (e.g., chemical oxidation, thermal degradation, catalytic combustion, ozonation, and chlorination). Some treatment processes, e.g., wet scrubbing, involve both a physical and chemical treatment phase. These traditional odor control methods can produce the desired result; but they also create secondary wastes which require treatment. Additionally, they are expensive to construct and operate.

The alternative to these traditional controls is biological treatment of odorous waste gases using a biofilter. The term "biofilter" is defined as a packed bed with microorganisms attached to a suitable media. The waste gases pass through the packed bed and the microorganisms metabolize the contaminants. A biofilter can effectively remove a wide spectrum of odor causing compounds (e.g., H_2S, mercaptans, terpenes, amines, etc.) from waste gas streams. Additionally, biofilters have the advantages of being inexpensive, reliable, and environmentally compatible.

The following sections of this paper review the theory of biofilter operation, present examples of treatment efficiencies, discuss alternative biofilter designs, and detail cost considerations.

THEORY AND PRINCIPLES

The theory behind biofilter operation is roughly analogous to the aerobic biological treatment processes (e.g., trickling filters, activated sludge units, etc.) used for waste water. In both cases, bacteria are provided with both a hospitable environment (in terms of oxygen, temperature, nutrients, and pH) and a carbon source for energy. The bacteria utilize these favorable conditions to metabolize the carbon source to its primary components (i.e., carbon dioxide, water, inorganic salts). The result is "clean" water or, in the case of biofilters, air stream.

Biological treatment for waste waters has been an established practice for the past century. Using biofilters for the treatment of air streams is a relatively recent development. Occasional references to biofilter technology can be found in scientific and engineering literature prior to the 1950s. Research during the 1960s and 1970s centered on evaluating the types of air borne contaminants amenable to biofilter treatment. Several researchers[1,2] documented biofilter effectiveness in the removal of sewage related odors. Soil beds also have the capacity for removing sulfur containing gases (e.g., CH_3SH) and serve as a sink for hydrocarbon gases[3].

An early discovery in the research and operation of biofilters was that the moisture content in the filter bed plays a crucial role in determining removal efficiency. Moisture in the biofilter serves both as a habitat for the microbes and as a transport mechanism of substrate and nutrients in the filter media. Other filter bed characteristics also influence biofilter treatment efficiencies. Based on the data provided by several researchers, the following range of operational parameters for biofilters has been presented[4]:

Parameter	Recommended Range
Retention Time	>15 seconds
Filter Bed Temperature	15 - 45 °C
Filter Bed pH	7 - 8
Filter Bed Moisture Content	50% - 70% (weight)
Influent Gas Humidity	80% - 100%
Filter Media Loss on Ignition	60% - 80%
Filter Media Dry Density	0.40 - 0.45 g/cc
Filter Media Porosity	80% - 90%

It has been determined that the most active microbes in a biofilter are the heterotrophic and chemoorganotrophic groups[5]. One of the more prevalent groups of heterotrophic bacteria is the Actinomycete spp. Actinomycete spp. are able to utilize a variety of organic substrates and are known to be able to degrade even stable compounds such as phenols, tannins, and long chain polymers. Eitner (1984) also noted that Actinomycete spp. are able to exploit a wide variety of substrates and are reported to kill pathogens under certain conditions. Only a few organoheterotrophs have been isolated from biofilters; some of these include the Pseudomonas and Micrococcus.

The distribution of microbes in the biofilters has been described as being vertically segregated[5,6]. In the lower portions of the biofilter bed (closest to the gas diffusion stream), the total microbial population density is the highest. It has been assumed that the microbes at the bottom of the biofilter preferentially metabolize the more readily degradable influent compounds and utilize less of the nutrients in the filter material. Conversely, the less degradable compounds are more readily assimilated in the upper portion of the biofilter.

Because the biofilter is a biological system, it is of interest to know how long it takes for the microbial population to achieve optimal removal efficiencies and their sensitivity to system shut downs. In the case of adaptation time for the biofilter, data has been presented[5] which indicate a significant reduction of the hydrocarbon concentrations will be achieved in approximately 1 week from start-up and maximum removal rates achieved within 1 month. The survivability of the microbial organisms in biofilters which were not actively being operated (i.e., no flow of air) has also been investigated[6]. These experiments showed minimal loss of microbial activity after suspending biofilter operations for 14 days.

In summary, the following statements can be made regarding the theory and principals of biofilter operation:

1.   Biofilters have been demonstrated to remove a wide spectrum of odor causing compounds from air streams.

2. Maintaining proper moisture content in the biofilter media is crucial to achieving and maintaining maximum removal efficiencies.

3. The microbial community in a biofilter is diverse and vertically stratified resulting in different compounds being removed at different levels in the biofilter.

4. A biofilter will become acclimated to the waste gas stream in about one week with maximum efficiency beginning after approximately one month. Once acclimated, a biofilter can be operated cyclically (versus continuously) without a loss in performance.

BIOFILTER TREATMENT EFFICIENCIES

The effectiveness of biofilters for treating odorous waste gases in a variety of applications is well documented in the literature. For sewage treatment plant applications[7], odors reduced by approximately 98%. Treatment efficiencies of 96% for a composting facility and 93% at a rendering plant have been observed[8]. Also, removal rates of 99.9% have been reported[9] for odors from a animal rendering facility.

In all the above cases, the waste gases treated were heterogeneous streams containing a variety of odor causing and other organic gases. Several researchers have demonstrated the effectiveness of biofilters at removing specific odor compounds and selected VOCs. A summary data of this research[4] includes the following compound specific results:

* Hydrogen sulfide   >99% removal
* Dimethyldisulfide   >91% removal
* Terpene   >98% removal
* Organo-sulfur gases   >95% removal
* Ethylbenzene   >92% removal
* Tetrachloroethylene   >86% removal

It is important to note that the filter media used in the biofilter can have a significant impact on the removal efficiencies experienced by the filter. It has been shown[10] that composted municipal refuse can be a superior filter media when compared to composted leaves or peat moss. The reasons for the differences in performance from different media are probably related to media quality (in terms of pH, available nutrients, etc.) and substrate diversity.

Given the above summary of information on biofilter performance, the following conclusions can be made regarding treatment efficiencies.

1. Biofilters can provide odor removal rates exceeding 95% (and, in some instances 99%) for primary applications such as: sewage treatment plants, composting facilities, and rendering plants.

2. The high removal efficiencies for H_2S (>99%) and organo-sulfur gases (>95%) make biofilters a potential treatment alternative at pulp and paper or other wood processing facilities.

3. Biofilters have the capacity to treat a broad spectrum of waste gas contaminants and are not limited to a few parameters.

COMMONLY USED BIOFILTER DESIGNS

Biofilter design is inherently simple. The major components of a biofilter include:

1. An air blower - to push the collected, contaminated air through the filter media;

2. An air distribution system - to provide a uniform air flow to the filter media;

3. A moisturizing system - to maintain proper moisture content in the filter media;

4. Filter media - to provide the habitat for the microorganisms; and

5. A drainage system - to remove excess water (e.g., rainfall and/or condensate) from the filter bed thus preventing waterlogging of the media.

Given these major components, there are two basic biofilter design configurations; the perforated pipe filter and biotowers. Both of these designs can be effectively utilized for treating contaminated air streams. A brief discussion of the designs follows.

The perforated pipe filter, or traditional biofilter, typically consists of perforated pipe buried in gravel (for air distribution) covered with either native soils or an organic substrate (e.g., peat, compost, etc.) as the filter media. Moisture content of the filter media is typically maintained by means of placing a sprinkler on top of the filter media. This type of biofilter has been used for three decades with favorable results. Its limitations tend to be logistical (e.g., very low loading rates, large area required, manual labor for O&M activities) but poor air distribution and inefficient moisturizing can also lead to operational problems. Traditional biofilters are not proprietary technology and are the most commonly used biofilter design in the US.

The design of biofilters is in many respects, more of an art than a science. This is especially true in the case of the traditional biofilter which tend to be inefficient in terms of air distribution (i.e., the air ports, regardless of conduit design, create conical air flow channels which may overlap or miss portions of the filter media) and the highly variable nature of the "native" soils which are typically used as the filter media. The EPA Odor Control Manual[11] provides some general design criteria for these system but no algorithm for calculating biofilter size based on the inlet air flow rate or odor concentration.

Biotowers resemble activated carbon units in general structure except that an organic media is substituted for the carbon (and serves as the biofilter media) and water spray nozzles are added for moisturizing. Biotowers can support very high loading rates, eliminating the space limitations inherent with traditional designs. The limitation of biotowers are related to the relatively low biofilter media to air flux volume ratio in the design. Under these conditions, operating parameters (e.g., temperature, moisture, pH, etc.) require strict operator control to minimize the potential for upsets. Also, the need for controls and tower design eliminate some of the cost advantages afforded by traditional designs. Biotowers can be "homemade" by reworking a carbon tower; proprietary biotowers are available in Europe but currently not in the U.S. Biotowers enjoy limited usage in both the US and Europe.

For the biotower design, some kinetics research[6,12] has been done to provide design methodology for sizing of the unit. This design methodology takes into account the inlet air flow rate and concentration(s) but the literature provides kinetics coefficients for the algorithms for only a relatively small number of volatile organic compounds. It should also be noted application of that these kinetics coefficients may be restricted to evaluating biotowers or other design configurations which have similarly controlled temperature conditions.

THE ALTERNATIVE BIOFILTER DESIGN

The alternative design to the above two configurations is the aerated rigid planum biofilter, or the BIKOVENT® System. This design is similar to the traditional design in that a single, on-grade filter is used. The design is used extensively in Europe and has recently been introduced to the US by BioFiltration, Inc.

The BIKOVENT® System consists of an aerated plate that provides uniform air distribution, integrated air supply and drainage ducts, a driveable surface for mechanized installation and maintenance of the filter media, and a non-clogging venturi configured air escape port; see Figure 1 for details. The engineering of the air escape ports is such that escape velocities prevent biological build-ups or physical clogging. Additionally, the venturi design minimizes the potential for incoming material to clog the opening. Figure 2 presents a comparison example of the air flow distribution as measured at the surface of a BIKOVENT® System biofilter and a traditional biofilter design. As seen in the Figure, the BIKOVENT® System provides relatively uniform air velocity across the plane while the traditional biofilter design exhibits considerable velocity gradients and resulting preferential flow pathways.

The prefabricated, interchangeable components of the BIKOVENT® System support multi-bed installations, thereby allowing for continued biofilter operation during maintenance periods. A schematic design of a typical BIKOVENT® System installation is shown in Figure 3. As seen in Figure 3, air supply can be made to either biofilter section through controls at the Humidification Chamber. The Humidification Chamber also serves as the air moistening process and the collection point for excess drainage from the biofilter. Excess water from the biofilter is collected by the air supply ducts and then routed to the Humidification Chamber via the U-Channels; this water can be used for raw gas moistening. The moistening of the raw gas prior to entering the filter media provides a relatively constant humidity both horizontally and vertically in the filter bed. This constant moisture content throughout the bed allows for uniform, optimal biological activity and thus, maximum biofilter treatment efficiencies.

A design methodology for the BIKOVENT® System was developed[7] specifically for odors at sewage treatment plants. The methodology considers inlet air flow rate and odor concentration (in odor units per cubic meter) as well as some biofilter design characteristics (e.g., filter media depth, retention time, etc.) and the ability of the filter media to degrade the odor compounds. The ability of the filter media to degrade the compounds (referred to as the specific odor coefficient load) will typically range from 10,000 to 60,000 OU/m^3 of filter media per hour (OU/fm^3-hr). The air loading rates with the BIKOVENT® System are typically in the 5 to 10 cfm/ft^2 range, and can approach 15 cfm/ft^2 for certain applications.

Figure 4 shows the relationship between biofilter air loading and odor level at various odor concentrations using this design method. Assuming an average odor concentration for a typical sewage treatment plant is 200 OU/m^3 and the specific odor coefficient load is 35,000 OU/fm^3-hr, the design air loading rate for a BIKOVENT® System would be about 10 cfm/ft^2. A hypothetical air stream (and proposed BIKOVENT® biofilter) with the above characteristics and an air flow rate of 10,000 cfm would require a minimum active biofilter surface area of 1,000 ft^2.

In instances where raw gas odor data is not available but H_2S concentrations of the raw are (e.g., the Kanapaha Waste Water Treatment Plant in Gainesville, Florida); sizing of the BIKOVENT® System can be accomplished by either reviewing past system designs or by pilot testing. At Kanapaha, the exhaust gas from the grit chamber has H_2S levels averaging nearly 200 ppmv with peaks estimated to be substantially higher. Given this air stream, a BIKOVENT® System was designed and installed with the general design parameters and removal characteristics shown in Table 1. Also presented in Table 1 are the design parameters for a traditional biofilter (at Moerawa, New Zealand). As noted in the Table, both biofilters provide excellent control of the raw gas (i.e., 99.9% H_2S removal). However, the BIKOVENT® System has a considerably higher air loading rate and requires a substantially lower air pressure to move the gas stream through the biofilter.

COST INFORMATION

Biofilter technology in general, and the BIKOVENT® System in particular, represents the most cost effective means of controlling odors. A comparison of cost data was prepared for several odor control technologies using a common comparison base. The comparison base is the cost data presented in the EPA Odor Control Manual.

In this publication, EPA developed cost data for several alternative odor control technologies. The design basis used was an air flow rate of 10,000 cfm with an inlet H_2S concentration of 20 ppm and an outlet concentration of <1 ppm. The capital costs estimates for all treatment alternatives included all site preparation, construction, equipment and labor necessary to have an operational unit. The capital costs for the EPA estimates (i.e., all alternatives except the BIKOVENT® System) were escalated to 1988 dollars using the Engineering News Record (ENR) Cost Index. Capital costs for the BIKOVENT® System were estimated using 1988 equipment prices.

The O&M costs for the various alternatives were obtained from several sources. Chemical and/or materials costs were obtained from the EPA Odor Control Manual for traditional biofilters, wet scrubbers and activated carbon; 5% of capital costs were added to account for routine maintenance. For catalytic and thermal incinerators, a fuel supplement cost (natural gas feed at 10% of the air flow rate) and the 5% maintenance cost were included.

Ozonation O&M costs were estimates[13] include materials, energy, labor, and maintenance. BIKOVENT® System O&M costs were obtained from operational data.

The cost comparison is presented in Table 2. As seen in this table, the BIKOVENT® System is the least expensive system to construct and operate. Indeed, the annualized cost of the BIKOVENT® System (i.e., capital recovery plus O&M) is less than the annual maintenance costs alone of the physical and chemical processes (i.e., ozonation, wet scrubbing, thermal incineration, catalytic incineration and activated carbon).

SUMMARY

The biofilter process is a proven technology for the control of odor problems in a wide variety of applications. There are several alternative biofilter configurations which can be successfully employed for odor treatment. An alternative and effective (in terms of both removal efficiencies, size, and ease and cost of operation) biofilter design is the BIKOVENT® System. The BIKOVENT® System design is a patented filter bed system which provides uniform air distribution to the filter media, thus greatly increasing air to filter volume loading rates while achieving high odor removal efficiencies (approaching 99%, depending on the compound). Given this new design and superior performance, the BIKOVENT® System represents an efficient alternative control technology for the treatment of odorous waste gases.

REFERENCES

1. Carlson, D.A. and C.P. Leiser, 1966, "Soil Beds for the Control of Sewage Odors", J. Water Pollution Control Federation, Vol. 38, No. 5, pgs. 829-840, May 1966.

2. Pomeroy, R.D., 1982, "Biological Treatment of Odorous Air", J. Water Pollution Control Federation, Vol. 54, No. 12, pgs. December 1982.

3. Smith, K.A., J.M. Bremmer and M.A. Tatabai, 1973, "Sorption of Gaseous Atmospheric Pollutants by Soils", Soil Sci, pgs. 313-319, April 1973.

4. Hartenstein, H., 1987, "Assessment and Redesign of an Existing Biofiltration System", Masters Thesis presented to the Department of Environmental Engineering Sciences, Univ of Fla, Gainesville, 1987.

5. Eitner, D., 1984, "Untersuchungen uber Einsatz und Leistungsfahigkeit von Konpostfilteranlagen zur biologischen Abluftreiningun im Bereich van Klaranlagen unter besonderer Berucksichtigung der Standzeit" (Investigations of the Use and Ability of Compost Filters for the Biological Waste Gas Purification with Special Emphasis on the Operation Time Aspects), GWA, Band 71, TWTH Aachen, West Germany, 1984.

6. Ottengraf, S.P.P., and A.H.C. Van Den Oever, 1983, "Kinetics of Organic Compound Removal from Waste Gases with a Biological Filter", Biotechnology and Bioengineering, Vol. XXV, pgs. 3089-3102, 1983.

7. Eitner, D. and H.G. Gethke, 1987, "Design, Construction and Operation of Bio-Filters for Odour Control in Sewage Treatment Plants", Paper Presented at the 80th Annual Meeting of the Air Pollution Control Association in New York, NY, June 1987.

8. VDI, 1984, "Biologische Abluftreininung - Biofilter (Biological Waste Air Purification - Biofilters), VDI Handbuch Reinhaltung der Luft, VDI No. 3477, Band 6, Dec 1984, West Germany.

9. Prokop, W.H., and H.L. Bohn, 1985, "Soil Bed System for Control of Rendering Plant Odors", J. Air Pollution Control Association, Vol. 35, No. 12, pgs. 1332-1338, December 1985.

10. Bohnke, B. and D. Eitner, 1983, "Vergleichende Utersuchung verschiedener Komposte in einen mobilen Biofilter (Investigation and Comparison of Different Kinds of Compost in a Mobile Biofilter)", Gutachten i.A. der AG Kompostabstaz NW, Aachen, West Germany, 1983.

11. EPA, 1985, Odor and Corrosion Control in Sanitary Sewerage Systems and Treatment Plants, Design Manual, EPA/625/1-85-018, October 1985.

12. Van Lith, C., 1989, "Design Criteria for Biofilters", Paper Presented at the 82th Annual Meeting of the Air Pollution Control Association in Anaheim, CA, June 1989.

13. Harris, J.C., 1978, "Ozonation", presented in Unit Operations for Treatment of Hazardous Industrial Wastes, D.J. DeRenzo, editor, Noyles Data Corporation, Park Ridge, New Jersey

14. Allen, E.R., Y. Yang, H.U. Hartenstein, "Final Project Report to Provide Technical Assistance in the Design and Installation of an Odor Control Biofilter System at the Kanapaha Wastewater Treatment Plant", University of Florida Department of Environmental Engineering Sciences, Gainesville, Florida, March 1989.

Table 1. Comparison of Gainesville and Meorawa Biofilters

Item	Units	Gainesville	Moerawa
Size of Biofilter	ft^2	1,076	460
Air Flow Rate	cfm	3,400	530
Loading Rate	cfm/ft^2	3.2	1.2
Static Air Pressure	inches of H_2O	6.7	14.4
Average H_2S Removal Rate		99.9%	99.9%

Notes: 1.   Performance data from the Gainesville BIKOVENT® System biofilter is from Allen (1989).
         2.   Design and performance data from the Moerawa biofilter is from EPA (1985).

Table 2. Comparison of Costs for Various Odor Control Systems

Treatment Alternative	Capital Cost	O&M Cost per Year	Annualized Cost
BIKOVENT® System Biofilter	$68,800	$5,110	$17,630
Traditional Biofilter	$97,300	$7,870	$25,750
Ozone	$81,180	$20,300	$38,780
Wet Scrubber	$84,180	$25,530	$45,850
Catalytic Incinerator	$76,860	$31,280	$51,900
Thermal Incinerator	$79,370	$45,100	$69,720
Activated Carbon	$139,940	$59,480	$97,690

Notes: 1.   All costs in December 1988 dollars.
         2.   O&M costs increase at 6.0% per year.
         3.   Estimated project life for all alternatives is 10 years.
         4.   Interest rate is 10.0% per year.
         5.   O&M labor included only for BIKOVENT® and Ozone.

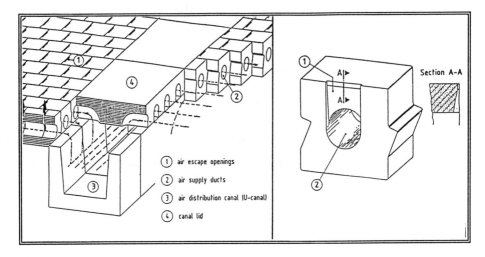

Figure 1. Details of BIKOVENT® System design.

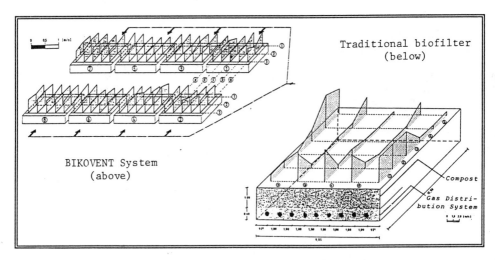

Figure 2.   Comparison of air distribution between BIKOVENT® System and traditional biofilter design.

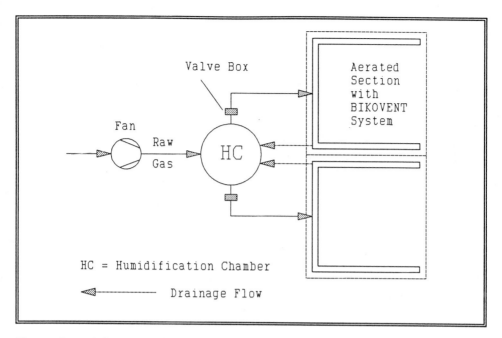

Figure 3.   Schematic of typical BIKOVENT® System installation.

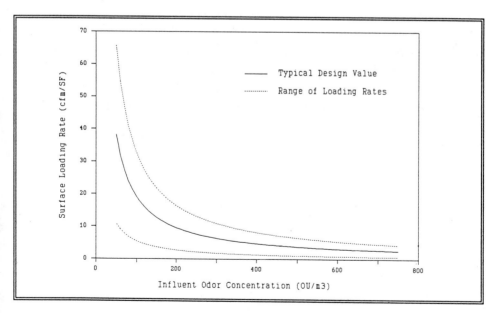

Figure 4.   BIKOVENT® System typical air loading rates for different inlet
odor concentrations.

VAPOR PHASE REACTIONS AND THEIR USE IN CONTROLLING ORGANIC ODORS

Kenneth J. Heller
NuTech Environmental Corp.
5350 N. Washington St.
Denver, CO. 80216

Odor control is an area of growing concern for all types of people; from the neighbors who complain, to the engineers who are required to solve the problem. Traditional methods of odor control have proven to be effective in most applications, but their cost can prove to be prohibitive. NuTech Environmental Corp. introduced "vapor phase" reaction products over five years ago in an attempt to provide inexpensive and effective alternatives.

The NuTech line of products and systems can be used to solve a wide variety of problems, as well as improve the performance and lessen the cost of many existing systems. This can include; increasing the life of carbon beds, replacing traditional oxidizing agents in wet scrubbers, providing simple systems for lift stations, and designing custom systems for the more difficult applications.

History

NuTech Environmental Corp., incorporated in 1982, is now in its seventh year of operations. Our company deals exclusively with odor control. The basis for the company's technology centers around the rights to certain patents and trade secrets which were purchased from Dr. James Cox. Dr. Cox's book "Odor Control and Olfaction"[1] served as a reference for us as well as the Water Pollution Control Federation's MOP 22 on odor control. Dr. Cox conducted significant research into the use of chemicals to control odors. We found, and have successfully demonstrated over the past six years, that this type of approach can successfully mediate odor complaints.

My brother, Jon Heller, and I are the principals of the firm. We were introduced to Dr. Cox while involved with a former business, The Denver Rendering Company. The company was one of two rendering plants in the Denver area. The rendering business provided our introduction to odors and odor control technology. Sadly, we also learned about unhappy neighbors, odor complaints and even lawsuits. That unfortunate occurrence however, did afford us the opportunity to learn about the use and operation of wet scrubbing equipment. In fact, we operated packed tower scrubbers and we were one of the early owners of a mist scrubber. We also learned about thermal oxidation of odors as we incinerated some of our high intensity odors in our boiler's firebox.

Dr. Cox did some consulting for us and introduced us to his technology. We were impressed enough to purchase the rights to the technology. I must say that now almost ten years later, I am still impressed each time we start up a new system.

Neutralizing Odorous Organics with Vapor Phase Reactions

The benefit of the NuTech approach is the ease of treatment. I'm sure most of you are all familiar with the use of masking agents. Please try for a moment, to separate your opinion of them from your knowledge of how they are applied. Everyone knows that masking agents are easy to apply. Our chemicals apply with the simplicity of masking agents, but offer effectiveness closer to that of the oxidizing agents. The solution to the odor control problem involves finding the best way to apply the product to the odorous air. We have developed a line of simple application equipment that address most application problems. We have also found existing wet scrubbing equipment to be quite satisfactory for our chemical's application. However, the higher vapor pressure of our neutralizing products lends itself to less costly pressure spray and air atomizing spray type systems.

Application Systems

The delivery systems we have developed and used successfully are based on simplicity. For such a system we use a small reservoir for the diluted product, an automatic dilution device, a chemical feed pump and an atomizing type nozzle. The systems range from small single or dual nozzle systems, to similar but larger systems with multiple nozzle capacity. We also have even larger units with blowers that operate in principle like a mist scrubber but are much smaller in size because the required contact time is only two seconds. This reduced contact time is due to the fact that we utilize construction and combining reactions rather than the more traditional oxidation reactions.

The smaller systems work by strategically placing the pressure atomizing nozzle in a place where the mist from the nozzle can achieve maximum contact with the odorous air. Now let me emphasize that the intent is to treat the odorous air at its point origination or in an area of containment, not just to spray chemicals in the air. This means that we set up nozzles in the head space of a tank or wet well. We can also use wide angle nozzles to treat the air just above the liquid level in a tank. We also set up the systems to treat existing ventilation ducts or air discharge points. When ventilation streams exceed 1500 cfm we recommend using air atomizing nozzles. This requires a compressed air supply and different brackets to mount the nozzles. All of our brackets are designed for easy access to the nozzle for maintenance checks.

NuTech Products

Now, lets take a look at the chemicals we use. The technology we purchased from Dr. Cox deals with odor neutralizers. The basis of the technology is a combination of ingredients that react in the vapor phase. The high vapor pressure of these products has demonstrated an ability to quickly mix into an air stream and reduce odors. This approach by-passes the requirement for more expensive absorption hardware required with traditional oxidation reactions. The formulations we use depend upon the odors we need to treat. This is because the different products have special components to increase their ability to react certain types of odors.

We explain the chemistry of our reactions as being a combination of construction and combining reactions. These reactions build the odorous molecules into a heavier less volatile form, or in effect create a more complex molecular structure which is odorless. This process requires a very short reaction time because it takes place in the vapor, rather than liquid state and does not require the odors to be solubilized into the scrubbing medium to be effective. This advantage allows for application equipment that is much less elaborate and costly than conventional wet scrubbing equipment. Now let me point out at this time that the NuTech approach is designed to react the organic component of the odor stream. The organic odors are comprised of amines, acids and thiols. We cannot chemically react hydrogen sulfide with the "vapor phase" technology but we can control low level H_2S odors using interference reactions which I'll explain later.

Due to the multitude of reactions involved it has been cost prohibitive to do research to validate the actual chemical reactions. The premise of our approach is that the amine odors can be reacted by both the acid and aldehyde components in our formulation. The fatty acid odors will form esters when reacted with the alcohol component of the formulation and the thiols can be polymerized. There is sufficient organic chemistry to support these types of reactions. In terms of quantitative tests, we can document reductions in total sulfides when an air stream is treated. There is equipment available to measure amines and total sulfides. I wish it were as easy to measure the total mercaptan and acid components.

In choosing a NuTech product, we determine if the subject air is more heavily laden with thiols or amines. If thiols are the primary problem, we would use our NuTralite™ or Chi-X® Odor Eliminators. If the amines are predominant, such as in composting operations, we would use our DeAmine™ Odor Eliminator which has a greater aldehyde component.

Now let me explain the interference component of our formulations. This gives the products the additional ability to counteract the odors of hydrogen sulfide up to 6 or 8 ppm in the air. If the levels are higher, complete odor control cannot be achieved, as the H_2S odors will begin to bleed through. This process involves the use of Zwaardemaker pairing which is discussed in the MOP #22[2].

I also want to mention that we recently added a straight counteractant to our line, our OCA-21™ product. This was in response to the demand for a non-corrosive, non-toxic product that could be easily handled with no special safety precautions. This product has demonstrated its ability to neutralize odors and eliminate complaints. However, due to its lower vapor pressure, it must be air atomized and thoroughly contact the odorous air in order to be effective. Lets look at the relative effectiveness of the different chemical approaches.

Common Chemical Approaches

There are four common groups of chemicals that are currently being used to control airborne odors; masking agents, odor counteractants, odor reactants, and oxidizing agents. Each of the chemicals is either atomized or used in a wet scrubber to reduce the airborne odors. Figure 1 shows graphically how each group interacts with the odor. How each chemical works in a different way to reduce the severity of the odors is discussed below:

Masking Agents. Masking agents attempt to overpower the malodor with another odor which is perceived to be more pleasant. In many cases, the end result is a fragrance flavored version of the original malodor. Masking always increases the total odor load. There is no chemical reaction and the individual constituents of the odor remain unchanged. The main advantage of masking agents is the low cost and their non-hazardous nature. The main disadvantage is the tendency for the agent to separate from the odor downwind. In addition, sometimes the fragrance can be as offensive as the original malodor. Masking agents are only effective on very mild odors and should be utilized only when no other practical solutions exist.

Counteractants. Counteractants are chemicals that work through a process that interferes with the malodor. While a counteractant does not chemically react with the malodor, it does reduce the odor level by eliminating the objectionable characteristics of the malodor. As can be seen in the illustration in Figure 1, this is different from masking in that the total odor level is actually reduced rather than increased. Odor counteractants offer an effective way of dealing with a wide variety of problems. Counteractants usually have a neutral pH, are easy and safe to handle, and are only moderately more expensive than masking agents. One disadvantage with counteractants is that due to their chemical characteristics, they require fine atomization or fogging. This is because they require intimate contact with the odorous air in order to be effective.

Reactants. NuTech offers a line of odor eliminating chemical products that actually react with the odorous molecules to remove them from the airstream. The reactive chemicals work through the process of polymerization and esterification (construction and combining) which we discussed previously. In addition, these products can be used to increase the performance of existing odor control systems that are operating ineffectively. If properly applied, the operating cost of these reactive chemicals can actually be less than masking agents or counteractants. The disadvantage is that the reactive chemicals have corrosive properties which must be addressed in the application.

Oxidizers. Oxidizing agents are an accepted means of dealing with many odor problems. Common oxidizing agents include hydrogen peroxide, sodium hypochlorite, potassium permanganate, ozone and chlorine. These chemicals work through the oxidation process to tear the odor molecule down into more elemental compounds which have no odor. One disadvantage to using oxidizing agents is that they must react in the liquid state (with the exception of ozone) which requires longer contact times. When used in scrubbers, the odorous molecule must first be absorbed into the scrubber fluid. Although oxidizers are very effective on inorganic H_2S, some of the organic odors are not as soluble in the scrubber water and can pass through the scrubber untreated.

Each of these chemicals have distinct advantages and disadvantages which must be addressed when trying to find the best solution for each individual odor control problem. In many cases a combination of chemicals, process changes and/or other odor control methods must be used concurrently to cost effectively solve the problem. There are also a number of other products such as enzymes, polymers and monomers that have not been discussed in this paper. At this time, we do not have enough information about these products to discuss them at any length. It is possible that they would also offer a potential solution.

Typical Applications

While no two odor control installations are identical, there is enough similarity that our typical applications can be divided into several common areas. These areas are: small headworks and wet wells or enclosed head spaces of small tanks; ducted process air or room ventilation exhaust streams; open tanks, such as gravity thickeners, equalization tanks, and sludge holding tanks; and belt press dewatering operations. In addition, the NuTech neutralization products can be used in existing wet scrubbing equipment. Lets take a closer look at these applications to give you an understanding of what we call our "Vapor Phase Approach".

Small Headworks and Wet Wells and Enclosed Head Spaces of Small Tanks

These applications are quite common and are relatively easy to treat with a Series 800 system. The nozzle is mounted by a special bracket in the wet well or tank. This bracket is designed to facilitate easy maintenance of the nozzle, and is placed so the atomized mist achieves maximum contact with the odorous air. It is also placed in a way that none of the low pH solution comes in contact with the walls or equipment in the wet well or tank.

In-duct Application Systems

Our Series 900 AT System is used in this type of application. It utilizes air atomization nozzles to create a finer mist of the diluted product for easier mixing in the air stream. At the Delta Diablo WWTP in California, a flow of 12,500 cfm is being treated with a single nozzle. The nozzle is mounted to spray co-current with the air flow. The presence of the mounting bracket helps to create turbulence and assists in mixing the diluted chemical with the air stream.

The Series 900 can also be used to run nozzles in multiple locations at the same time. An installation of this type can be found at the Naugatuck Treatment Co., in Naugatuck, CT. In this application, a Series 900 System is used for direct spray application of the NuTralite product onto a belt dewatering press, supplying two air atomizing nozzles to treat an 18,000 cfm air collection system and supplying two additional pressure atomizing nozzles at a gravity discharge of a remote weir location.

Application to Open Tanks Such as Gravity Thickeners and Equalization Tanks

Controlling odors from an open tank requires more application points and more chemical than it would take if the tank were covered and ventilated. There are two basic approaches to this situation. The first is to use the Series 900 System with wide angle spray nozzles mounted one to three feet above the high liquid level or surface of the tank. The number of nozzles depend upon the surface area, and the amount of freeboard available in the tank. With a fair amount of freeboard, the tank tends to act as a contact chamber and fewer nozzles are required. For example in one application where a 10' diameter gravity thickener is being treated, two nozzles are required because there is minimal freeboard. This gives us one nozzle for 39' sq.ft. of surface area. On an aerobic digester application where there is more than 6' of freeboard, the ratio can be increased to one nozzle per 327 sq.ft. of surface area.

Keeping this in mind, lets look at our other approach to open tanks. This approach utilizes air atomizing nozzles mounted on the wall or bridge of the tank. This approach is used only when the freeboard in the tank is enough to direct the atomized mist down into the tank. For large diameter tanks, this approach can be combined with a blower for wider distribution of the atomized chemical product. This will reduce the number of nozzles and amount of chemical required.

384

Wet Well Ventilation and Odor Control with a NuTech Series 1000 System

The Series 1000 System is designed to be used where there are space restrictions and a limited budget. The system is a 1000 cfm spray chamber unit built from PVC. Typical application involves placing the unit inside the pump station, usually right over the wet well. The air from the wet well is pulled through the spray chamber and mist packing and exhausted.

Ventilation and Odor Control With A Series 1500 System

The Series 1500 System is a versatile, multi purpose odor control system that can treat between 500 and 5000 cfm. The Series 1500 differs from the Series 1000 system in several ways. First, it is a skid mounted system with a built in chemical dilution and delivery system, as opposed to the the Series 1000 system which utilizes the Series 900 system for its chemical delivery. In addition, the Series 1500 system utilizes a centrifugal blower which can supply up to 13" static pressure as opposed to the "weaker" tube axial fan design of the Series 1000 unit. The 1500 series has a complete control panel including starter and disconnect. This system is also more flexible, as its design capacity can be increased or decreased by changing the relative sizes of the blower and the contact chamber. Ideal uses for this system, which costs about one-half that of wet scrubbing equipment include:

- Providing ventilation with an odor control option for wet wells, headworks, or small process buildings.
- For the removal and treatment of air from a launder cover.
- Treatment of the air from small trickling filters, biotowers or aeration tanks.
- Essentially, this system can be considered in most applications suited to a carbon absorption or permanganate pellet system.

Successes and Failures With the "Vapor Phase" Approach

Within the past four years, we have provided one hundred and forty-two installations. Of these, fifty-nine are Series 800 Systems treating lift stations and small headwork structures. Forty-one involve the use of the larger Series 900 or Series 900 AT System. These are used in situations such as small plants with multiple odor sources, open large diameter thickeners, digesters and equalization tanks and others we have already discussed. Seven of our Series 1000 systems have been installed on wet wells. Four of our new Series 1500 Systems have been shipped, and five mist scrubber operators have replaced hypochlorite with our NuTralite or Chi-X products. In the past four years, only seven systems have been taken out of service. Of these, only two were removed because the operator's standards for odor removal were not being met. The others involved factors such as corrosion, replacement with compost filters and maintenance neglect.

Significant Successes

We have experienced a high percentage of success with our "Vapor Phase" approach to odor control. Our greatest success has been in retrofitting the Virotrol, "S" shaped, open spray chamber scrubbers which were heavily specified and installed in the late 1970's and early 1980's. At that point in time, performance was evaluated on H_2S reduction only, as H_2S was thought to be the only source of the odor problem. Unfortunately, the design of this system did not incorporate enough contact time to react the organic components of the odor stream. As a result, many of these units were removed from service. The much faster "Vapor Phase" reactions of the NuTech products can take place within the designed contact time the Virotrol Units allow. As a result, 12 of these units have been retrofitted with NuTech feed systems and chemical products, allowing them to continue in service and provide satisfactory odor reduction.

We have also had success with our ability, to treat odorous air streams in-duct. The significance of this is that existing single stage scrubbing systems can be easily converted to a two stage systems for only a minimal additional capital investment. For air streams where only organic acids, aromatic amines, and thiols are present, treatment can be achieved through in-duct application alone or with contact chambers that provide a second or two of residence time.

Specific Installations

I would like to mention six specific installations, as I believe their successes are significant:

Delta Diablo WWTP - Pittsburg-Antioch, CA.   I mentioned this installation earlier. Delta Diablo has four biotowers for which covers and odor control systems were determined necessary. While preliminary testing showed some promise for NuTech's "vapor phase" approach, the final design called for four mist type scrubbers to treat the 12,500 cfm air flow being generated by each individual biotower. However, by agreement with both the district's engineer and the consulting engineer, the NuTech approach was to be evaluated prior to the purchase and installation of the scrubbers. The test results were so positive that the mist scrubbers were not ordered and NuTech was used to treat the odors in the existing ductwork. This saved the municipality a great deal of money.

Wilmington, DE.   Our success in treating the process exhaust gas from this in-vessel, co-composting facility presented an interesting odor control option. This facility composts a mixture of municipal sludge and refuse. On site experimentation with our DeAmine Odor Eliminator along with outside odor evaluation assistance, led to the successful approach. This involves the dilution of 10,000 cfm of process exhaust air with 40,000 to 50,000 cfm of ambient air. This diluted air stream is then being treated by adding 2 gph of a 100 to 1 dilution of our DeAmine solution. This is accomplished with a NuTech Series 900 System using pressure atomized nozzles. The contact area is a chamber created on the discharge side of the 6' diameter roof mounted exhaust fan. The atomized spray of chemical is being introduced on the discharge side of the fan. Reductions of 75% to 86% in the number odor units are being achieved. The chemical cost for this application is less than $.50 per hour.

Naugatuck Treatment Co.   The significance of the Naugatuck application was in proving the ability to control dewatering odors by pressure atomization of "Vapor Phase" neutralizers at the belt press. The Naugatuck plant is owned and operated by the Uniroyal Co. The plant processes municipal and industrial waste water. In addition, it contracts for the dewatering and incineration of municipal sludge from nearby waste water treatment plants.

The application is a 1.2 M Aris-Andrus belt filter press. The plant's management purchased a Series 900 System for product application. Experimentation with nozzle placement helped to determine the number, sizing, and location of the pressure atomizing nozzles. The arrangement consists of three, 1 gph nozzles located on a spray header positioning the tips of the nozzles 6" above the press's gravity section. They mist vertically down on the surface of the sludge. Two additional nozzles are positioned in the drain area of the press which treat the odors from the filtrate. Treating the filter press in this manner has substantially reduced the odors generated from the dewatering operations. In addition, the odors from the dewatered sludge being conveyed to the incinerator feed pump are eliminated. The total amount of diluted product applied was 7.5 gallons per hour. The dilution rate of the chemical for this particular application is 120 to 1 resulting in 8 oz./hour usage of the NuTralite product. The product has a cost of $.17/oz., which gives an operating cost of $1.36 per hour. It is important to note that the ambient hydrogen sulfide levels generated prior to treatment were less than the 6 to 8 ppm maximum levels required for a successful NuTech installation.

Supplementing The Treatment Of A Carbon System At Harlingen, TX.   The municipality of Harlingen, TX was experiencing difficulties with its 3000 cfm carbon absorption unit installed on a primary aerated grit chamber. The carbon bed was completely full and breakthrough was occurring. Their MSA detector indicated that the air stream contained H_2S inlet levels in the area of 50 to 70 ppm. Total sulfide readings ranged from 1500 to 2000 ppm on the discharge side of the carbon, as measured with a Sensidyne total sulfide meter. Sulfide break through was occurring every 45 days requiring the carbon to be regenerated or completely replaced.

A NuTech Series 800 System was installed for use in conjunction with the carbon bed. The diluted Chi-X product was atomized at a rate of 1.2 gph, in the ductwork upstream of the carbon system. The "Vapor Phase" neutralizer is being diluted 100 to 1 resulting in a chemical usage of 2.5 oz./hr of concentrate and an hourly operating cost of approximately $0.60 per hour. The NuTech system provided an organic sulfide emission reduction in excess of 99%. Thus, after the carbon was regenerated, its useful life was extended from 45 days to over 6 months.

The City of Rockland , ME WWTP. The Rockland Waste Water Treatment Plant is located in the heart of the city on the bay. Odor complaints became prevalent in the early 1980's and an odor study was done by the Boston based engineering firm of Whitman and Howard. This study included odor measurement and modeling through TRC Environmental of Hartford, CT. The sludge thickeners and sludge storage tank were considered to be among the primary odor sources. The study called for covering these sources and ducting them to a wet scrubber for treatment. In 1984, this installation would have cost Rockland over $200,000. Due to a lack of funding, nothing was done.

In June of 1989, Whitman and Howard recommended that a NuTech Series 900 System be installed on a rental basis for a one month evaluation. The NuTralite product was used as the "Vapor Phase" reactant. The installation provided for treatment of the two open 10' diameter sludge thickeners and the enclosed 20' diameter sludge storage tank. The thickeners were treated by placing 2 gph wide angle pressurized atomizing nozzles spraying down vertically from a point 18" above the liquid surface and at a point equidistant from the units bridge and outside wall. The sludge storage tank was treated by placing a 0.6 gph nozzle under each of the tanks two access hatches. The nozzles were located 24" beneath the top of the tank with a purpose of treating any air being emitted from the tank. The plant manager considered the test extremely successful. The system was purchased and expanded to treat the vacuum filter exhaust and several other emission points in the sludge processing building. The chemical is being diluted at a rate of 225 to 1 resulting in a cost of approximately $1.00 per hour. Also, the city purchased an additional Series 900 System to treat their headworks building.

System Failures

NuTech has experienced seven failures. All of these were with the Series 800 System in lift station or small headworks applications. There are three primary causes for these failures:

1. Corrosion.
2. Wide fluctuations in odor loads.
3. High levels of hydrogen sulfide.

Corrosion can occur from improper nozzle placement in the wet well or channel. This can cause obvious corrosion to concrete and metal after six months to one year of operation. Also, improperly designed ventilation systems can sometimes allow treated air into areas not designed for corrosive air. This combined with over treatment of the chemical will allow residual chemical to be discharged from an exhaust stack and pulled into an adjacent ventilation system. These possibilities must be reviewed and corrective measures taken where necessary.

In other instances, systems were installed before hydrogen sulfide readings were taken. When troubleshooting why systems were not performing correctly, it was discovered that the H_2S levels were well above the 8 ppm maximum that we can effectively handle. Widely fluctuating odor loads also make it very difficult to balance the chemical feed. In these applications, an alternative method of odor control may be required. The adequacy of the NuTech approach can easily be evaluated by using our 30 day rental/purchase option.

Odor Panel Testing

Odor panel testing is one indication of the success or failure of an odor control system. In several of the NuTech installations, this type of testing has confirmed that NuTech is a viable solution with removal efficiencies ranging from 65% to 85%. Other testing has yielded results that are less impressive, but the overall results have been enough to solve the odor problem at hand.

In the particular installation at the Delta Diablo WWTP, odor panel testing was conducted by taking samples and sending them to the IITRI lab in Chicago for analysis. The samples were taken according to standard procedures. Actual samples were taken by a representative of Camp Dresser & McKee and the results are given in Table I. It should be remembered that these tests were run with less than 1 second of contact time due to preexisting design constraints. It should also be pointed out that the tests were run using increasingly strong concentrations of the product. The system was well balanced in the first two tests, but as more chemical is injected, the odor units increase. This is due to the glacial acetic acid component of the Chi-X product which can show up as odor units, if too much product is used.

The results of the odor panel tests carried out by Odor Science and Engineering at the Wilmington, Delaware Composting Facility are shown in Table II. The results shown are self-explanatory and show reductions from 79% to 86%. The results for the tests run in Digester C with one nozzle, can not be explained. OS&E feels that something went wrong with the testing equipment.

Keep in mind the fact that NuTech's chemicals are not breaking down the odors into their component parts but building them up into less odorous forms. In the combining reactions used, esters are formed. Esters are not malodorous, but have a more pleasant odor character. This approach will not give as complete odor removal as other treatments (i.e. 99% removal). Yet the 66% to 86% reduction we have documented has been more than enough to solve the odor problem. At Delta Diablo we were able to eliminate the need for approximately $1/2 million in wet scrubbing equipment.

System Specifications

All of the NuTech systems are designed to be utilized on a retrofit basis. As such, they require a minimum of space. Table III gives the space requirements for our Series 800 through Series 1500 odor control systems. As can be seen, all of the systems require a minimal amount of space. The Series 800 and Series 900 are even portable so that they can be moved to various locations on an as needed basis.

A system is sized and chemical chosen according to the requirements of each particular application. Determining factors include:

- Existing ventilation systems, if any.
- Construction materials of the ventilation equipment.
- The nature and source of the odors to be treated.
- The area of the facility requiring treatment.
- Any physical constraints that may be applicable.

Operating Costs and Maintenance Requirements

All the systems were designed to utilize the benefits of NuTech's vapor phase odor neutralizers. In addition, they were designed with user in mind, so operation and maintenance costs are minimized.

Operating Cost

There are three components to the operating cost of the NuTech systems: electricity, water and chemical costs. The smaller systems require only a standard 110 V breaker service. The largest system uses a 3 HP blower motor and has roughly 2 amps of other 110 V electrical requirements. The water requirements are based on either airflow or area treatment needs. The smaller systems use as little as 15 gallons of water per day. Treating a 12,500 cfm air stream with an in-duct spray system could require as much as 60 gallons of potable water per day. Chemical cost is a function of the air requiring treatment. Higher airflows and high odor airstreams require a larger quantity of chemical for proper treatment. Less chemical is required to treat a contained odor stream (ie. in-duct) than an unconfined area such as an open tank. A general rule of thumb is that it costs 1¢ to treat 100 cfm of air for one hour.

As you can calculate, the annualized operating costs of the NuTech systems are very reasonable. The total annual cost will also vary with the operating time required. For instance, our systems can be turned off when the weather turns cold and easily restarted when the weather again warms.

Maintenance Requirements

All of the NuTech systems are constructed of simple, easy-to-use, efficiently designed components. This minimizes the systems maintenance costs. Our years of experience operating an animal by-products rendering facility gave us a great appreciation for the need to keep the maintenance requirements of any system to a minimum. We have tried to carry this principle into our manufacturing business. Preventative maintenance for our systems requires only 10 to 15 minutes per week. If a repair is required, the modular construction of the units permits easy component replacement.

References

1. J.P. Cox, Odor Control and Olfaction, Pollution Sciences Publishing Co., Lynden, WA. 1975, pp. 67-103.

2 Water Pollution Control Federation, Odor Control for Wastewater Facilities-Manual of Practice No. 22, Washington, DC.,1979. p. 4., p. 70.

Table I. Odor panel test results - Delta Diablo, CA.

SAMPLE ID	CHEMICAL USAGE/HR.	ED$_{50}$ VALUE	BUTANOL EQUIVALENT UNDILUTED SAMPLE	SLOPE
ambient air	----	19	N/A	N/A
#1 inlet	4.5 ozs.	100	198	0.52
#1 outlet	4.5 ozs.	44	N/A	N/A
# 2 inlet	7.8 ozs.	150	166	0.51
#2 outlet	7.8 ozs.	67	N/A	N/A
# 3 inlet	12.3 ozs.	225	417	0.54
# 3 outlet	12.3 ozs.	213	173	0.52
# 4 inlet	18.8 ozs.	149	294	0.61
# 4 outlet	18.8 ozs.	170	243	0.58
# 5 inlet	24.6 ozs.	78	N/A	N/A
# 5 outlet	24.6 ozs.	101	197	0.48

N/A: odor in sample not strong enough to measure

Table II. Odor panel test results - Wilmington, DE.

Source	Odor Level No DeAmine (D/T)	Odor Level (D/T)	Airflow (ACFM)	Control Efficiency Percent
One Spray Nozzle				
Digester A	610	128	36756	79%
Digester B	722	158	18,378	78%
Digester C	625	1000	42,411	-56%
Digester D	910	158	31,101	83%
Two Spray Nozzles				
Digester A	610	128	36756	79%
Digester B	722	100	18,378	86%
Digester C	625	128	42,411	80%
Digester D	910	128	31,101	86%

Table III. System size requirements.

System	Floor Space Req.	Height Req.
Series 800 & 800AT	2'x 2'	30"
Series 900 & 900AT	2'x 2'	42"
Series 1000	2'x 5'	8'
Series 1500	3'8" x 6'6"	4'10"

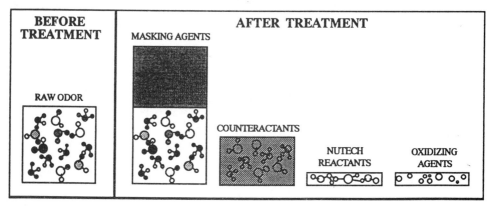

Figure 1. How chemical groups react with odor.

SUBJECT INDEX

AUTHOR INDEX